Sense the Real Change: Proceedings of the 20th International Conference on Near Infrared Spectroscopy

Xiaoli Chu · Longhai Guo ·
Yue Huang · Hongfu Yuan
Editors

Sense the Real Change: Proceedings of the 20th International Conference on Near Infrared Spectroscopy

Springer

Chemical Industry Press Co., Ltd.

Editors
Xiaoli Chu
Analytical Research Department
Sinopec Research Institute
of Petroleum Processing
Beijing, China

Longhai Guo
College of Materials Science
and Engineering
Beijing University of Chemical Technology
Beijing, China

Yue Huang
College of Food Science & Nutritional
Engineering
China Agricultural University
Beijing, China

Hongfu Yuan
College of Materials Science
and Engineering
Beijing University of Chemical Technology
Beijing, China

ISBN 978-981-19-4883-1 ISBN 978-981-19-4884-8 (eBook)
https://doi.org/10.1007/978-981-19-4884-8

Jointly published with Chemical Industry Press
The print edition is not for sale in China (Mainland). Customers from China (Mainland) please order the print book from: Chemical Industry Press.

This Springer imprint is published by the registered company Springer Nature Singapore Pte Ltd.
The registered company address is: 152 Beach Road, #21-01/04 Gateway East, Singapore 189721, Singapore

Introduction

ICNIRS 2021 Proceedings

As a member of the spectral family, the NIR spectroscopy with the wavelengths between the UV–visible and mid-infrared spectra has been widely applied in many fields as a rapid and online analysis technique since the late 1960s under the leadership of Dr. Karl Norris. Especially in the past two decades, near-infrared spectroscopy has made great progress in agriculture, food, medicine, chemical industry, and other fields. The reason why NIR spectroscopy has become a powerful analysis technique is that it has completely changed the traditional analysis method of taking a sample back to laboratory for further characterization and realized the in situ and online analysis in many fields. The measurement of NIR spectrum is very flexible and convenient. It can directly and non-destructively analyze a variety of products that are closely related to our lives through different special measurement accessories without any sample pretreatment. Moreover, the NIR spectrum can be transmitted through a silica fiber, which means that the spectrometer can be used in measurement sites with complex working conditions, so it can be used for online analysis of large-scale installations such as petrochemicals to measure the composition and chemical properties of materials in real time. Nowadays, the application of NIR spectrometer covers almost all aspects of human life and plays an increasingly important role in scientific research, industry, agriculture, commerce, and other fields.

The biennial meeting of the International Council for NIR Spectroscopy is undoubtedly an academic event for the researchers on NIR spectroscopy. However, unfortunately, due to the impact of the global epidemic of COVID-19, in order to effectively protect the health of the participants, NIR2021 International Conference was finally opened as a form of an online conference. It has also become a historical event so far, the first international conference on near-infrared spectroscopy held online. A total of 316 delegates from 29 countries of six continents attended online the NIR2021 International Conference held in Beijing, and they presented us with the wonderful lectures, including two award presentations, five keynote lectures, and 66 oral presentations as well as 111 post presentations. Although the delegates could not attend this academic event by a face-to-face form, they must have also felt the powerful and wide development of NIR spectroscopy through these wonderful reports, just like the slogan of NIR2021, “Sense the Real Change”. All the participants sensed the various spectral information of NIR, the development of spectral theory and chemometrics, and the enhanced performance of instruments and measurement accessories. We also believe that every participant could personally sense the real changes of NIR spectroscopy in China, especially for its wide application in different field. The theme of the NIR2021 is indicated by the logo, which is “Rainbow: Diversity, Optimization, and Inspiration”. The rainbow in logo represents the spectrum, as well as a dragon of Chinese element. The dancing dragon means that NIR is taking off in the international technological arena, playing an increasingly important role in agriculture, food, chemical industry, and people’s daily lives. We strongly hope that the NIR2021 can make a great contribution to accelerate advancing NIR technology through constructing a strong human network on NIR spectroscopy in the world. We also have to sincerely thank all 15 sponsors, especially the two platinum sponsors ABB and REEMOON, who economically supported NIR2021 even without the live exhibition.

The scientific program was divided into seven sections: (1) Spectroscopy Theory and Chemometrics, (2) Instrument, Accessory, and Experimental Technology, (3) Agriculture, (4) Pharmaceutical and Chemistry, (5) Biomedicine, Environment, and fNIR; (6) PAT and Imaging, and (7) Aquaphotomics. In each section, one keynote lecture and around four to eight oral presentations were selected. Around 100 poster presentations were performed at the conference. The keynote lectures were as follows: (1) Roumiana Tsenkova from Kobe University gave an overview on the development and application of NIR spectroscopy in Aquaphotomics; (2) Heinz Siesler from the University of Duisburg-Essen talked about the test and application of miniaturized handheld vibrational spectrometers over the last ten years; (3) Richard Crocombe from Crocombe Spectroscopic Consulting introduced the New Technologies and Applications of Ever-Shrinking Spectrometer; (4) Da-Wen Sun from University College Dublin talked about the Advances in Hyperspectral Imaging Technology for Food Quality and Safety Detection and Control; (5) Peiwu Li from Chinese Academy of Agricultural Sciences talked about the Application of NIRS Technology in the Development of Green Animal Husbandry. Yukihiro Ozaki, as the winner of the Karl Norris Award from Kansai Gakuin University, gave a lecture with the title: “NIR Spectroscopy-What a Wonderful World!”. Vincent Baeten, as the winner of Tomas Hirschfeld Award from Walloon Agricultural Research Center, gave a lecture with the title: “The treachery of NIRS applications: authentic or not?”. These programs can be seen at the homepage of NIR2021 (www.nir2021.com).

The selected papers in NIR2021 Proceedings have been professionally reviewed by the Scientific Committee. Although there are only about 30 papers in the NIR2021 Proceedings, these papers cover almost all the research fields and the practical applications of NIR spectroscopy, from which we can clearly see the hotspots and frontiers of NIR research and application in recent years. We hope that you will enjoy reading the NIR2021 Proceedings and these original researches could promote the investigation and the further development of NIR spectroscopy.

Finally, we hope that the COVID-19 epidemic will end soon. Looking forward to seeing you all at the NIR2023 Innsbruck. Warmest regards.

Hongfu Yuan
Xiaoli Chu
Longhai Guo

Organization

Organizing Committee

Chair

Hongfu Yuan

Secretary-General

Xiaoli Chu

Vice Secretary-General

Longhai Guo

Committee Members

Hengchang Zang
Huihua Yang
Jingzhu Wu
Lian Li
Lihui Yin
Nanning Cao, USA
Shungeng Min
Tao Pan
Xihui Bian
Xudong Sun
Yande Liu
Yiping Du
Yonghuan Yun
Yue Huang
Zengling Yang
Zhisheng Wu
Jian Ye
Xiaoshi Zhang

Scientific Committee

Akifumi Ikehata
Ana Garrido-Varo
Anders Larsen
Carrie Vance
Celio Pasquini
Christian Huck
Cristina Malegori
David Highton

Contents

Selected Articles

Agriculture, Food and Forestry

Biomedicine, Environment, and fNIR

Instrument, Accessory and Experimental Technology

PAT and Imaging

Pharmaceutical and Chemistry

Spectroscopy Theory and Chemometrics

Others

Selected Articles

How Can We Unravel Complicated NIR Spectra? –Challenges of the Ozaki Group for the Last 30 Years–

Yukihiro Ozaki(✉)

School of Biological and Environmental Sciences, Kwansei Gakuin University, Sanda, Hyogo 669-1330, Japan
ozaki@kwansei.ac.jp

Abstract. This review consists of two parts. The first part is concerned with the outline of the spectra analysis methods in NIR spectroscopy. In this part conventional spectra analysis methods such as the analysis based on group frequencies and calculations of difference spectra and second derivatives are explained. The second part describes big efforts of the Ozaki group in unraveling complicated NIR spectra for last three decades. Our studies using two-dimensional correlation spectroscopy (2D-COS) are mentioned first. The analysis of the temperature-dependent spectra variations of water using difference spectra, 2D-COS, principal component analysis (PCA), and self-modeling curve resolution (SMCR) is reported next. Our proposals of new chemometrics algorisms and 2D-COS algorisms are introduced. Moreover, our recent investigations of quantum chemical calculation studies of NIR spectra are discussed.

Keywords: Spectral analysis · Chemometrics · Derivative spectra · Quantum chemical calculation · Two-dimensional correlation spectroscopy

1 Introduction

NIR spectroscopy is a spectroscopy of overtones and combinations, and in NIR spectra a number of bands due to overtones and combinations overlap severely (NIR spectroscopy is electronic spectroscopy as well as vibrational spectroscopy, but this review is concerned only with the latter) [1]. Therefore, it is not easy to analyse NIR spectra even spectra of simple compounds. Figure 1 shows experimental and calculated NIR spectra of low concentration (0.005 M) methanol [2]. Note that even the simple compound yields many bands derived from the overtones and combinations in the 7500–4000 cm^{-1} region. Reliable bands assignments become possible only with the aid of quantum chemical calculation. I will explain in more detail about the band assignments and quantum chemical calculation of the NIR spectrum of methanol.

The purpose of this review is to outline the spectral analysis methods of NIR spectra first, and then, describe how the Ozaki group has challenged the spectral analysis of NIR spectra for the last 30 years or so.

X. Chu et al. (Eds.): ICNIR 2021, *Sense the Real Change: Proceedings of the 20th International Conference on Near Infrared Spectroscopy*, pp. 3–16, 2022.
https://doi.org/10.1007/978-981-19-4884-8_1

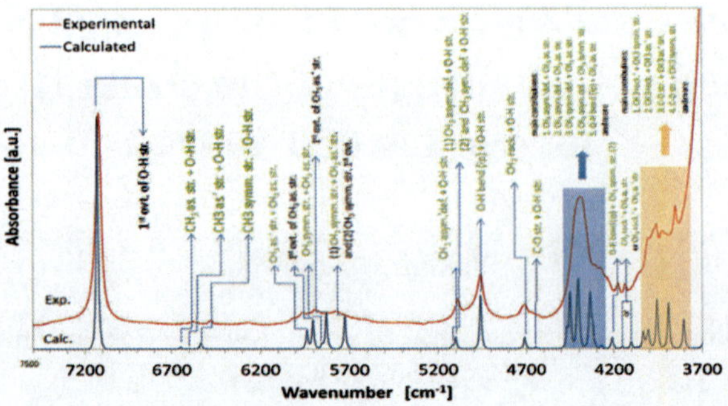

Fig. 1. Experimental (5×10^{-3} M CCl_4) and simulated NIR spectra of diluted methanol. Reproduced from Ref. [2] with permission.

2 Spectral Analysis in the NIR Region

As in the case of other spectroscopy band assignment is the base for spectral analysis in NIR spectroscopy [1, 3]. However, the band assignment in the NIR region is, in general, not easy because of severe overlaps of many overtones and combinations bands. In some cases, bands due to combinations including an overtone such as $\nu_1 + 2\nu_2$ appear, and for such cases it is very difficult to make solid band assignments. Moreover, Fermi resonance gives rise to very complicated spectral patterns, making band assignments almost impossible. However, even in such cases it may be possible to know to which functional group a band belongs.

Conventional spectral analysis methods such as spectral analysis based on group frequencies and spectra-structure correlations are used as the bases for the spectral analysis of NIR spectra [1, 3, 4]. However, these are just a starting point and far from enough, and thus, various other spectra analysis methods such as calculations of second derivatives and difference spectra and comparison with spectra of similar compounds are employed. Chemometrics is also very useful for spectra analysis, and it also allows one to extract rich information from NIR spectra [5, 6]. Nowadays, in addition to the conventional spectral analysis methods and chemometrics, quantum chemical calculations, such as density function theory (DFT) calculations [1, 7, 8], and two-dimensional correlation spectroscopy (2D-COS) [1, 9] have become important in the analysis of NIR spectra. Using quantum chemical calculations one can calculate the intensities and frequencies of overtones and combination bands [1, 7, 8].

3 Conventional Spectral Analysis Method

The following conventional spectral analysis methods are used in NIR spectroscopy [1, 3].

(1) Spectral analysis based on group frequencies:

Each functional group has group frequencies [1, 4]. In NIR spectroscopy group frequencies due to the first and second overtones of CH, CH_2, CH_3, OH, and NH_2, stretching modes, the second overtone of the C = O stretching mode, and some combination modes such as combinations of amide A and amide II (or III) are very important. Group frequency tables in the NIR region can be found in a few NIR textbooks [4, 10].

(2) Spectra-structure correlations:

A comparison of the NIR spectrum of a compound with those of similar compounds is useful for the band assignment of the spectrum. For example, NIR measurements of a series of alcohols allow one to make assignments of bands due to OH, CH_2 and CH_3 groups.

(3) Calculations of derivative spectra and difference spectra:

Both are useful to unravel overlapping bands and to find out a weak feature hidden by a strong band. One good example of difference spectra will be shown later.

(4) Spectral analysis based on perturbation:

Temperature-dependent, pH-dependent, and concentration-dependent spectra changes, often provide important information about the band assignments.

The above four are particularly important for spectral analysis, and the following methods are also sometimes useful.

(5) Comparison of an NIR spectrum with the corresponding IR and/or Raman spectra:

This is very important to understand the relation between a band due to fundamental and bands derived from its corresponding first and second overtones and the relation between bands originating from fundamentals and those due to their combinations.

(6) Curve fitting:

Before one performs curve fitting, the number of bands used for the curve fitting must be determined precisely [11]. One can use the second derivative, for example, for this purpose.

(7) Self-Modeling Curve Resolution (SMCR), Multivariate Curve Resolution with Alternating Least Squares (MCR-ALS):

SMCR and MCR-ALS are used to predict the pure component spectra and pure concentration profiles from a set of NIR spectra [12].

(8) 2D-COS:

We may be the first in applying (6), [11] (7), [12] and (8) [13] in NIR spectroscopy. I will explain the examples of the uses of SMCR and 2D-COS in NIR spectroscopy later.

(9) Spectral interpretation by polarization measurements:

This method has been used, for example, for the determination of the molecular orientation of solid-oriented compounds such as uniaxially stretched polymers [14, 15].

(10) Isotope exchange experiments:

Deuterium exchange experiment is useful for some special case.

(11) Quantum chemical calculations:

I will explain in some detail quantum chemical calculation using one independent session.

4 Challenges of the Ozaki Group in the NIR Spectral Analysis for the Last 30 Years

In the following sections I will describe some examples of challenges of the Ozaki group in the analysis of NIR spectra for the last three decades. The following topics will be reported. 1) The use of 2D-COS. 2) Analysis of the NIR spectra of water using the difference spectra, PCA, 2D-COS and SMCR. 3) Proposals of some chemometrics and 2D-COS algorithms. 4) Applications of quantum chemical calculations to NIR spectra.

4.1 The Use of 2D-COS

2D-COS was originally proposed by Noda in 1986, and later developed into more applicable generalized 2D correlation spectroscopy by the same author [9] In 2D-COS one can obtain synchronous and asynchronous correlation spectra; the former represents the simultaneous or coincidental changes of spectral intensities at ν_1 and ν_2, whereas the latter spectrum represents sequential or unsynchronized variations. 2D-COS enables enhanced spectra resolution by spreading the spectral data over the second dimension. By use of 2D-COS the correlation between the different bands and intermolecular interactions can be investigated.

We together with Noda introduced 2D-COS to NIR spectroscopy; we first investigated temperature-dependent NIR spectra variations of oleyl alcohol [13]. It was also applied to NIR spectra of various molecules such as alcohls [16], *N*-methylacetamide [17], nylon 6 [18], polymers [19, 20], and proteins [21]. 2D-COS provided not only useful information about the band assignments but also information about hydrogen bonding, intermolecular interactions, hydration, and so on [9]. Here, I introduce one interesting example of 2D-COS studies of proteins. Wu et al. carried out an investigation on secondary structure and hydration of human serum albumin (HSA) by 2D-COS NIR spectroscopy [21]. They measured FT-NIR spectra of HSA in aqueous solutions with the concentrations of 1.0, 2.0, 3.0, 4.0, and 5.0 wt % over a temperature range of 45–80 °C in the regions of 7500–5500 and 4900–4200 cm^{-1} (Fig. 2(A)) [21]. 2D-COS spectra generated by the temperature-dependent spectra of HSA can decompose NIR spectra of HSA with complex and heavily overlap features into spectral components by spreading peaks along the second dimension [21]. To examine the correlation between the temperature-induced secondary structural changes and hydration of HSA, power and slice spectra were calculated, respectively, from 2D synchronous and asynchronous correlation spectra. Figure 2(B) and 2(C) show the power and slice spectra in the 4900–4200 cm^{-1} region, respectively. The corresponding power and slice spectra in the 7500–5500 cm^{-1} region are shown in Fig. 2(D) and 2(E), respectively. It is noted that in the power spectra in the 4900–4200 cm^{-1} region, a band at around 4600 cm^{-1} derived from the combination of amide B and amide II (amide B/II) modes shows a shift by 5 cm^{-1} between 58 and 60 °C, indicating that the secondary structural changes of HAS occur in this temperature range (Fig. 2(B)). Both the power and slice spectra in the 7500–5500 cm^{-1} region also show remarkable changes near 60 °C (Fig. 2(D) and 2(E)). Note that in this region a board band due to the combination of OH symmetric and antisymmetric stretching modes of water dominates. Comparison of these temperature-dependent frequency shifts between the band due to amide B/II mode of HSA around

4600 cm^{-1} and the combination band derived from water near 7000 cm^{-1} indicates that the protein unfolding occurs almost in parallel with the change in the protein hydration. This study has demonstrated the usefulness of 2D-COS NIR spectroscopy in monitoring subtle changes in protein dynamics, especially for hydration [21].

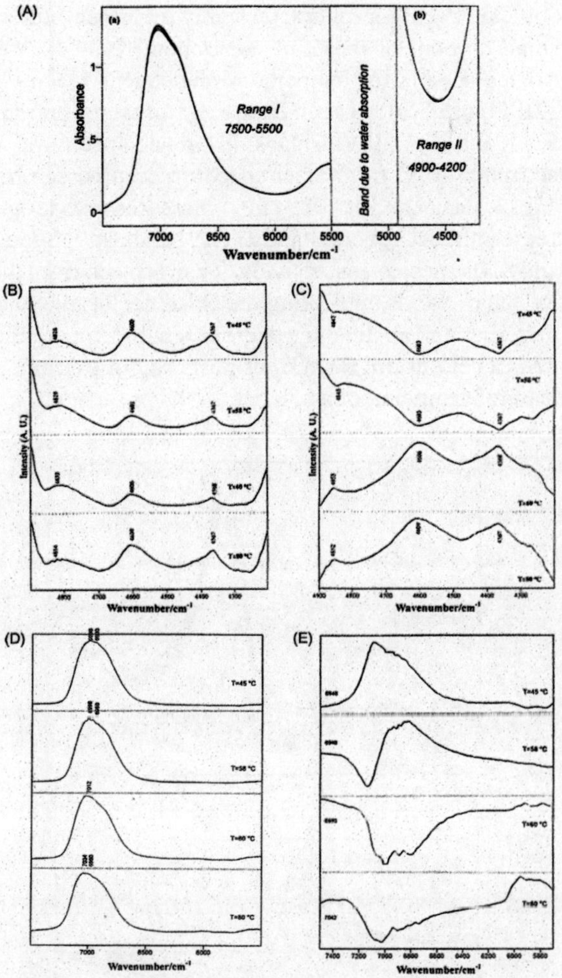

Fig. 2. (A) FT-NIR spectra of HSA in aqueous solutions with the concentrations of 1.0, 2.0, 3.0, 4.0, and 5.0 wt% measured in a temperature range of 45–80 °C. (B) Power and (C) slice spectra of the synchronous and asynchronous 2D correlation spectra, respectively, in the 4900–4200 cm^{-1} region calculated from the spectra shown in (A). (D) The corresponding power and (E) slice spectra in the 7500–5500 cm^{-1} region. Reproduced from Ref. [21] with permission.

4.2 Analysis of the NIR Spectra of Water by Difference Spectra, 2D-COS, PCA and SMCR

We analyzed temperature-dependent NIR spectral variations of water using difference spectra, 2D-COS, PCA and SMCR [22–24]. All the spectra analysis methods provided the same conclusion regarding water structure; liquid water consists of two major species, strongly hydrogen-bonded (SHB) species and weakly hydrogen-bonded (WHB) species but there is very small amount of the third component [22–24]. We first carried out calculation of difference spectra for temperature-dependent spectra variations of NIR spectra of water [22]. Figure 3(a) shows NIR spectra of water measured in a temperature range of 5–85 °C, and Fig. 3(b) displays difference spectra of water obtained by subtracting the spectrum at 5 °C from other spectra measured at various temperatures [22]. It is noted in Fig. 3 that a band at 7089 cm^{-1} increases with temperature while that at 6718 cm^{-1} decreases almost concomitantly. The bands at 7089 and 6718 cm^{-1} are assigned to WHB and SHB species, respectively. It can be seen from the difference spectra that there are two major bands, indicating the existence of two major water species; one water species increases and another one decreases with temperature. The calculation of difference spectra is a very useful for various purpose, for example, the absorption of a solute may be separated from that of a solvent.

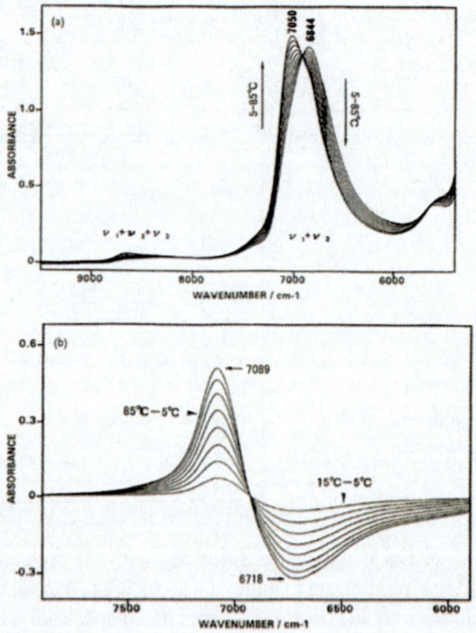

Fig. 3. (a) NIR spectra of water collected in a temperature range of 5–85 °C. (b) Difference spectra of water calculated by subtracting the spectrum at 5 °C from other spectra measured at various temperatures. Reproduced from Ref. [22] with permission.

We also tried PCA and 2D-COS to investigate the structure of water and its temperature dependence [23, 24]. A synchronous 2D correlation map and PCA shown in

Fig. 4(A) and (B), respectively, were generated from the temperature-dependent spectral variations of water measured over a temperature range of 6–80 °C at 2 °C increments. The 2D synchronous spectrum gives two peaks at 1412 and 1491 nm (7089 and 6718 cm^{-1}), corresponding to the WHB and SHB species, respectively, and in the PCA results the wavelengths at 1412 and 1491 nm account for more than 99% of the spectral variations. Thus, based on these results Segtnan and Šašić et al. [23, 24] concluded that water can be portrayed as a quasi-two-component mixture.

Šašić et al. [24] also carried out a SMCR study to investigate the 1300–1600 nm region of the water spectrum. They used a SMCR method named Simplisma. Without any prior knowledge about the system or supplementary experimental data, they were able to demonstrate that two species are enough to represent more than 99% of the original data with at least one additional spectroscopically identifiable component. Figure 5 shows the calculated spectra (A) and concentrations profiles (B) of SHB and WHB species [24]. It can be seen from Fig. 5 that the spectra of water can be decomposed clearly into two species and that the SHB species decreases while WHB species increases with temperature. SNCR plays an important role in band decomposition in various spectra including NIR spectra. For example, Šašić et al. performed a SMCR analysis of on-line NIR spectra of monitoring of the melt-extrusion transesterification of ethylene/vinyl acetate copolymer.

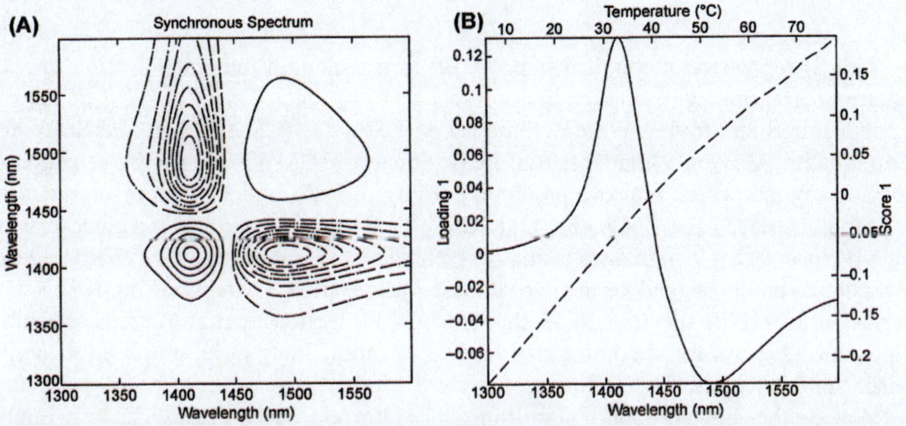

Fig. 4. (A) A synchronous 2D correlation map constructed by the temperature-dependent spectral variations of water measured in a temperature range of 6–80 °C at 2 °C increments. (B) Loadings and scores of the PCA from the same data as those for (A). Reproduced from Ref. [23] with permission.

4.3 Proposal of New Chemometrics and 2D-COS Algorisms

4.3.1 Chemometrics

The Ozaki group proposed several important ne chemometrics algorisms. As for chemometrics [25–30]. Jiang et al. [25] proposed Moving-Window Partial Least Squares

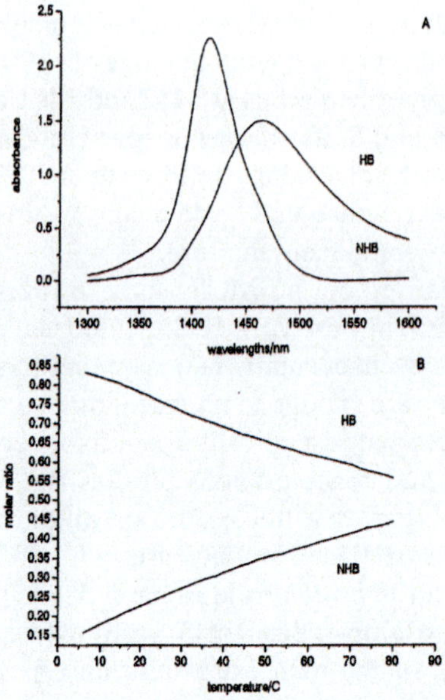

Fig. 5. The component spectra (A) concentration profiles (B) of HB (SHB) and NHB (WHB) species obtained by SMCR. Reproduced from Ref. [24] with permission.

(MWPLS) regression method. It is powerful in wavelength interval selection in least squares regression with applications to spectroscopy data. Du et al. [26] conceived Changeable Size Moving Window Partial Least Squares (CSMWPLS) and Searching Combination Moving Window Partial Least Squares (SCMWPLS) for the selection of spectral regions which improve prediction ability of PLS models. Kasemsumran et al. [27] applied MWPLS, CSMWPLS, and SCMWPLS for noninvasive blood glucose assay by NIR spectroscopy. Shinzawa et al. [28] created several interesting new chemometrics algorithms; one is bagged kernel partial least squares (KPLS) and boosting KPLS with applications to NIR spectra and another is sample selection by multi-objective genetic algorithm [29]. As for chemometrics for classification Jiang et al. proposed Principal Discriminant Variate (PDV) [30].

Among the above proposed algorithms here I introduce MWPLSR. The very important point for developing the best calibration models using PLS is to choose informative NIR regions where one can obtain an optimized calibration model for them. In general, wavelength selection can significantly improve the performance of full-spectrum calibration techniques, such as PLS, and various wavelength or wavenumber selection methods have been developed. Jiang et al. proposed a novel chemometrics algorithm for wavelength interval selection named moving window partial least squares regression (MWPLSR) [25]. The aim of MWPLSR is to search for informative spectral regions that contain useful information for PLS model building. MWPLSR develops a series of PLS models for all PLS factor numbers (LVs) in a window that moves over the full spectra, and then locates relevant spectral intervals in terms of the least complexity of PLS models reaching a desired error level [25]. Moreover, the selection of spectral intervals in terms of the least model complexity enables one to reduce the size of a calibration sample set in calibration modeling.

Figure 6 depicts scheme for explaining of MWPLSR, where a spectral window starting at the ith spectral channel, and ending at the $(I + h - 1)$ th spectral channel is constructed, where h is the window size. There are $(n - h + 1)$ windows over the whole spectra, each window corresponding to a subset of the original spectral $\boldsymbol{X}$ ($m \times n$ matrix; m samples and n spectral channels).

The PLS models with different numbers of LVs can then be built to relate the spectra in the window to the concentrations of the analyte, as follows:

$$\boldsymbol{y} = \boldsymbol{X}_i \boldsymbol{b}_{i,k} + \boldsymbol{e}_{i,k} \tag{1}$$

where $\boldsymbol{b}_{i,k}$ ($h \times 1$ vector) is the regression coefficients vector, estimated by PLS with k PLS components, and $\boldsymbol{e}_{i,k}$ is the residue vector obtained with k PLS components. The window is moved over the whole spectral region. At each position, PLS models with varying PLS component number are built for the calibration samples, and the log of the sums of squared residues (log (SSR)) are calculated with these PLS models and plotted as a function of the position of the window.

Figure 7 displays residue lines obtained by MWPLSR for the NIR spectra of skin (This result was obtained for noninvasive blood glucose assay by NIR spectroscopy) [27]. A representative informative region should show low values of the SSR, and often shows the shape of an upside-down peak, corresponding to a band in the same region. Thus, one can easily select the beginning and end points of the region.

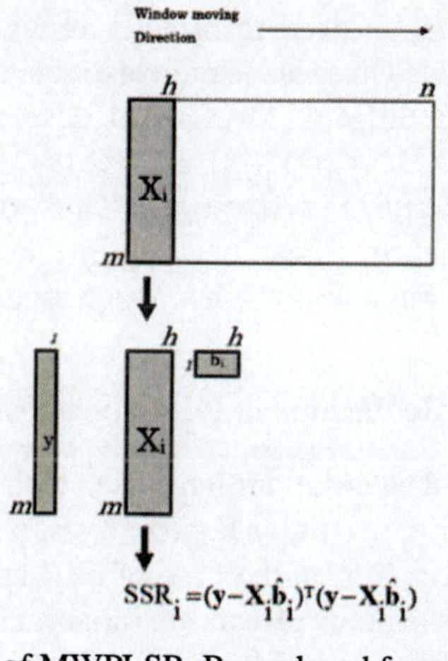

Fig. 6. Scheme for explaining of MWPLSR. Reproduced from Ref. [25] with the permission.

4.3.2 2D-COS

In 2000's we proposed several unique 2D-COS algorithms. Šašić et al. designed sample-sample 2D-COS [31] and statistical 2D-COS [32]. In sample-sample correlation spectra different from conventional wavelength – wavelength 2D-COS spectra, the rows

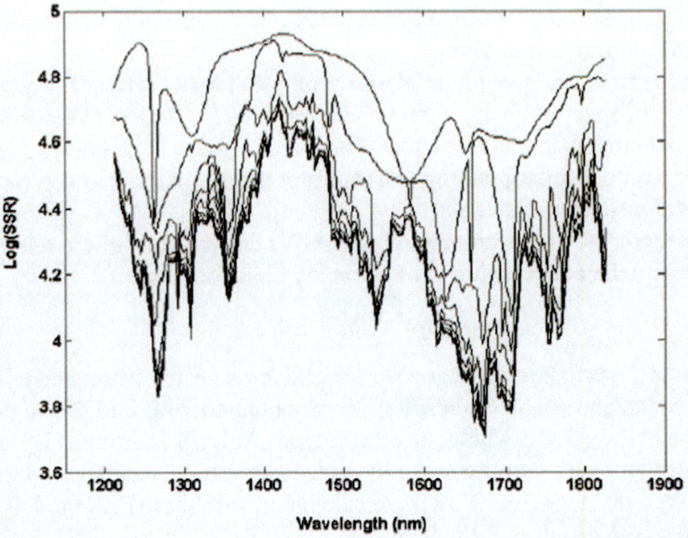

Fig. 7. Residue lines obtained by MWPLSR for the NIR spectra of skin. Reproduced from Ref. [27] with the permission.

and columns of the experimental matrix are exchanged; that is, the spectral data set is arranged.

In rows during the construction of the 2D matrices. The resultant 2D correlation map, having sample axes, can be used to directly reflect concentration dynamics of different chemical species [31]. Thus, one can investigate a correlation between sample using sample-sample 2D-COS. Shigeaki Morita et al. developed Perturbation Correlation Moving-Window 2D-COS (PCMW2D) [33]. In PCMW2D a 2D map between spectra and perturbation (e.g., temperature) is developed. Thus, one can explore perturbation-dependent spectra changes clearly. These new algorithms and methods are, of course, can be used not only for NIR but also for other spectroscopies.

4.4 Quantum Chemical Calculations in NIR Spectroscopy

Since 1990s quantum chemical calculations have been extensively used in IR and Raman spectroscopy because harmonic oscillator approximation is a straightforward simplification of a molecular vibration. Yet, in the case of NIR spectroscopy, the situation has been different because the anharmonic effects are inherent in NIR spectroscopy, and thus complex theoretical approaches are involved in the quantum chemical calculations of NIR spectra [7, 8]. Normal modes of vibrations can be approximately described by harmonic oscillator approximation. This happens because within fundamental transitions the harmonic approximation of vibrational potential resembles adequately the shape of true vibrational potential. Therefore, the prediction of IR and Raman spectra can be handled with relative ease nowadays.

We may be the first in carrying out quantum chemical calculations of frequencies and overtones of overtones and combinations of simple organic compounds such as alcohols and fatty acids [2, 7, 8, 34–38].

In their investigations of NIR spectra of methanol [2], ethanol [2], 1-propanol [2], and butanols [34]. Beć et al. evidenced that very accurate reproductions of the NIR spectra became possible, thus opening doors for further studies in NIR region. I already show the comparison of the experimental (5×10^{-3} M CCl_4) and simulated NIR spectra of diluted methanol in Fig. 1. Good agreement has been obtained between the experimental and calculated spectra including the reproduction of minor bands and the detailed band assignments have been made. A very intense band at 7120 cm^{-1} is due to the first overtone of stretching mode of free OH group. Most of bands below 6700 cm^{-1} derive

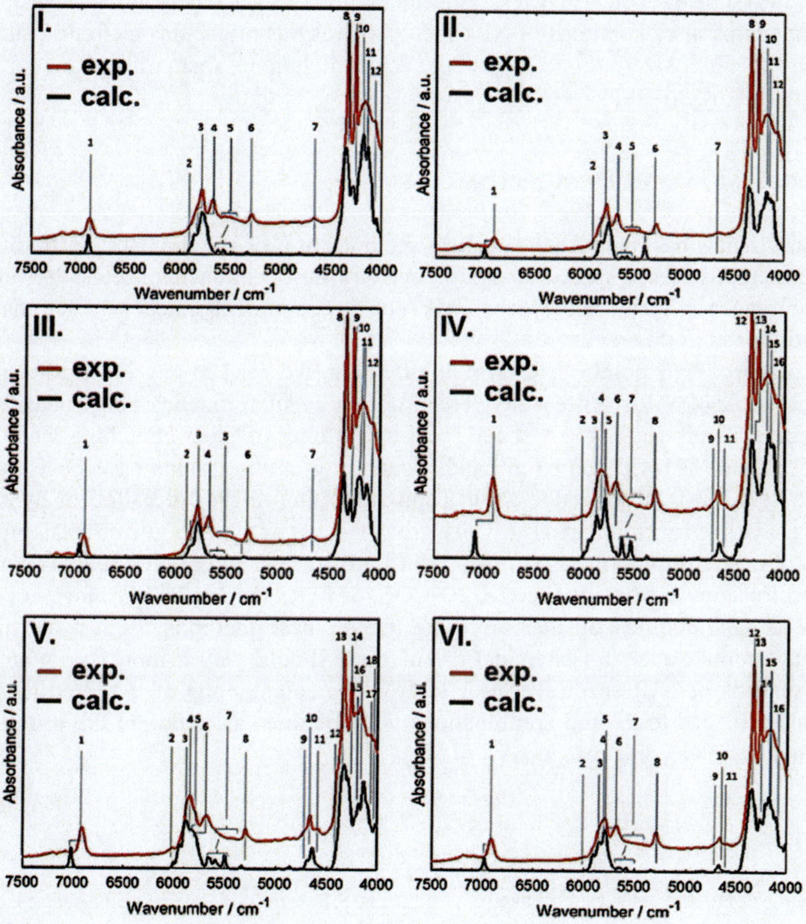

Fig. 8. Experimental and simulated (DVPT2) NIR spectra of long-chain fatty acids in solution phase (CCl_4); I. arachidic acid, II. palmitic acid, III. stearic acid, IV. linoleic acid, V. α-linolenic acid, VI. oleic acid. Reproduced from Ref. [35] with permission.

from combination modes except for a few bands due to the first overtones in the 6000–5750 cm^{-1} region.

Grabska et al. [35] investigated experimental and simulated spectra of long chain fatty acids in relation to application of NIR spectroscopy to biomedical analysis. They used the potential of DVPT2/GVPT2 method for successful anharmonic treatment of considerable complex molecules. Figure 8 compares experimental and simulated (DVPT2) NIR spectra of long-chain fatty acids in solution phase (CCl_4); I. arachidic acid, II. palmitic acid, III. stearic acid, IV. linoleic acid, V. α-linolenic acid, VI. oleic acid. A similarity between the experimental NIR spectra of long-chain fatty acids is apparent. This reduces the structural specificity of NIR spectroscopy towards long chain fatty acids. However, the theoretical spectra correctly reproduced the fine differences between these systems, improving one's ability discriminate between long chain fatty acids in their NIR region.

Beć and Grabska et al. also investigated quantum chemical calculations of carried out quantum chemical calculations of NIR spectra of various molecules including butanol, small and medium size fatty acids, nucleic acids, melamine, rosmarinic acid, caffeine, and thymol [7, 8, 36–38].

5 Conclusion and Perspective

This short review has reported the analysis methods of NIR spectra. Besides traditional analysis methods such as spectra analysis based on group frequencies, second derivatives, and chemometrics, various new techniques such as quantum chemical calculations have been introduced in this review.

As perspective, I point out several points. First, we need to develop further group frequency table of NIR spectroscopy. Recently, new group frequencies have been found, for example, bands due to the first and second overtones of C≡N stretching mode have been identified [39]. Combination bands containing amide groups have been studied in more detail [40]. Knowledge about band shifts due to the formation of hydrogen bonding have been accumulated [1]. Thus, group frequency table should be reestablished. Studies of new group frequencies may expand further research of NIR spectroscopy. In NIR spectroscopy difference spectra, 2D-COS, SMCR, and MCR-ALS have not often been used. One should consider more the use of these methods. At last, but not at least, anharmonic quantum chemical calculations should play a more important role in the analysis of NIR spectra in near future. The calculations of the intensities and frequencies of overtones and combinations will become easier due to the progress in algorisms, software, and computers.

References

1. Ozaki, Y., Huck, C.W., Tsuchikawa, S., Engelsen, S.B. (eds.): Near-Infrared Spectroscopy—Theory, Spectral Analysis, Instrumentation, and Applications. Springer, Singapore (2020). https://doi.org/10.1007/978-981-15-8648-4
2. Beć, K.B., Futami, Y., Wojcik, M.J., Ozaki, Y.: Phys. Chem. Chem. Phys. **18**, 13666 (2016)
3. Ozaki, Y., McClure, W.F., Christy, A.A.: Near-Infrared Spectroscopy in Food Science and Technology. Wiley-Interscience, New York (2007)

4. Ciurczak, E.W., Igne, B., Workman Jr., J., Burns, D.A.: Handbook of Near-Infrared Analysis, pp. 45–70. CRC Press, Boca Raton (2021)
5. Næs, T., Isaksson, T., Fearn, T., Davis, T.: A User-Friendly Guide to Multivariate Calibration and Classification. NIR Publications, Chichester (2002)
6. Martens, H., Martens, M.: Multivariate Analysis of Quality; An Introduction. Wiley, Chichester (2001)
7. Beć, K.B., Grabska, J., Huck, C.W., Ozaki, Y.: A Quantum Chemical Approach. In: Ozaki, Y., Wojcik, M.J., Popp, J. (eds.) Molecular Spectroscopy, vol. 2, pp. 353–388. Wiley-VCH, Weinheim (2019)
8. Czarnecki, M.A., Beć, K.B., Grabska, J., Hofer, T.S., Ozaki, Y.: In: Near-Infrared Spectroscopy Theory, Spectral Analysis, Instrumentation, and Applications. Ozaki, Y., Huck, W.C., Tsuchikawa, S., Engelsen S.B. (eds.), pp. 297–330. Springer, Singapore (2021). https://doi.org/10.1007/978-981-15-8648-4
9. Noda, I., Ozaki, Y.: Two-Dimensional Correlation Spectroscopy-Applications in Vibrational and Optical Spectroscopy. Wiley, Chichester (2004)
10. Workman, J., Jr., Weyer, L.: Practical Guide and Spectral Analysis for Interpretive Near-Infrared Spectroscopy, 2nd edn. CRC Press, New York (2012)
11. Katsumoto, Y., Adachi, D., Sato, H., Ozaki, Y.: J. Near Infrared Spectrosc. **10**, 85 (2002)
12. Šašić, S., Kita, T., Furukawa, T., Watari, M., Siesler, H.W.: Ozaki, Monitoring the melt-extrusion transesteriffication of ethlene/vinyl acetate copolymer by self-modeling curve resolution analysis of on-line near infrared spectra. Analyst **125**, 2315 (2000)
13. Noda, I., Liu, Y., Ozaki, Y., Czarnecki, C.W.: J. Phys. Chem. **95**, 3068 (1999)
14. Bokobza, L.: Spectroscopic Techniques for Polymer Characterization, Methods, Instrumentation, Application, pp. 75–105. Wiley-VCH, Weinheim (2021). Ozaki, Y., Sato, H. (eds.)
15. Bokobza, L., Buffeteau, T., Desbat, B.: Appl. Spectrosc. **54**, 360 (2000)
16. Czarnecki, M.A., Ozaki, Y.: Phys. Chem. Chem. Phys **1**, 797 (1999)
17. Liu, Y., Ozaki, Y., Noda, I.: J. Phys. Chem. **100**, 7326 (1996)
18. Ozaki, Y., Liu, Y., Noda, I.: Macromolecules **30**, 2391 (1997)
19. Noda, I., Story, G.M., Dowrcy, A.E., Reeder, R.C., Marcott, C.: Makromol. Chem. Macromol. Symp **119**, 1 (1997)
20. Ren, Y., Murakami, T., Nishioka, T., Nakashima, K., Noda, I., Ozaki, Y.: J. Phys. Chem. B **104**, 679 (2000)
21. Wu, Y., Czarnik-Matusewicz, B., Murayama, K., Ozaki, Y.: J. Phys. Chem. B **104**, 5840 (2000)
22. Maeda, H., Ozaki, Y., Noda, Y., Mimura, Y., Nakai, T., Tani, T.: J. Near Infrared Spectrosc. **3**, 43 (1995)
23. Segtnan, V.H., Šašić, S., Isaksson, T., Ozaki, Y.: Anal. Chem. **73**, 3153 (2001)
24. Šašić, S., Segtnan, V.H., Ozaki, Y.: J. Phys. Chem. A **106**, 760 (2002)
25. Jiang, J.H., Berry, R.J., Siesler, H.W., Ozaki, Y.: Anal. Chem. **74**, 3555 (2002)
26. Du, Y.P., Liang, Y.Z., Jiang, J.H., Berry, R.J., Ozaki, Y.: Anal. Chim. Acta **501**, 183 (2004)
27. Kasemsumran, S., Du, Y.P., Murayama, K., Huehne, M., Ozaki, Y.: Anal. Chim. Acta **512**, 223 (2004)
28. Shinzawa, H., Jiang, J.H., Ritthiruangdej, P., Ozaki, Y.: J. Chemometrics **20**, 436 (2006)
29. Shinzawa, H., Li, B., Nakagawa, T., Maruo, K., Ozaki, Y.: Appl. Spectrosc. **60**, 631 (2006)
30. Jiang, J.H., Tsenkova, R., Wu, Y., Yu, R.Q., Ozaki, Y.: Appl. Spectrosc. **56**, 488 (2002)
31. Šašić, S., Muszynski, A., Ozaki, Y.: J. Phys. Chem. A **104**, 6380–6388 (2000)
32. Šašić, S., Ozaki, Y.: Anal. Chem. A **73**, 2294 (2002)
33. Morita, S., Shinzawa, H., Noda, I., Ozaki, Y.: Appl. Spectrosc. **60**, 398 (2006)
34. Grabska, J., Beć, K.B., Ozaki, Y., Huck, C.W.: J. Phys. Chem. A **121**, 1950 (2017)

35. Grabska, J., Beć, K.B., Ishigaki, M., Huck, C.W., Ozaki, Y.: J. Phys. Chem. B **122**, 6931 (2018)
36. Grabska, J., Ishigaki, M., Beć, K.B., Wojcik, M.G., Ozaki, Y.: J. Phys. Chem. A **121**, 3437 (2017)
37. Beć, K.B., Grabska, J., Kichler, C.G., Huck, C.W.: J. Mol. Liq. **268**, 895 (2018)
38. Kichler, C.G., et al.: Analyst **142**, 455 (2017)
39. Beć, K.B., Karczmit, D., Kwasniewicz, M., Ozaki, Y., Czarnecki, M.A.: J. Phys. Chem. A **123**, 4431 (2019)
40. Ishigaki, M., et al.: Anal. Chem. **93**, 2758 (2021)

The Ever-Shrinking Spectrometer: New Technologies and Applications

Richard Crocombe(✉)

Crocombe Spectroscopic Consulting, 30 Thornberry Road, Winchester, MA, USA
racrocombe@gmail.com

Abstract. Spectrometers, especially those operating in the near-infrared and visible, are today so small and such low cost that they can be embedded in consumer goods or sold directly to the public. This paper outlines what is available today in portable NIR spectroscopy, and how these instruments can be categorized; emerging applications of miniature spectrometers, especially those in consumer goods using embedded spectrometers and marketed directly to consumers; and caveats on these direct-to-consumer instruments.

Keywords: Miniature · Portable · Embedded · Consumer · Applications

1 Introduction

Over the past twenty years we have seen near-infrared and Raman spectrometers shrink in size from 2 ft (0.75 m) square beige boxes to devices a little bigger than a pack of playing cards. These spectrometers can have very respectable performance, and can be completely self-contained with battery, display, user controls, sample interface, databases and calibrations. This combination critically enables their use in the field, and the sample to be taken to the spectrometer, changing the game and significantly improving efficiency of the process in question. Portable optical spectroscopy was reviewed in 2018 [1], and the field of portable spectroscopy as a whole is now so large that it has taken a two-volume book to review its instrumentation and applications [2, 3]. In some cases these spectrometers, especially those operating in the near-infrared and visible, are so small and such low cost that they can be embedded in consumer goods or sold directly to the public. However, these possibilities raise their own issues, which are explored in this paper.

This paper is therefore divided into three major parts: what is available today in portable NIR spectroscopy, and how can we categorize these instruments; emerging applications of miniature spectrometers, especially those in consumer goods using embedded spectrometers; and caveats on direct-to-consumer instruments.

X. Chu et al. (Eds.): ICNIR 2021, *Sense the Real Change: Proceedings of the 20th International Conference on Near Infrared Spectroscopy*, pp. 17–31, 2022.
https://doi.org/10.1007/978-981-19-4884-8_2

2 Today's Portable Near-Infrared Spectrometers

2.1 Technologies

In the near-infrared, a wide variety of technologies are available, and have been applied, for portable instruments. Commercially available portable spectrometers were recently reviewed by Beć et al. [4] Fig. 1, and research on 'photonic', 'plasmonic' and 'computational' devices was reviewed by Yang et al. [5] Fig. 2. In addition to the devices covered in those reviews, instruments based on discrete filtered detectors, patterned filters on 2D sensors, and thin film organic semiconductors are becoming available [6]. Further miniaturization may also be possible via metamaterials [7] and photonic integrated circuits [8–10], with developments to extend operation into the mid-infrared [11]. The lowest cost devices, available for a few dollars each [12], appear to be multispectral sensors (as opposed to full spectrometers), with a small number (e.g., 6–16) of discrete detectors each having a different thin-film bandpass filter, or implemented via organic semiconductors [13]. This raises the prospect of multi-spectral sensors being incorporated into the next-generation smartphones. We may also see analytical spectroscopy in this region benefit from advances in the LIDAR and optical coherence tomography (OCT) fields, especially if photonic integrated circuit technology is applied in the automotive sector. It should be noted that miniaturization of hyperspectral instruments is similarly advanced; coverage of them is beyond the scope of this paper, but is briefly covered in elsewhere [14]. Cameras in smartphones are essentially RGB sensors [15] which can be converted into a visible-region spectrometer with ancillary optics [16] and applied to colorimetric, fluorescence [17], biomedical and clinical analyses [18, 19]. The underlying sensors, being based on silicon, are sensitive to wavelengths as long as 1000nm, but a filter is employed to block the ~700–1000 nm region. It should be noted that the optical resolution for condensed phase *molecular* spectroscopy in this region are quite modest; typically several nm. However, *elemental* spectrometers (e.g., laser-induced breakdown spectroscopy or LIBS) can require a resolution of up to 0.1 nm combined with wide wavelength coverage, and therefore designs of portable LIBS instruments feature either multiple spectrometers, or employ an echelle-based design.

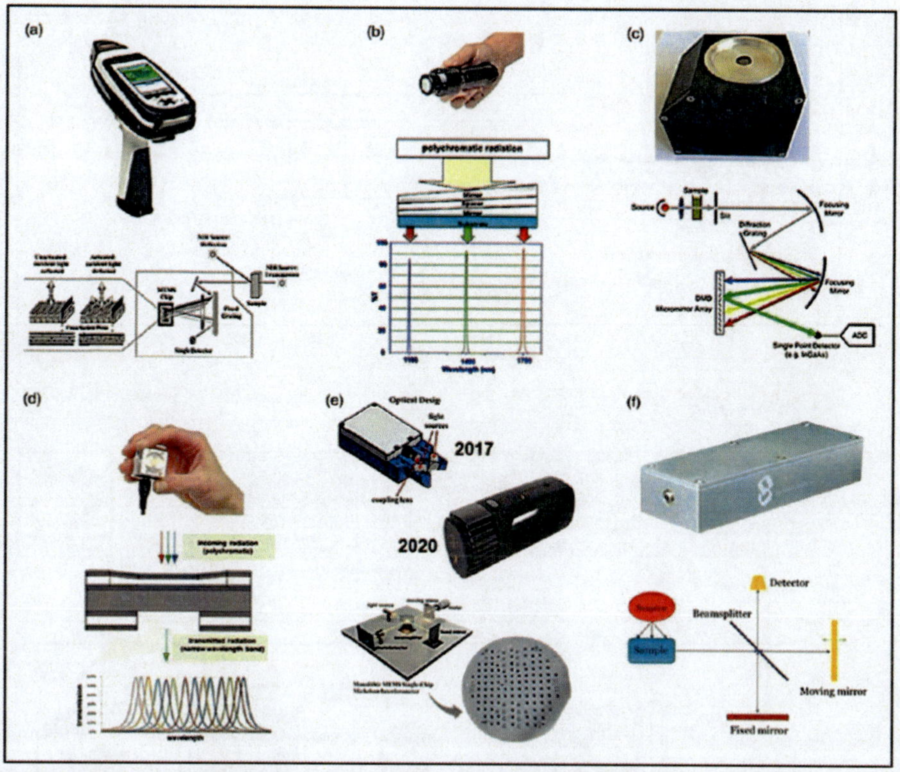

Fig. 1. Principles of wavelengths selectors built into different handheld NIR spectrometers: (a) MEMS Hadamard mask – microPHAZIR, Thermo Fisher Scientific, Waltham, USA; (b) LVF – MicroNIR Pro ES 1700, VIAVI, Santa Rosa, USA; (c) MEMS DMD – implementation of DLP NIR scan module, Texas Instruments, Dallas, USA; (d) MEMS Fabry–Perot interferometer – NIRONE Sensor S, Spectral Engines, Helsinki, Finland; (e) MEMS Michelson interferometer – NeoSpectra, Si-Ware, Cairo, Egypt; (f) MEMS Michelson interferometer with a large mirror – nanoFTIR NIR, SouthNest Technology, Hefei, China. ADC: analog-to-digital converter; InGaAs: indium–gallium–arsenide; MEMS: micro-electro-mechanical system. [Reproduced by permission from SAGE: Krzysztof B Beć, Justyna Grabska, Heinz W Siesler, Christian W Huck, "Handheld near-infrared spectrometers: Where are we heading?", NIR News, https://doi.org/10.1177/0960336020916815, Vol. 31(3–4), 28–35 (June 1, 2020). Published by SAGE. https://journals.sagepub.com/doi/full/10.1177/0960336020916815].

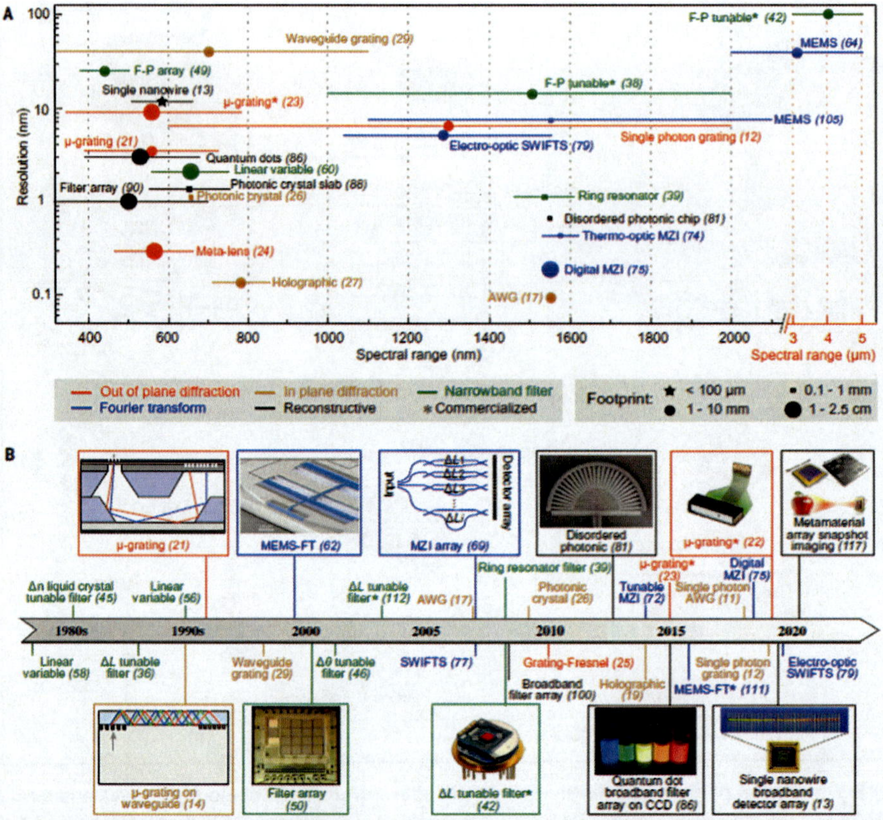

Fig. 2. The field of miniaturized spectroscopic devices. (A) Plot comparing the resolution, operational spectral range, and footprint for selected device demonstrations in the literature and those that are commercially available (indicated by asterisks), as categorized into their respective subfields (see color key). Footprint encompasses those elements of the device that are active in resolving and detecting light, and does not include accessory components such as the readout electronics or packaging. (B) Timeline illustrating the emergence of different technological platforms for microspectrometer systems from the 1980s to the present day, sorted by subfield as displayed in the color key in (A). [Reproduced, with permission from American Association for the Advancement of Science, from Z. Yang, et al., "Miniaturization of Optical Spectrometers", Science 371, eabe0722, (2021). © American Association for the Advancement of Science]

3 Miniature Spectrometers in the "VNIR" Region

The region from 700–1000 nm has been dubbed "VNIR", and there it is possible to detect the 3rd and 4th overtone vibrational bands from molecules with C-H, N-H and O-H bonds, and also some chromophores (i.e., electronic transitions). These higher overtone bands are weak, but that also implies a large penetration depth into the sample, and this region has been used for many years for the quantitative analysis of grain [20].

This region is characterized by the availability of bright sources (e.g., miniature tungsten-halogen light bulbs and broad-band light emitting diodes [21]) with sensitive array detectors, and commercial miniature instrument designs are therefore dominated by simple grating-based spectrographs. The facile access to this region has attracted a lot of commercial attention, and a large number of products are available. Because the spectrographs can be very small, more than one spectrometer can be built into a 'mouse-sized' device, covering an extended range from 400–1700 nm, with a smartphone used as the data system [22].

4 Miniature Spectrometers in the Region Beyond 1000 nm

At wavelengths longer than 1000 nm, indium-gallium-arsenide (InGaAs) detectors are commonly employed, with 'extended' InGaAs detectors operating to around 2500 nm. In addition, a new generation of lead salt detectors are also available and used in portable spectrometers [23]. InGaAs detectors are significantly more expensive than those based on silicon, and for best operation they also require cooling. Therefore, miniature and portable spectrometers operating in this region are significantly more expensive than those using silicon detectors, and this places them out of the range of consumer products at the present time.

5 Raman Spectroscopy

In the realm of miniature and portable molecular spectroscopy, Raman devices are the main competition for near-infrared spectrometers, and it's therefore worth briefly examining that competition. A remarkable trend in portable Raman instruments is that their size has diminished, while the performance (signal-to-noise ratio – SNR - for the same collection time and resolution) has increased. The first generation portable Raman instruments were based on a reflective Čzerny-Turner design with fiber-coupled components. Second generation instruments eliminated fiber coupling, are significantly smaller, and are more tightly integrated, using free-space optical coupling. These instruments improved the SNR (with the same range and resolution) by about a factor of 5 over their predecessors. Now, there are even smaller instruments, a little larger than a pack of playing cards, with yet improved SNR, possibly as much as a factor of 10, made possible via transmission grating designs [24]. This leads to the possibility of even smaller instruments in the future – today, a Raman instrument about $1'' \times 1'' \times 1''$ (a 2.5 cm cube) in size is possible.

The development of portable Raman instruments, with 785 nm excitation, started competition between Near-IR and Raman spectroscopy in the material identification sector, and in the chemical/pharmaceutical QA/QC area in particular (raw material identification – RMID). Here, Raman has been increasingly used due to a combination of lower sensitivity to particle size and moisture, and more specific spectroscopic information (fundamentals in Raman spectroscopy vs. overtones and combinations in the near-IR). For similar reasons, portable Raman instruments also dominate in the hazardous material and narcotics identification applications, despite the possibility of fluorescence interference. An intriguing possibility is a combined portable NIR-Raman instrument, which is certainly within the current state-of-the-art. A miniature, low cost, VNIR spectrometer, or an NIR multispectral sensor, could be added to a portable Raman instrument in a straightforward manner. In that way an RMID instrument could not only identify the material, but could also take advantage of near-infrared's sensitivity to particle size and moisture, which could be critical-to-quality attributes in a pharmaceutical manufacturing process.

Classical comparisons between mid-infrared, near-infrared and Raman spectroscopies are given in many textbooks, but Table 1 looks at this from the point of view of portable instruments and their sampling techniques.

Table 1. Comparing the attributes of portable mid-infrared, Near-Infrared and Raman spectrometers.

	Mid-IR	Near-IR	Raman
Sampling	Historically complicated, now much easier using diamond ATR for solids and liquids	Very straightforward – covering a quartz window (lab), or point-and-shoot (portable); non-invasive	Very straightforward – point-and-shoot; non-invasive
Sampling	Cannot interrogate through glass or transparent plastic	Can interrogate through glass (& quartz) containers, and transparent plastics	Can interrogate through glass (& quartz) containers, and transparent plastics
Sampling	Fiber optic probes difficult to use and fragile	Fiber optic probes easy to use	Fiber optic probes easy to use
Observed bands	Fundamentals - narrow	Combinations and overtones – broad, overlapping	Fundamentals – narrow
Chemical specificity	High	Medium	High
Sensitive to moisture/water	High	Medium	Low
Sensitive to particle size	Low with ATR	High – can be corrected (MSC, etc.)	Low
Problem samples?	Inorganics	Inorganics	Dark, colored and fluorescent materials
Sample area interrogated	Typically ~1–2 mm diameter by ATR	Depends on sampling arrangement, but can be several cm in diameter	Typically 300 microns to 1mm in diameter Some systems raster the beam. Larger area probes available

(*continued*)

Table 1. (*continued*)

	Mid-IR	Near-IR	Raman
Sampling depth	A few microns by ATR	Wavelength dependent, but can vary from <1 mm to several mm at short wavelengths	A confocal surface technique, but spatially offset Raman spectroscopy (SORS) can probe through translucent barrier materials, and into bulk material
Suitable for pure materials	Yes	Yes	Yes
Suitable for heterogeneous samples	No	Yes	With a larger sampling area or by 'transmission Raman'

6 Portable Spectrometer Taxonomy

In order to understand applications development and the challenges for miniature optical spectrometers, it is useful to categorize the instruments and their typical applications. This is presented in Table 2, and two areas that have emerged recently, enabled by very low-cost devices, are spectrometers embedded in consumer goods and those marketed directly to consumers. The remainder of this paper will examine these emerging fields.

Table 2. A taxonomy for portable spectrometers. The emerging classes are those embedded in consumer goods and those marketed directly to consumers.

Category	Maturity	Applications development	Approx. cost
Completely integrated instruments, often from major analytical instrument companies	Well-established, and also includes portable Raman instruments used by hazardous material technicians and portable XRF instruments used by scrap metal dealers	Well-defined and controlled by the instrument company	>$20,000

(*continued*)

Table 2. *(continued)*

Category	Maturity	Applications development	Approx. cost
Miniature spectrometers and spectral sensors embedded in consumer goods	An emerging field in 'white goods' (washing machines, refrigerators, vacuum cleaners) and personal care & fitness products	Well-defined, and often single purpose, controlled by the consumer goods manufacturer	A few $ to $1,000
Wearable or implantable spectrometers and sensors for biomedical applications	An emerging field	Done by medical device companies. Note that this is a highly regulated area	?
Stand-alone spectrometers or spectrographs for scientific use	Very familiar area. Spectrographs and components with well-defined specifications sold for lab use by established companies	Produce spectra used by researchers. Can be integrated into complete instruments	~$1,000
Spectrometers marketed directly to consumers	An emerging area, with a number of start-up companies active	More problematic – who develops and validates those applications, and does the end user understand the limitations of the instrument and of the calibration?	<$1,000
Miniature process spectrometers	Emerging	No-moving-parts and MEMS-based spectrometers can be integrated into small spaces and high vibration environments	Depends on the level of integration

7 Emerging and Future Applications

With the development of the low-cost spectrometers and multispectral sensors the number of possible applications has grown rapidly: applications drive the development of instrumentation, and in turn, instrumentation drives the development of applications. Internal R&D typically develops a 'platform', maybe designed for a single applications,

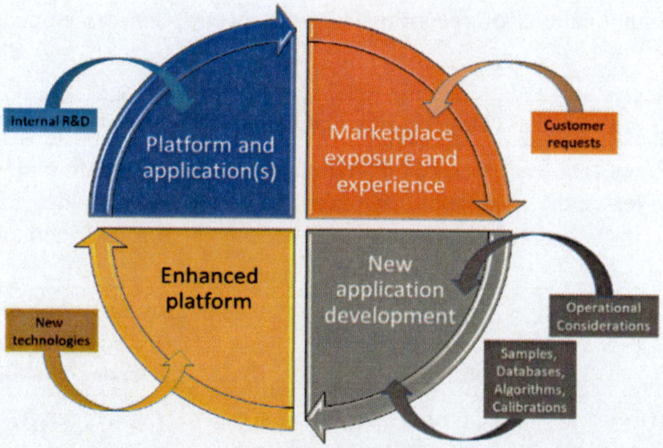

Fig. 3. The cyclic nature of portable instrumentation and application development. Reproduced, with permission from John Wiley, from Ref [25]. © 2021 John Wiley & Sons, Ltd.

but once that reaches the market applications requests emerge and can drive development of future products [25]. This is illustrated in Fig. 3.

For the completely integrated instruments two growth areas are the identification of street narcotics, especially fentanyl and its derivatives, and detection of counterfeit pharmaceuticals. Table 3 summarizes possible applications of spectrometers and spectral sensors embedded in consumer goods, and Table 4 possible applications of 'personal spectrometers', marketed directly to the general public. Biomedical applications using wearable or implantable spectrometers and spectral sensors is an early field and progress to commercial products may be slow as this is a highly regulated area. Nonetheless, there is considerable ongoing research in this area [26]. The 'holy grail' here is a spectroscopic non-invasive blood glucose monitor, but history has shown that this is an exceedingly difficult application [27].

8 Embedded Applications

Apart from an altruistic desire to improve products, and a move towards the 'smart home', there are two possible other motivations for large companies to incorporate spectroscopic devices into consumer products: a 'halo' effect, enhancing the image of a product; and the 'razorblade' strategy of selling hardware at a low cost, and making money on personal care consumables. The latter strategy is used, for instance, with inkjet printers. In the personal care area two products have recently been released, a smart cosmetics applicator (Opté [28]) and a personalized hair care system (Henkel/Schwartzkopf [29]). Other products have been described on the web, but not formally released as products (e.g., Bosch Home Connect X-Spect [30]). In what might be a surprising area, the Japanese company, Toto, has explored the concept of a 'smart toilet' [31], a recent research paper studied personalized health monitoring via the analysis of excreta [32], and smart sensors have been proposed for monitoring sanitation infrastructure [33]. Current generation sports-watches can monitor hydration, and incorporate heartrate and blood oxygenation monitors (i.e., pulse oximetry).

Table 3. Possible applications of spectrometers and spectral sensors embedded in consumer goods.

Application area	Desired information	Comments
White goods (refrigerators, ovens, washers, dryers, etc.)	Optimal operation Use of consumer preferences Spoilage alerts	E.g., stain removal E.g., cooking times E.g., refrigerator
Kitchen appliances	Optimal operation Use of consumer preferences	Smaller appliances: toasters, rice cookers, slow cookers, fryers
Other appliances (robot vacuum cleaners)	Optimal operation – adjustment to different surfaces	E.g., carpet vs. wood vs. tile
Indoor air quality	Carbon monoxide, methane, formaldehyde, etc	Note that radioactive radon is a problem in some locations, but will not be detected by molecular spectroscopy
Fitness	Hydration Oxygenation	Smartwatch and sportswatch capability
Personal care	Which cosmetics to use Smart cosmetic application Which hair products to use UV protection sufficiency	Disguising imperfections Optimal treatment Sunburn warning
Personal health	Brushing guide	Smart toothbrush
Smart toilets	Analysis of urine and stool	Detection of blood and other indicators

9 Consumer Applications

As component costs are reduced to the level of a few dollars, consumer-based drivers influence the development of portable spectroscopic instruments and their applications. In addition to the normal commercial drivers noted above, we now see the direct influence of consumer desires and the potential to enhance consumer products (Fig. 4).

The possible application areas for spectroscopic devices for consumers are summarized in Table 4, and it should be noted that each of these would need its own validated calibration and spectral database.

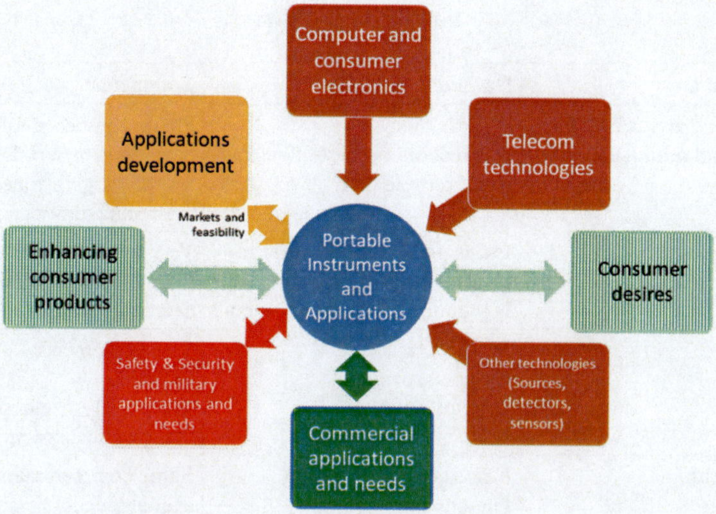

Fig. 4. As component costs are reduced, consumer-based drivers influence the development of portable spectroscopic instruments and their applications. Reproduced, with permission from John Wiley, from Ref [34]. © 2021 John Wiley & Sons, Ltd.

Table 4. Possible applications of 'personal spectrometers', marketed directly to the general public [34]. [Adapted from Ref. [34], with permission from John Wiley. © 2021 John Wiley & Sons, Ltd.]

Application area	Desired information	Comments
Food nutrition	What is it? Amount of fat, protein, carbohydrate? Calorie content? Is it fresh?	Heterogeneous samples!
Food fraud	Is it genuine?	Probably possible by optical spectroscopy (NIR or Raman) for manufactured products, but note that speciation for meat or fish will typically require DNA-type analyses
Food adulteration	Is it safe? Is it contaminated?	Note that desired detection levels for pathogens, allergens, pesticide and herbicide residues, etc. are orders of magnitude below that achievable by optical spectroscopy

(*continued*)

Table 4. (*continued*)

Application area	Desired information	Comments
Identification and verification of household products	What is this? Is it real or counterfeit? Can I recycle this?	Pharmaceuticals, fabrics, hazardous materials, recycling, gemstones, other high-value products, etc
Environmental	Is this good for me and/or my children? Is it good for the environment?	Very wide range of materials to be identified, and trace detection desirable
Cannabis	Potency Hemp vs. cannabis CBD oil analysis	Some commercial instruments available Trace analyses not possible using optical spectroscopy
Color matching	Names of paint colors, Pantone numbers	Some commercial instruments available

10 The Reality of Analytical Spectroscopy for Direct-to-Consumer Products

There is a mismatch between direct-to-consumer aspirations and the reality of analytical spectroscopy, especially for low-cost instruments marketed to and operated by non-scientists. This may seem to be an obvious statement to make, but some devices marketed to the general public seem to have been developed by people unfamiliar with fundamental concepts of analytical chemistry like surface vs. bulk sampling and analysis, homogeneous vs. heterogeneous samples, and limits of detection, etc. For instance, some trace detection requirements, which a consumer could desire, can only be met by sophisticated laboratory instruments like ICP-MS and GC/MS/MS.

A critical point is how the sample is presented to the spectrometer by a non-scientist operator. Sampling issues in solid phase NIR using laboratory instruments were addressed a long time ago with the use of large spot sizes, sample rotators and integrating spheres, etc. It's interesting to note that a commercial handheld NIR spectrometer, designed for specifically for quantitative grain analysis, does incorporate an integrating sphere [35], and a recent paper discusses this issue and describes an approach for portable instruments [36].

But a portable or handheld spectrometer, marketed to the general public at a low price, is likely to have a small spot size, and very unlikely to incorporate a sample rotator or integrating sphere. These portable instruments, if used improperly, could generate spectra contaminated with considerable amounts of stray light, leading to, at a minimum, problems in quantitative analysis [37, 38]. Would instruments marketed directly to consumers warn the operator about poor quality spectra? Would the customer expect foolproof 'point-and-shoot' operation? It is clear that what is needed is a front-end discriminator that will warn the operator if they have not collected a good spectrum, as well as a second step to confirm that the spectrum falls within the operating range (sample space) of the chosen analysis. The operator (consumer) will also have to be educated regarding heterogeneous samples, and what portion (area, volume) of the sample is actually being interrogated.

11 Applications Development

For a spectrometer to be useful to a non-scientist, and especially a general consumer, spectra must be processed to give meaningful results. Qualitative analysis will involve a spectral database or 'library', along with algorithms that give confident results, while quantitative methods require a validated calibration, with concentration data in the calibration samples obtained from a reference chemical method. The question for instruments marketed directly to consumers is who will develop the applications?

Years of experience in the near-infrared community have shown the need for careful and validated calibrations, to eliminate the possibility of accidental correlations when limited sample sets are used, mostly on one type of sample. But, some of these new vendors, targeting general consumers, seem to be relying on 'crowd-sourced' data, with the dual questions of the quality of the spectroscopic measurement, and the accuracy of the metadata (identification, or quantitative information). The lowest cost instruments, utilize the silicon detector region from ~400–1050 nm, and will there be appropriate information in this spectral region? You can obviously match colors, and you can certainly examine classes of compounds like lycopenes and carotenoids in fruits and vegetables, but the (more expensive) 1000–1700 nm region carries much more specific molecular information.

With some vendors, there seems to be a lack of understanding about the possible detection limits of optical spectroscopy in the condensed phase, with claims to detect trace amounts – e.g., allergens, pathogens, pesticide and herbicide residues. It's generally accepted that in condensed phase near-infrared spectroscopy the limit of detection is in the percent range, and a recent paper examining adulteration of milk powder determined a 5% level for the reliable detection and identification of the adulterants using portable near-infrared instruments [39]. Finally, there is the question of what portion of the sample is being interrogated: just the surface, what portion of the surface, or the bulk?

With all the issues noted above, and the lack (most likely) of rigorous reference analyses, the potential for 'garbage-in/garbage out' analysis is extremely high. This list of caveats can go on and on, but suffice it to say this is an area of concern, with the potential to mislead the general public, even with fatal consequences.

12 Conclusions

Good quality near-infrared spectrometer hardware and portable instruments are now widely available, and there are numerous applications of these instruments described in the refereed literature. Given the low cost of some spectrometers and spectral sensors, we will see spectrometers embedded into consumer products, and the very low cost of hardware in the "VNIR" region enables spectrometers that can be marketed directly to the public, and are reasonably affordable. But application development for these devices remains the major question: is the hardware ahead of application development? Do the developers of these devices really understand what is required to develop validated and rugged analyses (qualitative or quantitative) that will not give misleading information? Could problems with analyses lead consumers to believe that 'spectroscopy doesn't work' or is 'unreliable'? Are many consumers' application desires impractical?

Acknowledgements. This paper was presented at the ICNIRS Conference in October 2021. The author thanks Ellen Miseo, Pauline Leary and Brooke Kammrath for helpful discussions.

References

1. Crocombe, R.A.: Portable spectroscopy. Appl. Spectrosc. **72**, 1701–1751 (2018)
2. Crocombe, R.A., Leary, P.E., Kammrath, B.W. (eds.) Portable Spectroscopy and Spectrometry. Technologies and Instrumentation, vol. 1. Wiley, Chichester and Hoboken (2021)
3. Crocombe, R.A., Leary, P.E., Kammrath B.W (eds.): Portable Spectroscopy and Spectrometry. Applications, vol. 2. Wiley, Chichester and Hoboken (2020). ISBN 978-1-119-63642-7
4. Beć, K.B., Grabska, J., Siesler, H.W., Huck, C.W.: Handheld near-infrared spectrometers: where are we heading? NIR News **31**, 28–35 (2020)
5. Yang, Z., et al.: Miniaturization of optical spectrometers. Science **371**, eabe0722 (2021)
6. Zobenica, Z., van der Heijden, R.W., Petruzzella, M., et al.: Integrated nano-opto-electromechanical sensor for spectrometry and nanometrology. Nat. Commun. **8**, 2216 (2017)
7. Wei, J., Ren, Z., Lee, C.: Metamaterial technologies for miniaturized infrared spectroscopy: light sources, sensors, filters, detectors, and integration. J. Appl. Phys. **128**, 240901 (2020)
8. Chandrasekar, R., Lapin, Z.J., Nichols, A., Braun, R., Fountain, A.W., III.: Photonic integrated circuits for department of defense-relevant chemical and biological sensing applications: state-of-the-art and future outlooks. Opt. Eng. **58**(2), 020901 (2018)
9. Blumenthal, D.J.: Photonic integration for UV to IR applications. APL Photon. **5**, 020903 (2020)
10. Chang, L., Cole, G.D., Moody, G., Bowers, J.E.: CSOI: beyond silicon-on-insulator photonics. Opt. Photonics News **33**, 34–41 (2022)
11. See, for instance, Rockley Photonics. https://rockleyphotonics.com/platform/. Accessed 31 Dec 2021
12. See, for instance: AMS-OSRAM. https://ams.com/en/spectral-sensing. Accessed 31 Dec 2021
13. Senorics. https://www.senorics.com/. Accessed 1 June 2022
14. Crocombe, R.A.: The future of portable spectroscopy. In: Crocombe, R.A., Leary, P.E., Kammrath, B.W. (eds.) Portable Spectroscopy and Spectrometry. Applications, vol. 2. Wiley, Chichester and Hoboken, April 2020. ISBN 978-1-119-63642-7
15. Burggraaff, O., et al.: Standardized spectral and radiometric calibration of consumer cameras. Opt. Express **27**(14), 19075–19101 (2019)
16. Scheeline, A., Smartphone technology – instrumentation and applications. In: Crocombe, R.A., Leary, P.E., Kammrath, B.W. (eds.) Portable Spectroscopy and Spectrometry. Technologies and Instrumentation, vol. 1. Wiley, Chichester and Hoboken, April 2021
17. Oskolok, K.V., Monogarova, O.V., Garmay, A.V.: Molecular optical analyzers based on smartphones for high school and universities. J. Chem. Edu. **98**(6), 1937–1945 (2021)
18. Hussain, I., Bowden, A.K.: Smartphone-based optical spectroscopic platforms for biomedical applications: a review. Biomed. Opti. Expr. **12**, 1974–1998 (2021)

19. Peveler, W.J., Algar, W.R.: Toward clinical applications of smartphone spectroscopy and imaging. In: Crocombe, R.A., Leary, P.E., Kammrath, B.W. (eds.) Portable Spectroscopy and Spectrometry. Applications, vol. 2. Wiley, Chichester and Hoboken, April 2020. ISBN: 978-1-119-63642-7
20. Williams, P.: Near Infrared Technology: Getting The Best Out Of Light. Sun Press (2019)
21. See, for instance, OSRAM. https://www.osram.com/os/applications/mobile-competence/mobile-competence-spectroscopy.jsp. Accessed 31 Dec 2021
22. See, for instance, BaySpec Breeze. https://www.bayspec.com/spectroscopy/breeze-smart-palm-spectrometer/
23. Trinamix. https://trinamixsensing.com/ir-detectors/. Accessed 31 Dec 2021
24. Rathmell, C., Bingemann, D., Zieg, M., Creasey, D.: Portable Raman spectroscopy: instrumentation and technology. In: Crocombe, R.A., Leary, P.E., Kammrath, B.W. (eds.) Portable Spectroscopy and Spectrometry. Technologies and Instrumentation, vol. 1. Wiley, Chichester and Hoboken, April 2021
25. Crocombe, R.A., Leary, P.E., Kammrath, B.W.: The Role of Applications in Portable Spectroscopy. In: Crocombe, R.A., Leary, P.E., Kammrath, B.W. (eds.) Portable Spectroscopy and Spectrometry. Applications, vol. 2. Wiley, Chichester and Hoboken, April 2020. ISBN 978-1-119-63642-7
26. Ramella-Roman, J., Gandjbakhche, A.H., Kanick, S.C., Shadgan, B., Tromberg, B.J.: Wearable, implantable, mobile, and remote biomedical optics and photonics. J. Biomed. Opt. **26**(6), 062701-1–062701-4 (2021)
27. Smith, J.L.: The pursuit of non-invasive blood glucose: hunting the deceitful turkey, 7th edn. https://www.nivglucose.com/The%20Pursuit%20of%20Noninvasive%20Glucose%207th%20Edition.pdf. Accessed 3 Jan 2022
28. Opté. https://opte.com/. Accessed 1/3/2022
29. SalonLab Smart Analyzer. https://www.salonlab-server.de/en-GB/. Accessed 3 Jan 2022
30. Bosch Home Connect. https://www.home-connect.com/mt/en/connected-household/x-spect#/Togglebox=5030115-5030583-1/. Accessed 3 Jan 2022
31. See, for instance. https://asia.nikkei.com/Business/Health-Care/How-s-your-health-Toto-s-smart-toilet-will-keep-tabs-for-you. Accessed 3 Jan 2022
32. Park, S.-M., Gambhir, S.S., et al.: A mountable toilet system for personalized health monitoring via the analysis of excreta. Nat. Biomed. Eng. **4**, 624–635 (2020)
33. Rary, E., Anderson, S.M., Philbrick, B.D., Suresh, T., Burton, J.: Smart sanitation—biosensors as a public health tool in sanitation infrastructure. Int. J. Environ. Res. Public Health **17**, 5146 (2020)
34. Crocombe, R.A.: The future of portable spectroscopy. In: Crocombe, R.A., Leary, P.E., Kammrath, B.W. (eds.) Portable Spectroscopy and Spectrometry. Applications, vol. 2. Wiley, Chichester and Hoboken, April 2020. ISBN: 978-1-119-63642-7
35. Grainsense. https://grainsense.com/. Accessed 3 Jan 2022
36. Pasquini, C., Hespanhol, M.C.: A rotational-linear sample probing device to improve the performance of compact near-infrared spectrophotometers. Microchem. J. **170**, 106747 (2021)
37. Cook, R.B., Jankow, R.: Effects of stray light in spectroscopy. J. Chem. Ed. **49**, 405–408 (1972)
38. Slavin, W.: Stray light in ultaviolet, visible and near-infrared spectrophotometry. Anal. Chem. **35**, 5611–6566 (1963)
39. Delatour, T., et al.: Handheld spectral sensing devices should not mislead consumers as far as non-authentic food is concerned: a case study with adulteration of milk powder. Foods **11**, 75 (2020)

The New Avenue – Theoretical Simulation of NIR Spectra and its Potential in Analytical Applications

Krzysztof B. Bec(✉), Justyna Grabska, and Christian W. Huck

Institute of Analytical Chemistry and Radiochemistry, University of Innsbruck, Innrain 80-82, 6020 Innsbruck, Austria
Krzysztof.Bec@uibk.ac.at

Abstract. Quantum mechanical calculations are routinely used as a major support in mid-infrared (MIR) and Raman spectroscopy. In contrast, practical limitations for long time formed a barrier to developing a similar synergy between near-infrared (NIR) spectroscopy with computational chemistry. Recent advances opened the pathway to modeling NIR spectra of various molecules reaching the size of long-chain fatty acids. Theoretical NIR spectra unveil staggering number of overlapping vibrational bands that create complex spectral patterns in many cases hindering our comprehension of NIR spectra but also revealing rich information in NIR spectra.

This review article summarizes the most recent accomplishments in the emerging field. Simulation studies of NIR spectra of a variety of compounds are discussed, with particular attention given to those that are significant from the point of view of physical and analytical chemistry. The examples range from basic molecules (alcohols, nitriles, carboxylic acids) through complex molecules with importance to biophysical science (fatty acids, nucleobases) and analytical chemistry (natural drugs, polyphenols, food adulterants) to organic polymers including a set of standard materials used in various industries. Innovation into analytical routines includes e.g. interpreting the chemical information in the performance profiles of miniaturized spectrometers, or explaining the matrix effects in NIR spectra. These advances lead to intelligent design of the analysis with proper comprehension of chemical information. This review article is based on the topic of an invited oral talk presented at NIR-2021 Conference that took place on 18th–21st October 2021 as an online event organized and coordinated from Beijing.

Keywords: *Ab initio* spectra simulation · Spectra interpretation · Overtones · Combination bands · Anharmonic methods

1 Introduction

During the last four decades, near-infrared (NIR) spectroscopy (800–2500 nm; 12,500–4000 cm^{-1}) has become one of the most attractive and used methods for analysis for the following reasons: it represents a non-destructive analytical tool allowing a fast and

X. Chu et al. (Eds.): ICNIR 2021, *Sense the Real Change: Proceedings of the 20th International Conference on Near Infrared Spectroscopy*, pp. 32–46, 2022.
https://doi.org/10.1007/978-981-19-4884-8_3

simultaneous qualitative and quantitative characterization of a wide variety of samples [1, 2]. In many aspects it is the most rapidly advancing analytical technique nowadays, gaining recognition in research laboratories and in various practical applications. Mainly known as an analytical tool useful for quantitative analysis, NIR spectroscopy is essential in various other fields, e.g., NIR imaging techniques in biophotonics, medical applications, pharmaceutical industry based on natural products, or in qualitative analysis and material characterization [3–5]. Its contribution to basic science and physical chemistry should be noted as well, e.g., in exploration of the nature of molecular vibrations or intermolecular interactions [6, 7].

With the advances in modern sensor technology and miniaturization, airborne (i.e. drone-mounted) spectrometers, rapid development of sophisticated data-analytics such as multi-block data fusion, deep learning, calibration transfer strategies and continuously refined more conventional approaches to chemometrics, even as a matured technique the contemporary NIR spectroscopy enjoys stimulating progress in multiple directions [1–12]. Regardless of the application, the intrinsic complexity of NIR spectra and the difficulty in their direct interpretation remains a limiting factor [14, 15]. The broad absorption lineshapes characteristic for NIR spectra underneath contain a high number of overlapped bands; often, a given functional group gives rise to multiple bands appearing in broader wavenumber region. The resulting lack of clear 'fingerprint' of chemical groups in many cases hinders a direct interpretation of NIR spectra [14].

Throughout its history of advancement, vibrational spectroscopy much benefitted from theoretical and computational chemistry; a brief sketch of this pathway is available in a recent perspective article [8]. Very good understanding of the fundamental bands that appear in mid-infrared (MIR) or Raman spectra is to significant extent achieved by the readily available quantum mechanical calculations, which have routinely been used in vibrational spectroscopy studies [13]. This was permitted by the availability of efficient double harmonic approximation, which enables predicting fundamental bands with often sufficient accuracy to provide rough interpretation of MIR or Raman spectra [8]. In contrast, NIR spectroscopy has not developed a similar synergy with computational chemistry because of a major limitation. Overtones and combination bands are not available in harmonic calculations, making it necessary to employ decisively more complex, in computational sense, anharmonic approaches to predict NIR bands [14–16].

Recent few years have witnessed considerable advances in the applicability of the methods of theoretical chemistry to NIR spectroscopy [8, 14–16]. Anharmonic calculations provide very accurate prediction and detailed interpretation of the spectra of basic molecules, e.g. alcohols [17, 18] phenol [19], or nitriles [20] and much improve one's understanding of the fine lineshape features appearing in NIR spectra. However, particular attention should be given to simulation of NIR spectra of practically meaningful molecules [15], ranging from vitamins [21], terpenes [22], alkaloids [23], nucleobases [24], through simple carboxylic acids [25, 26] to long-chain fatty acids [27] and polymers [28].

Accurate simulation of NIR spectra of reasonably large molecules decisively improves our comprehension of NIR spectra of complex materials. However, the information available from in silico modelling of spectra offer many opportunities to make a step beyond just the band interpretation [29]. Noteworthy examples include predictions of the effects of intermolecular interactions, including hydrogen bonding, but also general effects of chemical neighborhood constituting to important in analytical spectroscopy 'matrix effects'. Insights into the cooperative character of solute-hydration shell interaction deliver mechanistic understanding of the important effects occurring in aqueous solution firstly observed and described in analytical NIR study [30]. Finally, this detailed knowledge about how and at which wavenumbers a specific chemical information is expressed in NIR spectra permits interpreting instrumental differences between handheld and miniaturized spectrometers, which most often operate in selected, narrow wavenumber regions and do not possess the ability to measure the entire NIR spectrum [31, 32]. Thus, it becomes possible to understand how the availability of the chemical information influences the performance profile of miniaturized NIR spectrometers, and further anticipate the performance of a given sensor in a given analytical scenario [33].

This review article overviews the accomplishments made in the emerging area of in silico NIR spectroscopy. The interest of NIR community in the development and outcomes of this new avenue in NIR spectroscopy seems to be confirmed by the warm receipt of NIR-2021 lectures devoted to this direction of research, as well as a dedicated workshop presenting the theoretical and technical background also offered at the conference [34].

2 Applications

2.1 Interpretation of NIR Spectra

Physical backgrounds of NIR absorption spectra features key differences vs. fundamental (i.e. MIR or Raman) bands [35]. This implies specific differences between MIR and NIR spectroscopy in their physical nature that propagates to disparity in their applications. Anharmonic methods of quantum chemistry are necessary for being helpful NIR in spectroscopy. However, huge computational cost (i.e. computing time) required for calculating anharmonic spectra places particular importance on efficiency of the quantum approach. Accordingly, recent years demonstrated in silico NIR spectroscopy benefits much from the availability of efficient DVPT2/GVPT2 method [8, 15, 16].

The most straightforward benefit from the availability of simulated spectra is a much-improved interpretability of NIR absorption bands. However, unlike the case of MIR or Raman techniques, conventional methods of spectroscopy used for band identification (e.g. systematic studies of series of compounds, isotopic substitutions, etc.) provide only limited assistance in comprehending NIR spectra, in which numerous bands overlap creating complex, broadened absorption features. As such, the simulated NIR spectra provide a substantial amount of information that is not available from alternative sources.

Table 1. Band assignments in NIR spectrum of rosmarinic acid, based on DVPT2//B3LYP/N07D calculations. Reproduced with permission from Royal Society of Chemistry from Ref. [36].

	Wavenumber/cm^{-1}		Major contributions
	Exp.	Calc.	
1	6854.9	6853	2νOH (ar)
2	6767.2	6741	2νOH (ar)
3	~6680	6645	2νOH (carboxyl)
4	~6044	6056	2νCH (ar, aliph, in-phase)
5	5986.5	6001	2νCH (ar, aliph, opp.-phase); 2νCH (ar)
6	5929.7	5930	2νCH (ar); 2νCH (ar)
7	5752.5	5780	[$\nu_{as}CH_2$, νCH (ar)] + [ν_{as} CH_2, νCH (ar)];
8	5128.0	5126	[ν C = O, δ_{ip}OH (carboxyl)] + [νOH (carboxyl)]
9	5075.8	5027	[δ_{ring}, δ_{ip}OH (ar)] + [νOH (ar, para-)]; [δ_{ring}, δ_{ip}OH (ar)] + [νOH (ar, para-)]; [δ_{ring}] + [νOH (ar, para-)]
10	4994.9	4980	[δ_{ring}, δ_{ip}OH (ar)] + [νOH (ar, meta-)]; [δ_{ring}, δ_{ip}OH (ar)] + [νOH (ar, meta-)]
11	4923.8	4906	[δ_{ring}, δ_{ip}OH (ar)] + [νOH (ar, para-)]; [δ_{ring}, δ_{ip}OH (ar)] + [νOH (ar, para-)]
12	4860.0	4847	[δ_{ring}, δ_{ip}OH (ar)] + [νOH (ar, meta-)]; [δ_{ring}, δ_{ip}OH (ar)] + [νOH (ar, meta-)]
13	4788.3	4798	[νCC] + [νOH (ar, para-)]; [νCC] + [νOH (ar, para-)]; [νCC] + [νOH (ar, meta-)]; [δCCH (carboxyl)] + [νOH (carboxyl)]; [δCH (ar), δ_{i}OH (ar)] + [νOH (ar, para-)]; [δCH (ar), δ_{ip}OH (ar)] + [νOH (ar, para-)]
14	4701.0	4701	δCH (aliph)] + [νOH (ar, meta-)]; δCH (ar), δ_{ip}OH (ar)] + [νOH (ar, meta-)]
15	4629.4	4632	[δCH (ar), δ_{ring}, δ_{ip}OH (ar)] + [νOH (ar, meta-)]; [δCH (ar), δ_{ring}, δ_{ip}OH (ar)] + [νOH (ar, para-)]; [δ_{ip}OH (ar), δCH (ar), δ_{ring}] + [νOH (ar, para-)]

(continued)

Table 1. (*continued*)

	Wavenumber/cm^{-1}		Major contributions
	Exp.	Calc.	
16	4575.7	4757	[δCH (ar), δ_{ring}, δ_{ip}OH (ar)] + [νOH (ar, meta-)]; [δ_{ip}OH (ar),δCH (ar), δ_{ring}] + [νOH (ar, meta-)]
17	~4508	4465	[δ_{ring}, δ_{ip}OH (ar)] + [νsCH (ar)]; [δ_{ring}] + [νCH (ar)]; [δ_{ring}, δ_{ip}OH (ar)] + [νCH (ar)]; [δC-O (carboxyl), ν_{ip}OH (carboxyl)] + [νOH (carboxyl)]; [δ_{ring}] + [νCH (ar)]; [δ_{ring}, δ_{ip}OH (ar)] + [νCH (ar)]
18	4372.3	4360	[δ_{ring}] + [νCH (ar)]; [δ_{ring}] + [νCH (ar)]
19	4233.3	4237	[$\delta_{sciss}CH_2$] + [$\nu_{as}CH_2$, νCH (ar)]; [$\delta_{sciss}CH_2$] + [$\nu_{as}CH_2$, νCH (ar)]
20	4179.4	4194	[δCH (aliph)] + [νCH (ar, aliph, opp.-phase)]; [δ_{sciss} CH_2] + [$\nu_s CH_2$]; [δCH (aliph)] + [νCH (ar, aliph, in-phase)]

Abbreviations: aliph - moiety connected to aliphatic chain; ar - moiety connected to aromatic ring;

Detailed band identification of the analyzed chemical constituent in a complex sample finds direct use in designing analytical applications. For example, calculated NIR spectrum of rosmarinic acid, an important anti-oxidant presents in many plants, enabled exhaustive band assignments (Table 1) that allowed a comprehensive design of analytical method [36]. In that pioneering work, the chemical information on the characteristic wavenumbers of the targeted plant constituent was used in hyphenation with hetero-correlated two-dimensional correlation spectroscopy (2DCOS) to establish the relative sensitivity of different spectrometers.

On the other hand, contributions to NIR absorption lineshape can conveniently be estimated from integral spectral intensity [37]. As presented here on the example of caffeine, simulated NIR spectra unveil that the major contributions to the observed absorption intensity are stemming from the combination bands (Fig. 1) [23]. This shows that the combination bands are the most important for NIR spectra.

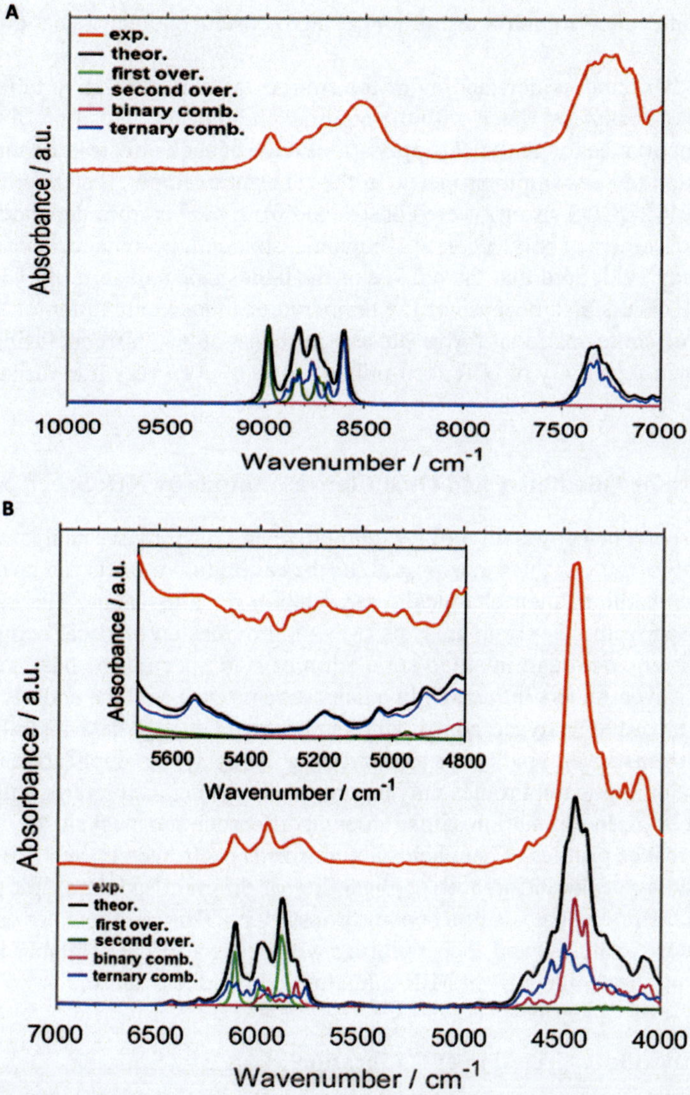

Fig. 1. Experimental and simulated NIR spectrum of caffeine. Additionally, presented are the contributions to absorption lineshape from the first and second overtones, as well as binary and ternary combinations. Reproduced (CC-BY 4.0 license) from Ref. [23].

2.2 Physical Chemistry

Substantial gains have been received from accurately simulated NIR spectra in physicochemical studies of molecular structures, interactions and dynamics. Besides enabling extremely detailed and systematic investigations of spectra-structure correlations in NIR region for numerous fundamental compounds [17, 19, 21, 25, 28, 38]. Mentioned should be here, e.g. effects of isotopic substitution, as an important pathway to band assignment [39–41].

On the other hand, understanding of the sources of fine spectral variations in NIR lineshape not directly correlated with the nominal chemical composition of the sample have great importance for analytical applications. One of such effects is the temperature-driven variation of the sample property. In the relevant example, the experimental and calculated NIR 2DCOS spectra were constructed from temperature-dependent spectral series for four isomers of butyl alcohol; 1-butanol, 2-butanol, *iso*-butanol and *tert*-butanol [18]. The study evidenced that the change of the bandshape and shift of OH stretching first overtone occurs also because of the temperature-induced alteration of the relative populations of conformational forms present in the sample. This conclusion confirms the well-known sensitivity of NIR absorption spectra to even very fine variations in the sample property.

2.3 Support for Qualitative and Quantitative Analysis by NIR Spectroscopy

NIR spectroscopy is a powerful tool for qualitative and quantitative analysis. However, while most often the variables meaningful for the calibration models can be determined using various methods, their chemical interpretation remains vague. The backbone of NIR spectroscopy, multi-variate analysis (MVA), provides no physical insight into the molecular system. Through in silico simulation of NIR spectra, the basic relationship between the wavenumbers influential in quantitative models, and the underlying molecular vibrations of the analyzed constituent can be established. The availability of such information opens several pathways to improving the analytical applications.

Three major application routes may be provided: (i) chemical interpretation of calibration models, (ii) interpretation of instrumental difference and mechanistic understanding of performance profiles of handheld/miniaturized spectrometers, and (iii) mechanistic understanding of the differences applicability of different spectroscopic techniques (e.g. NIR vs. MIR) in analysis of a specific constituent. This permits knowledge-based design of future analyses, and their tailoring with respect to the available instrument, e.g. selection of a specific NIR or MIR miniaturized/handheld sensor.

2.3.1 Chemical Interpretation of Calibration Models

Simulated NIR spectra open the possibility to interpret how the chemical information is expressed indirectly in the calibration models. Noteworthy here is the study focused on thymol, a prototypic polyphenol and monoterpene with prime importance for therapeutic properties of numerous medicinal plants [22]. As known, vibrations of OH group give rise to very intense and meaningful NIR bands. Thymol is a relatively simple molecule, however, simulated spectrum clearly evidenced that OH bands, not only the overtones but also combinations of νOH with other modes, are major contributors to the NIR absorption lineshape in this case as well. This helped to interpret the PLSR model calibrated for the thymol content in the dried extract used for the manufacture of a natural medicine. The meaningful vibrations were identified as C–H and CH3 stretching modes. However, the vibrations of OH group were found to not be dominant in the respective PLSR models.

In addition, the features of quantitative models obtained in analytical routines may be interpreted. Fundamental relationships with the basic factors may be established, e.g. how the sensitivity of the molecule to its chemical environment is reflected in the models, and thus an understanding of how these factors affect the analytical spectroscopy is obtained. Such frameworks were applied in various analytical scenarios, e.g. piperine analysis in black pepper [42], quantification of silymarin in Silybi mariani fructus [43], analysis of theanine and caffeine content in black tea (Thea nigrae) [44] or anthocyanin content in Sambucus fructus [45].

2.3.2 Chemical Interpretation of Instrumental Difference and Mechanistic Understanding of Performance Profiles of Handheld/Miniaturized Spectrometers

The contemporary NIR spectroscopy is largely defined by a rapid and continuing trend of the miniaturization of NIR spectrometers [31, 32]. The miniaturization required implementing a number of distinct engineering solutions. Different design philosophies make the performance and applicability of various miniaturized NIR spectrometers specific, governed by the operating spectral region, resolution and sensitivity. As the result, their analytical performance is often inferior in absolute terms to the benchtop laboratory instrumentation. These devices remarkably increase the flexibility of analysis; however, attention needs to be paid to the various factors affecting their performance in different scenarios. Currently, it is a focused and very active research direction to perform systematical evaluation studies of the analytical accuracy and reliability of various miniaturized spectrometers available at the market [31, 32].

The differences in the operational spectral regions of miniaturized spectrometers, which are often narrow and only partially overlapping between different devices, render only selective chemical information available to analysis. Studies oriented at elucidating this information and using it to improve NIR analysis were reported recently [33, 42]. For instance, PLSR calibration of piperine in black pepper involved spectral sets measured by VIAVI MicroNIR and ThermoFischer microPHAZIR handheld instruments, and a benchtop spectrometer Büchi NIRFlex N-500. Selective availability of the chemical information to each sensor is a meaningful factor influencing its performance (Fig. 2). In the case of piperine analysis, microPHAZIR can measure only the first overtones and binary combinations of C–H stretching modes of piperine, which subsequently contribute to the calibration model.

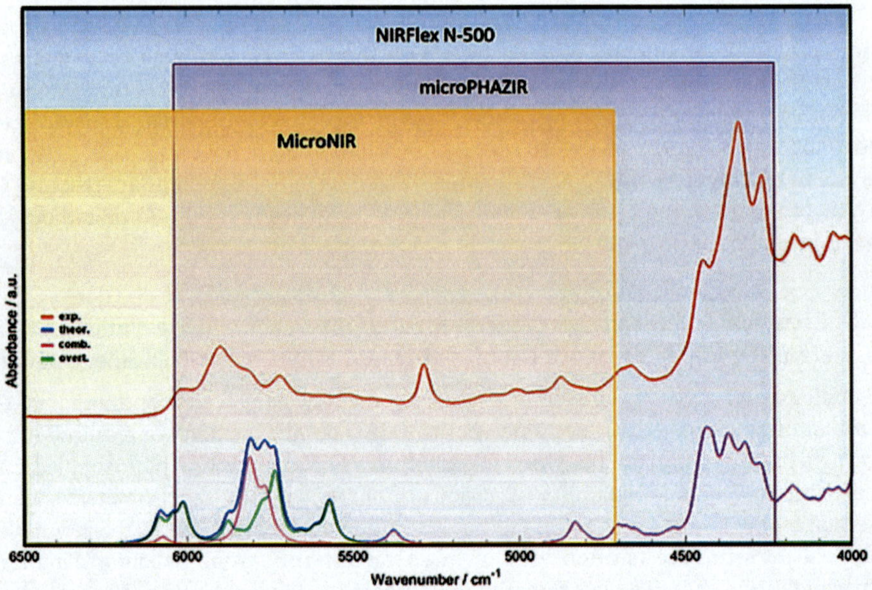

Fig. 2. Please Quantum mechanically simulated NIR spectrum of piperine compared with the experimental spectrum of a piperine standard (solution in CCl4). The summed contributions form the first overtones and combination bands are presented as well. Highlighted in color are the ranges that were recorded by the different spectrometers after spectra pre-treatment. Reproduced (CC-BY 4.0 license) from Ref. [42].

2.3.3 Insight into Analytical Performance of NIR vs. MIR Analysis

Detailed understanding of vibrational bands permits obtaining a deeper insight which of them are particularly meaningful for analysis by different spectroscopic techniques, i.e. NIR vs. MIR analysis. Thus, in-depth understanding of the differences of different spectroscopic techniques in a specific analysis becomes viable. The potential for such interpretation is given here by MIR and NIR examination of the vibrational bands of melamine [46]. In the recent study, MIR + NIR spectra simulation was carried out, and elucidated the chemical origin for the meaningful wavenumbers of melamine available for the calibration which is different between MIR and NIR spectra (Fig. 3) [46].

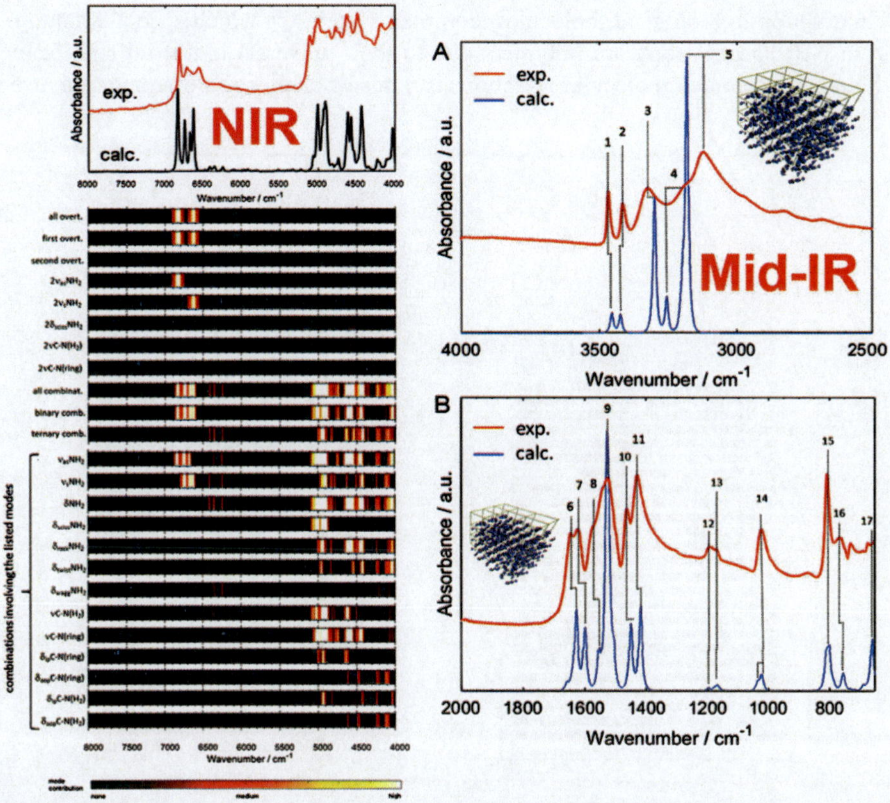

Fig. 3. Interpretation of chemical differences between analytical techniques (mid-IR vs. NIR) with respect to meaningful variables in calibration models. Reproduced (CC-BY 4.0 license) from Ref. [46].

2.4 NIR Fingerprinting of Specific Molecular Structures

Biomolecules are systems of universal importance but especially challenging for NIR spectroscopy. Because of structural complexity, their NIR spectra are strongly convoluted and particularly difficult for direct interpretation. Unlike well-studied MIR or Raman characteristic bands, it is much more difficult to determine well-defined wavenumber ranges in which the presence of peptides, fatty acids, or nucleic acids can be detected in NIR spectra.

Recently, several successful NIR studies of biomolecules aimed at establishing reliable spectra-structure correlations were performed. The absorption bands of short-, medium- and long-chain fatty acids [26, 27, 38], as well as nucleic acid bases (adenine, guanine, thymine and cytosine) [24] have been reconstructed and analyzed. For example, regions in NIR spectra specific to the saturation of the fatty acid chain were identified, demonstrating the future potential of NIR bio-spectroscopy supported by in silico simulation.

In addition to biological molecules, complex systems to which special attention is paid in NIR spectroscopy are polymers due to their universal industrial significance. Although the complex molecular structures can pose a challenge for simulation, use of

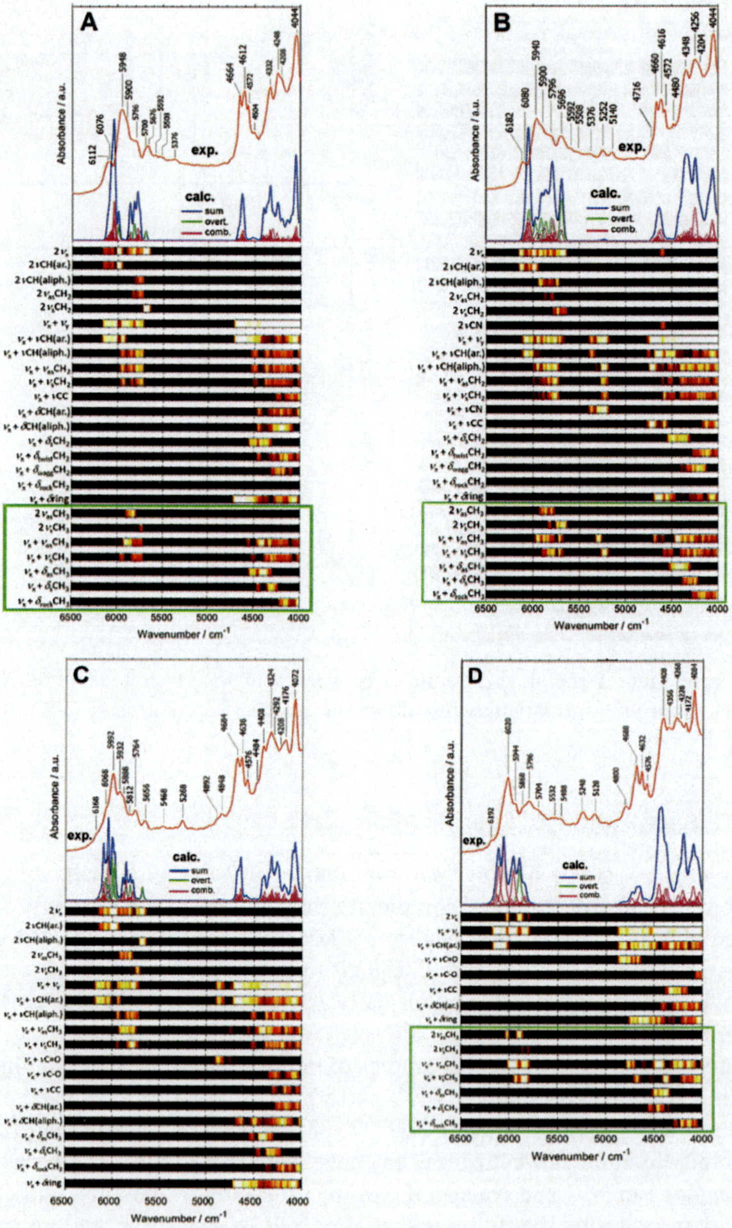

Fig. 4. Experimental and simulated (B3LYP/6-311+G(df,pd)) NIR spectrum of polymers: PS (A), ABS (B), PC (C) and PET (D). Reproduced (CC-BY 4.0 license) from Ref. [28].

simplified models that include the main structural motifs of larger molecular systems are particularly suitable for polymers due to their periodic structure. With this approach, NIR spectra of nine polymers were recently precisely simulated [28], as presented in Fig. 4. This approach achieved a successful identification of reliable structural markers, e.g. aromatic rings, present in polymers regardless of the rest of their molecular structure.

2.5 Understanding of the Intermolecular Interactions, Chemical Neighborhood and Matrix Effects Including Solute-Solvent System and Aqueous Solution

Vibrational coupling between carbohydrates and the hydration shell was unveiled as the underlying mechanism that improves wavenumber-selectively the carbohydrate discrimination performance by NIR spectroscopy. This effect has been previously identified as the 'water mirror effect' and evidenced by empirical studies and observations [30]. It has been concluded in the literature as the driving factor behind the improved ability to detect the presence of an analyte in aqueous solution, because of its interaction with the surrounding water molecules [30]. Its existence, however, was only indirectly observed through its impact on analytical performance of NIR spectroscopy. With the availability of in silico NIR spectroscopy, an attempt could be made to obtain independent evidence and detailed insights into this important phenomenon for fundamental science as well as applications of NIR spectroscopy [47].

Recently, the effect of water mirror was investigated with respect to the classification performance of six carbohydrates (fructose, glucose, mannose, ribose, xylose and sorbitol) in aqueous solution in different concentration levels. The results of MVA classification were subsequently interpreted by employing calculation of the NIR spectra of the investigated carbohydrates and their first hydration shells. The simulation unveiled that the water mirror phenomenon is vibration-selective and thus wavenumber-selective, and leads to an enhancement of the qualitative information contained in the specific spectral regions. The location of these regions and the related performance correspond fully to the appearance and magnitude of the cooperative vibration effect indicated by MVA [47].

3 Summary

A remarkable potential arises from the growing applicability of anharmonic quantum mechanical calculations to solving the problems, which arise in both basic NIR spectroscopy and its analytical applications. This review highlights an increased value of in silico NIR spectroscopy in relation to other kinds of vibrational spectroscopy. Much more complex spectra than those typical for MIR or Raman techniques, reduce the chemical specificity of NIR spectroscopy when supported only by conventional methods of spectral analysis. Advances in theoretical methods have paved the way for modeling NIR spectra of considerably large molecules and unlock hidden potential of NIR spectroscopy.

Ab initio calculated NIR spectra provide a significant improvement in our understanding of the NIR spectrum, enables research focused on physical chemistry as well as provide direct support in designing applications of NIR spectroscopy. Unveiling NIR 'fingerprint' of complex systems improves the much desired chemical specificity of

NIR spectroscopy. In MVA procedures, simulation allows interpreting the chemical contribution into the differences in performance observed between different miniaturized spectrometers. Furthermore, interpenetration of calibration models is permitted with the detailed information on the origins of NIR bands.

This review article is based in part on the topic of invite talk presented at the NIR Conference 2021, held from 18–21 October 2021 as an online event hosted and coordinated by Beijing.

Conflict of Interest. The authors declare no conflict of interest.

References

1. Ozaki, Y., Huck, C.W., Beć, K.B.: Near-IR spectroscopy and its applications. In: Gupta, V.P. (ed.) Molecular and Laser Spectroscopy. Advances and Applications, pp. 11–38. Elsevier, San Diego (2018)
2. Huck, C.W., Beć, K.B., Grabska, J.: Near infrared spectroscopy in natural product research. In: Meyers, R.A. (ed.) Encyclopedia of Analytical Chemistry: Applications, Theory and Instrumentation, pp. 1–29. Wiley, Hoboken (2020)
3. Beć, K.B., Grabska, J., Huck, C.W.: Near-infrared spectroscopy in bio-applications. Molecules **25**, 2948 (2020)
4. Beć, K.B., Grabska, J., Bonn, G.K., Popp, M., Huck, C.W.: Principles and applications of vibrational spectroscopic imaging studies in plant science: a review. Front. Plant. Sci. **11**, 1226 (2020)
5. Beć, K.B., Grabska, J., Huck, C.W.: NIR spectral analysis of natural medicines supported by novel instrumentation, methods of data analysis and interpretation. J. Pharm. Biomed. Anal **193**, 113686 (2020)
6. Beć, K.B., Grabska, J., Huck, C.W., Ozaki, Y.: Quantum mechanical simulation of near-infrared spectra. Applications in physical and analytical chemistry. In: Ozaki, Y., Wójcik, M.J., Popp, J. (eds.) Molecular Spectroscopy: A Quantum Chemistry Approach, vol. 2, pp. 353–388. Wiley-VCH, Weinheim (2019)
7. Czarnecki, M.A., Beć, K.B., Grabska, J., Hofer, T.S., Ozaki, Y.: Overview of application of NIR spectroscopy to physical chemistry. In: Ozaki, Y., Huck, C., Tsuchikawa, S., Engelsen, S.B. (eds.) Near-infrared spectroscopy, pp. 297–330. Springer, Singapore (2021). https://doi.org/10.1007/978-981-15-8648-4_13
8. Beć, K.B., Grabska, J., Huck, C.W.: Current and future research directions in computer-aided near-infrared spectroscopy: a perspective. Spectrochim. Acta A **254**, 119625 (2021)
9. Beć, K.B., Grabska, J., Plewka, N., Huck, C.W.: Insect protein content analysis in handcrafted fitness bars by NIR spectroscopy. Gaussian process regression and data fusion for performance enhancement of miniaturized cost-effective consumer-grade sensors. Molecules **26**, 6390 (2021)
10. Mayr, S., et al.: Challenging handheld NIR spectrometers with moisture analysis in plant matrices: Performance of PLSR vs. GPR vs. ANN modelling. Spectrochim. Acta A **249**, 119342 (2021)
11. Losso, K., et al.: Rapid discrimination of Curcuma longa and Curcuma xanthorrhiza using direct analysis in real time mass spectrometry and near infrared spectroscopy. Spetrochim. Acta A **265**, 120347 (2022)
12. Beć, K.B., Grabska, J., Huck, C.W.: Near-infrared (NIR) sensors in environmental analysis. In: Narayan, R. (ed.) Encyclopedia of Sensor Technology. Elsevier (2022)

13. Beć, K.B., Grabska, J., Huck, C.W.: Physical principles of infrared spectroscopy. In: Cozzolino, D. (ed.) Comprehensive Analytical Chemistry, vol. 95 (2020)
14. Beć, K.B., Grabska, J., Ozaki, Y.: Advances in anharmonic methods and their applications to vibrational spectroscopies. In: Wójcik, M., Nakatsuji, H., Kirtman, B., Ozaki, Y. (eds.) Frontiers of Quantum Chemistry, pp. 483–512. Springer, Singapore (2018). https://doi.org/10.1007/978-981-10-5651-2_20
15. Beć, K.B., Huck, C.W.: Breakthrough potential in near-infrared spectroscopy: spectra simulation. A review of recent developments. Front. Chem **7**, 48 (2019)
16. Ozaki, Y., Beć, K.B., Morisawa, Y., Yamamoto, S., Tanabe, I., Huck, C.W., Hofer, T.S.: Advances, challenges and perspectives of quantum chemical approaches in molecular spectroscopy of the condensed phase. Chem. Soc. Rev **50**, 10917–10954 (2021)
17. Beć, K.B., Futami, Y., Wójcik, M.J., Ozaki, Y.: A spectroscopic and theoretical study in the near-infrared region of low concentration aliphatic alcohols. Phys. Chem. Chem. Phys. **18**, 13666–13682 (2016)
18. Grabska, J., Beć, K.B., Ozaki, Y., Huck, C.W.: Temperature drift of conformational equilibria of butyl alcohols studied by near-infrared spectroscopy and fully anharmonic DFT. J. Phys. Chem. A **121**, 1950–1961 (2017)
19. Beć, K.B., Grabska, J., Czarnecki, M.A.: Spectra-structure correlations in NIR region: spectroscopic and anharmonic DFT study of n-hexanol, cyclohexanol and phenol. Spectrochim. Acta A **197**, 176–184 (2018)
20. Beć, K.B., Karczmit, D., Kwaśniewicz, M., Ozaki, Y., Czarnecki, M.A.: Overtones of νCN vibration as a probe of structure of liquid CH3CN, CD3CN, and CCl3CN: combined infrared, near-infrared, and Raman spectroscopic studies with anharmonic density functional theory calculations. J. Phys. Chem. A **123**, 4431–4442 (2019)
21. Beć, K.B., Grabska, J., Huck, C.W., Mazurek, S., Czarnecki, M.A.: Anharmonicity and spectra-structure correlations in MIR and NIR spectra of crystalline menadione (vitamin K3). Molecules **26**, 6779 (2021)
22. Beć, K.B., Grabska, J., Kirchler, C.G., Huck, C.W.: NIR spectra simulation of thymol for better understanding of the spectra forming factors, phase and concentration effects and PLS regression features. J. Mol. Liq. **268**, 895–902 (2018)
23. Grabska, J., Beć, K.B., Ozaki, Y., Huck, C.W.: Anharmonic DFT study of near-infrared spectra of caffeine. Vibrational analysis of the second overtones and ternary combinations. Molecules **26**, 5212 (2021)
24. Beć, K.B., Grabska, J., Ozaki, Y., Czarnecki, M.A., Huck, C.W.: Simulated NIR spectra as sensitive markers of the structure and interactions in nucleobases. Sci. Rep. **9**, 17398 (2019)
25. Beć, K.B., Futami, Y., Wójcik, M.J., Nakajima, T., Ozaki, Y.: Spectroscopic and computational study of acetic acid and its cyclic dimer in the near-infrared region. J. Phys. Chem. A **120**, 6170–6183 (2016)
26. Grabska, J., Ishigaki, M., Beć, K.B., Wójcik, M.J., Ozaki, Y.: Correlations between structure and near-infrared spectra of saturated and unsaturated carboxylic acids. Insight from anharmonic density functional theory calculations. J. Phys. Chem. A **121**, 3437–3451 (2017)
27. Grabska, J., Beć, K.B., Ishigaki, M., Huck, C.W., Ozaki, Y.: NIR spectra simulations by anharmonic DFT-saturated and unsaturated long-chain fatty acids. J. Phys. Chem. B **122**, 6931–6944 (2018)
28. Beć, K.B., Grabska, J., Badzoka, J., Huck, C.W.: Spectra-structure correlations in NIR region of polymers from quantum chemical calculations. The cases of aromatic ring, C=O, C≡N and C-Cl functionalities. Spetrochim. Acta A **262**, 120085 (2021)
29. Grabska, J.: Current frontiers in quantum chemical simulations of NIR spectra – polymers, biomolecules, aqueous matrix, and interpretation of instrumental difference of handheld spectrometers. NIR News **32**, 7–14 (2021)

30. Bázár, G., et al.: Water revealed as molecular mirror when measuring low concentrations of sugar with near infrared light. Anal. Chim. Acta **896**, 52–62 (2015)
31. Beć, K.B., Grabska, J., Huck, C.W.: Principles and applications of miniaturized near-infrared (NIR) spectrometers. Chem. Eur. J. **27**, 1514–1532 (2021)
32. Beć, K.B., Grabska, J., Siesler, H.W., Huck, C.W.: Handheld near-infrared spectrometers: where are we heading? NIR News **31**(3–4), 28–35 (2020)
33. Grabska, J., Beć, K.B., Mayr, S., Huck, C.W.: Theoretical simulation of near-infrared spectrum of piperine. Insight into band origins and the features of regression models. Appl. Spectr. **75**, 1022–1032 (2021)
34. Beć, K.B.: A Simple guide to complex world of overtone and combination bands: theoretical simulation and interpretation of NIR spectra – summary of the workshop at NIR-2021 Beijing Conference. NIR News **32**, 15–18 (2021)
35. Beć, K.B., Grabska, J., Hofer, T.S.: Introduction to quantum vibrational spectroscopy. In: Ozaki, Y., Huck, C., Tsuchikawa, S., Engelsen, S.B. (eds.) Near-Infrared Spectroscopy, pp. 83–110. Springer, Singapore (2021). https://doi.org/10.1007/978-981-15-8648-4_5
36. Kirchler, C.G., et al.: Critical evaluation of spectral information of benchtop vs. portable near-infrared spectrometers: quantum chemistry and two-dimensional correlation spectroscopy for a better understanding of PLS regression models of the rosmarinic acid content in Rosmarini folium. Analyst **142**, 455–464 (2017)
37. Huck, C.W., Beć, K.B., Grabska, J., Ozaki, Y.: Quantum chemical calculation of NIR spectra of practical materials. NIR News **28**(2), 13–20 (2017)
38. Grabska, J., Beć, K.B., Ishigaki, M., Wójcik, M.J., Ozaki, Y.: Spectra-structure correlations of saturated and unsaturated medium-chain fatty acids. Near-infrared and anharmonic DFT study of hexanoic acid and sorbic acid. Spectrochim. Acta A **185**, 35–44 (2017)
39. Grabska, J., Czarnecki, M.A., Beć, K.B., Ozaki, Y.: Spectroscopic and quantum mechanical calculation study of the effect of isotopic substitution on NIR spectra of methanol. J. Phys. Chem. A **121**, 7925–7936 (2017)
40. Beć, K.B., Grabska, J., Huck, C.W., Czarnecki, M.A.: Spectra–structure correlations in isotopomers of ethanol (CX3CX2OX, X = H, D): combined near-infrared and anharmonic computational study. Molecules **24**, 2189 (2019)
41. Beć, K.B., Grabska, J., Huck, C.W., Czarnecki, M.A.: Effect of conformational isomerism on NIR spectra of ethanol isotopologues. Spectroscopic and anharmonic DFT study. J. Mol. Liq **310**, 113271 (2020)
42. Mayr, S., Beć, K.B., Grabska, J., Schneckenreiter, E., Huck, C.W.: Near-infrared spectroscopy in quality control of Piper nigrum: a comparison of performance of benchtop and handheld spectrometers. Talanta **223**, 121809 (2021)
43. Mayr, S., et al.: Quantification of Silymarin in Silybi mariani fructus: challenging the analytical performance of benchtop vs. handheld NIR spectrometers on whole seeds. Planta. Med. **87**, 1–13 (2021)
44. Mayr, S., et al.: Theae nigrae folium: Comparing the analytical performance of benchtop and handheld near-infrared spectrometers. Talanta **221**, 121165 (2021)
45. Stuppner, S., et al.: Near infrared spectroscopy as a rapid screening method for the determination of total anthocyanin content in sambucus fructus. Sensors **20**, 4983 (2020)
46. Grabska, J., Beć, K.B., Kirchler, C.G., Ozaki, Y., Huck, C.W.: Distinct difference in sensitivity of NIR vs. IR bands of melamine to inter-molecular interactions with impact on analytical spectroscopy explained by anharmonic quantum mechanical study. Molecules **24**, 1402 (2019)
47. Beganović, A., Beć, K.B., Grabska, J., Stanzl, M.T., Brunner, M.E., Huck, C.W.: Vibrational coupling to hydration shell – mechanism to performance enhancement of qualitative analysis in NIR spectroscopy of carbohydrates in aqueous environment. Spectrochim. Acta A **237**, 118359 (2020)

Chemometric Studies in Near-Infrared Spectroscopy

Hongle An[1,2], Li Han[1,2], Yan Sun[1,2], Wensheng Cai[1,2], and Xueguang Shao[1,2](✉)

[1] Research Center for Analytical Sciences, Frontiers Science Center for New Organic Matter, College of Chemistry, Tianjin Key Laboratory of Biosensing and Molecular Recognition, State Key Laboratory of Medicinal Chemical Biology, Nankai University, Tianjin 300071, China
xshao@nankai.edu.cn

[2] Haihe Laboratory of Sustainable Chemical Transformations, Tianjin 300192, China

Abstract. Near-infrared (NIR) spectroscopy has been a powerful technique for both qualitative and quantitative analysis. Due to the highly overlapping of the spectral bands, however, it is difficult to extract structural and quantitative information from the spectral data. Therefore, chemometric methods have been widely applied to enhance the spectral resolution or extract the spectral information, including modeling techniques, spectral preprocessing, variable selection, outlier detection, modeling transfer, etc. These methods provided colorful approaches for improving the models in both quantitative and discrimination analysis, greatly enhanced the applicability of NIR spectroscopy. On the other hand, temperature-dependent near-infrared spectroscopy was developed for analyzing liquid mixtures or aqueous systems. Chemometric methods were also established to build the quantitative models and extract the temperature-induced spectral variations. The former provided powerful tools for predicting the temperature or the concentration of the components, and the latter provided efficient approaches for understanding the structures and the interactions in chemical and biological samples or processes.

Keywords: Near-infrared spectroscopy · Chemometrics · Qualitative analysis · Quantitative analysis

1 Introduction

Near-infrared (NIR) spectroscopy, as a technique for rapid and non-destructive analysis with little or no sample preparation, has been widely used in different fields, such as environment, food, agriculture and industries [1–3]. However, the weak absorption and overlapping of the NIR spectral peaks make it difficult to provide the required spectral information for qualitative and quantitative analysis. Chemometric methods, therefore, have been adopted to improve the qualitative analysis and quantitative prediction of the technique. The first work in our group was published in 2002, in which the models for predicting the chemical components of tobacco samples using NIR spectra were established [4]. Since then, chemometric methods for improving the models of NIR spectra

X. Chu et al. (Eds.): ICNIR 2021, *Sense the Real Change: Proceedings of the 20th International Conference on Near Infrared Spectroscopy*, pp. 47–58, 2022.
https://doi.org/10.1007/978-981-19-4884-8_4

were developed [5, 6]. Besides partial least square (PLS) regression, consensus or ensemble modeling, multi-block strategies, and non-linear methods were studied, which may provide better prediction and interpretability in specific cases. NIR spectra contain not only the information of the components in the samples but also the spectral interferences, such as varying background and noise signals. To eliminate the irrelevant information, spectral preprocessing methods were studied. It has been proved that spectral preprocessing can effectively eliminate the interferences and improve the calibration model. Continuous wavelet transform (CWT) has been a commonly used method in our works to remove the variant background and enhance the spectral resolution [7–9]. Moreover, an NIR spectrum is composed of hundreds or even thousands of variables. The uninformative variables may degrade the model. Therefore, variable selection methods were proposed to select informative variables for building the predictive models [10–12]. These methods have been proved helpful for building sparse models, particularly in the case that the number of calibration samples is not enough. Furthermore, methods for calibration transfer were also studied for practical applications of NIR spectroscopy in the case that more than one instruments are needed.

From 2010, temperature-dependent NIR spectroscopy was developed for analyzing the structures and interactions in aqueous or bio-systems using the spectral variation induced by temperature. A quantitative spectra-temperature relationship (QSTR) model between NIR spectra of water and temperature was established [13, 14], and chemometric methods to extract quantitative and structural information from the temperature-dependent spectra were developed [15, 16]. High order chemometric algorithms were employed to extract the information from the high-dimensional spectral data. To simplify the calculation, multilevel simultaneous component analysis (MSCA) was studied and mutual factor analysis (MFA) was proposed for analyzing the high-dimensional data with the two-dimensional algorithm. Besides, chemometric algorithms were proposed to improve the resolution of NIR spectra or separate the mixed spectra of multi-component systems. Works were concentrated on capturing the temperature-induced spectral change of water, which can reflect the structures and interactions of water and solutes in aqueous solutions. Through the variation of water structures, the structure and the structural transformation of functional molecules such as proteins and thermo-responsive polymers were analyzed, and the role of water in the chemical and biological processes was revealed.

2 Modeling Techniques

PLS is the most commonly used linear calibration method due to its practicability and versatility. To improve the reliability of the models, particularly in the case that the calibration samples are not enough, ensemble modeling methods based on PLS were developed for modeling the NIR spectra [17–20]. For modeling the data set with large number of variables and making the model more interpretable, multiblock partial least squares (MB-PLS) method was proposed [18]. In the method, the spectra were decomposed into blocks by discrete wavelet transform (DWT), and the relative importance of the blocks was estimated by both the super-weights and the block-weights determined by the prediction error of the sub-models in cross validation. Two NIR data sets of tobacco samples were investigated by the proposed method, and it was found that the weighted MB-PLS coupled with DWT gives a better predictive accuracy and interpretability compared with the ordinary PLS and MB-PLS methods. Besides, weighted-PLS regression method in multivariate calibration of NIR spectra was proposed [19]. In the approach, the spectra were decomposed into different scale blocks (or frequency components) by wavelet transform (WT) at first, and then PLS models were built with the decomposed components. A more precise prediction was obtained by the combined model compared with the ordinary PLS model. In weighted-PLS regression method, the spectra were split into groups of variables according to the statistic values of variables, i.e., the stability, which were used to evaluate the importance of variables in a calibration model [20]. Because the stability reflects the relative importance of the variables for modeling, these groups present different spectral information for the construction of PLS models. The two weighted-PLS methods were applied in modeling the NIR spectra of different complex samples. Compared with the results obtained with ordinary PLS method, the proposed methods are proved to be high-performance tools for multivariate calibration of NIR spectra.

In practical works, it was found that there is non-linear relationship between the NIR spectral response and the component content, which may make PLS model poor in predictive ability. To solve this problem, a local regression method based on WT was developed to improve the universality and prediction precision of the models [21]. In the algorithm, DWT was firstly utilized to compress the NIR spectra, and then, the calibration subsets were individually selected for each prediction sample according to the Euclidean distance in wavelet domain. The method was used to quantitative determination of chlorine in tobacco samples. The results obtained by the method are superior to that of the global PLS method and principal component analysis (PCA) based local regression method. Moreover, a consensus least squares support vector regression (LS-SVR) method was proposed based on the principle of consensus modeling [22]. In the proposed method, NIR spectra of tobacco lamina samples were firstly preprocessed using DWT for eliminating the spectral background and noise, then, consensus LS-SVR technique was used for building the calibration model. With an optimization of the parameters involved in the modeling, a satisfied model was achieved for predicting the content of reducing sugar in tobacco lamina samples.

3 Preprocessing

NIR spectra contain not only the information of the components to be analyzed but also the light scattering and fluctuating background. Therefore, appropriate spectral preprocessing is vital to eliminate the irrelevant information before multivariate calibration. Methods such as multiplicative scatter correction (MSC), standard normal variate (SNV), Savitzky-Golay (SG) smoothing and orthogonal signal correction (OSC), were developed to remove spectral interference [23–26]. The influence of scattering caused by uneven distribution of sample particles and different particle sizes can be effectively eliminated by these methods. In addition, WT is a useful signal processing method and widely applied in spectral analysis. WT with a wavelet function can be regarded as a smoothing and differentiation process. Owing to the characteristic of the double localization in position and frequency domains, WT can decompose a signal into localized contributions representing the information of different frequencies. Among these contributions, there are contributions which represent the high resolution signals because their frequency is higher than the lower frequency background and lower than the high frequency noise. Based on this property, WT can be often employed for data compression, data smoothing, baseline correction, resolution of multicomponent overlapping signals, etc. In our previous works, WT has been proved to be an efficient tool for removing the variant background and noise. For example, a hybrid algorithm was developed based on the utilization of multi-resolution, which is one of the most important advantages of WT [27]. The spectral signals were split into different frequency components, and the number of spectral data points remained unchanged. The method was applied to the simulated data and experimental NIR spectral data. It was found that the method can be used to remove the low-frequency background and the high-frequency noise simultaneously with the help of WT. CWT was also applied as the pretreatment method to eliminate the background in the discrimination of pharmaceutical products [28]. Furthermore, CWT was found to be an efficient tool due to its advantage in smoothing and flexibility for calculating high order derivatives, and the fourth order derivative was proved to be a good choice for improving resolution as well as reducing the noise and sidelobe effects [29].

4 Variable Selection

NIR spectral data matrix is often quite large and some variables may be irrelevant to the multivariate calibration, which reduce the quality of the model. Eliminating uninformative variables can simplify calibration modeling and improve the accuracy and robustness of prediction models. In order to obtain better quantitative calibration models, several variable selection methods for selecting characteristic wavelengths of NIR spectra have been developed, such as uninformative variable elimination (UVE), competitive adaptive reweighted sampling (CARS) and genetic algorithm (GA) [30–32].

A variable selection method, named as MC-UVE, was proposed based on the principle of Monte Carlo (MC) and UVE, which is applied in the quantitative analysis of NIR spectral data. Firstly, a large number of regression models were established with corresponding training subsets selected by MC technique, and then each variable was sorted based on the stability of the PLS coefficients in these models. The variables with poor stability were removed as uninformative variables. The method was used to analyze the contents of nicotine and sugar in tobacco samples. The results showed that this method can select important wavelengths and make the prediction more precise compared with UVE-PLS and conventional PLS [10]. Moreover, MC-UVE combined with successive projections algorithm (SPA) was proved to be an effective way to generate variable subsets by removing the uninformative variables from NIR spectra of complex lamina samples [33]. In addition, the silver substrate can reduce spectral interference and enhance the spectral response, and the detection performance can be further enhanced with the help of chemometric methods. To obtain satisfactory PLS model between the spectra and concentration of lysozyme solutions, MC-UVE method was used to eliminate uninformative variables [34].

An effective variable selection method named as RT-PLS was proposed based on randomization test for NIR spectral analysis. The importance of the variables can be evaluated by a statistic P, and the variable will be removed as an uninformative variable if P value is greater than the threshold. With applications of RT-PLS for the quantitative analysis of corn and tobacco samples, it was proved that RT may be a good alternative for multivariate analysis [11]. RT can also be employed as a gene selection method to deal with gene expression data [35]. Besides, latent projective graph (LPG) was adopted in variable selection for NIR spectral analysis [36]. The method was based on the assumption that collinear variables may have the same contribution to the modeling. An LPG was calculated by performing PCA on an NIR spectral data firstly, and then informative variables were selected from LPG. The method was applied in three NIR datasets of pharmaceutical tablets, blood and tobacco samples, and it was shown that accurate models are built by using only the variables at the inflections of LPG. Moreover, a method for variable selection was developed based on the detection of the influential variables (IVs) [37]. PCA was used to insight into the clustering of the models built with different combination of the spectral wavelengths. The IVs can be distinguished by the frequency number of variables. From the results of five datasets, it was proved that parsimonious and precise models can be obtained by the proposed method. Furthermore, a variable importance criterion named as C was developed to determine the importance of variables in the multivariate calibration model. The value of C was a measurement of the average contribution for a variable to the prediction error of a model. The multi-step shrinkage strategy can further improve the variable selection effect of the C value. The criterion was applied in the diesel fuel dataset and two blood datasets, and the results demonstrated that the value of C is an effective parameter for selecting the important wavelengths compared with MC-UVE, RT and CARS [12].

5 Outlier Detection

Outliers contained in the calibration data set are usually caused by the instrument, operation, and samples preparation and the outliers may interfere with the prediction accuracy of the multivariate calibration model in NIR spectral analysis. In order to detect outliers and improve the quality of the models, a large number of alternative approaches were proposed for outlier detection. A useful algorithm named as CWT-mIPOW-PLS for simultaneous outlier detection and variable selection was proposed for NIR spectra [38]. The mIPOW-PLS was proposed to remove both the useless wavelengths and the multiple outliers in CWT domain. The calibration models based on NIR spectra of sugar aqueous solutions and tobacco samples were built accurately by using this method. Moreover, an extension of the Kennard-Stone (EKS) algorithm was used to extract the optimal samples in the CWT domain, which can effectively reduce spectral interference and the number of samples for obtaining high-quality calibration model [39]. There is another algorithm to detect outliers in NIR spectra analysis. A large number of PLS models were constructed by random test cross validation firstly, and then the models were sorted by the prediction residual error sum of squares. The outliers can be detected by the plot of the accumulative probability which can be gotten from the sorted PLS models [40]. Applying this method to wheat, gasoline, corn and tobacco lamina samples datasets, the method was proved to be a more efficient method for detecting outliers compared with the conventional LOOCV method. In addition, an improved boosting

PLS was proposed to enhance the prediction ability of the quantitative model when the outliers existed [17]. A robust step was added to modify sampling weight for weakening the effect of the outliers in the models.

6 Model Transfer

Model transfer is essential for practical applications of NIR spectroscopy because differences may exist between the spectra measured using different instruments. Model transfer methods for correcting prediction error without using the spectra standard samples were developed. Based on the assumption that spectral differences of different instruments and the prediction error were linear, a useful method without standard samples was developed according to the linear relationship [41]. The linear model correction (LMC) method was proved to be an efficient way for transferring the models between different instruments in NIR spectra analysis when standard samples are not available [42]. The method assumes that the NIR spectra measured on different instruments are linearly correlated if the samples are similar. The coefficients of the master model can be applied in the slave model by using the constrained optimization method. Furthermore, a modified method based on LMC was proposed for improving the model transfer accuracy and computational efficiency [43]. This method was used to analyze the NIR spectra data of pharmaceutical tablets from different instruments. It was found that the method can achieve efficient transfer and the calculation can be simplified by the modification.

7 Discrimination Analysis

Discrimination analysis of samples, especially complex actual products, is a significant subject in many fields. PCA is one of the most important pattern recognition methods, which can reduce spectral data dimensionality and extract distinguishing features information from NIR spectra. In order to discriminate azithromycin tablets from four manufacturers, PCA was performed after CWT preprocessing and variable selection of NIR diffuse reflectance spectra [44]. The results showed that the classification of manufacture sites was acceptable. A method was developed to discriminate the Chinese patent medicines using NIR spectroscopy and principal component discriminant transformation [45]. The optimal set of orthogonal discriminant vectors was designed by maximizing Fisher's discriminant function. The discrimination models were very accurate for the discrimination of different types of medicines. To distinguish different brands of tobacco products, methods were proposed using the NIR spectra datasets. PCA combined with hierarchical cluster analysis (HCA) can accurately determine the brand of tobacco products [46]. In addition, the principal component accumulation (PCAcc) method was used for the multiclass problem of tobacco samples [47]. It was found that the method can discriminate different parts of tobacco leaves and different brands of cigarettes accurately. Moreover, PCA combined other chemometric methods is often applied in the rapid disease diagnosis using NIR spectra of human serum samples [48, 49]. For example, PCA, DWT, linear discriminant analysis (LDA) and partial least squares discriminant analysis (PLSDA) were used for discrimination of the sera from healthy and possible kidney patients, and satisfactory classification was obtained.

8 Extracting Information from Temperature-Dependent Near-Infrared Spectra

Temperature-dependent NIR spectroscopy is a technique for measuring the NIR spectra of a sample at different temperatures. Temperature effect can be considered as a source of information for multivariate spectroscopic analysis. A QSTR model between NIR spectra and the temperature was established based on PLS regression and applied to the quantitative determination of the compositions in aqueous solutions of methanol, ethanol, n-hexane, and their mixtures [13, 14]. The temperature-dependent NIR spectra measured at different conditions generally generate high-dimensional data. Thus, high order chemometric algorithms are attractive for dealing with such data. N-way principal component analysis (NPCA), parallel factor analysis (PARAFAC) and alternating trilinear decomposition (ATLD) were adopted to explore the spectral information from the temperature-dependent NIR spectra of a binary water-ethanol and a ternary water-ethanol-isopropanol mixtures [50]. The temperature- and concentration-induced spectral variations were obtained and the quantitative model was successfully built. In order to simplify the calculation, two-dimensional algorithms were developed for analyzing the high-dimensional data. Multilevel simultaneous component analysis (MSCA), which unfolds a data array into a two-dimensional matrix, was used to study temperature-dependent NIR spectra. A two-level MSCA model was employed to capture the temperature- and concentration-induced spectral variations of aqueous solutions and real serum samples [51, 52]. Moreover, quantitative analysis of the NIR spectra for aqueous proline solution under multiple perturbations was investigated by a three level MSCA [53]. The spectral changes induced by pH, concentration and temperature were discribed by the three level models, respectively. A new method, mutual factor analysis (MFA), was also developed [54, 55]. The method unfolds a high-dimensional data array into a combined data matrix and then extracts the common spectral feature contained in the spectra of different temperatures or different concentrations by PCA. The relative quantity of the extracted spectral feature can be used to build the quantitative model. The method was employed for the quantitative determination of glucose solutions and serum samples. A calibration model with a good correlation coefficient was obtained for the measurement of the glucose content.

The structure of water has been an interesting subject in chemistry and biology for decades due to the complexity and flexibility of hydrogen bonding. With the help of chemometric methods, the spectral features of different water structures can be obtained. To investigate the effect of the temperature on the NIR spectra of water, a method combined CWT and MC-UVE was proposed for the selection of the temperature-dependent variables (wavenumbers) from the NIR spectra measured at different temperatures [56]. Seven variables with a significant temperature dependency can be found and the variables for different solutions are not identical, indicating that the variables can be used to discriminate different solutions. To understand water structures in liquid water and aqueous solutions, Gaussian fitting was adopted to analyze the temperature-dependent NIR spectra of water [57]. The spectral components corresponding to the nine water structures with different hydrogen were obtained from the spectra of water and glucose solutions by Gaussian fitting with a knowledge-based genetic algorithm [58]. The knowledge about the variation of these water structures with temperature was included in the method. Through the structural variation with temperature and glucose concentration, the dissociation of the water clusters and the enhancement of tetrahedral water structure was observed, respectively. Alcohols and amines can be regarded as models for investigating the hydrogen-bonded interactions. The spectral features of eight alcohol structures were identified from the resolution-enhanced spectra calculated by CWT [59]. The stability of the aggregates was obtained from the temperature effect and a sequence of the stability was deduced. Independent component analysis (ICA) was adopted for analyzing the temperature effect on the spectra of primary aliphatic amines [60]. Three independent components (ICs) corresponding to the spectral information of the free, linearly and cyclically hydrogen-bonded NH groups, respectively, were obtained. With the reconstructed spectra from the ICs, the variation of the three forms of NH groups with concentration and temperature was observed.

The interaction of water and solutes is of great significance for understanding the properties of aqueous solutions or bio-systems. The interaction in alcohol-water mixtures has been studied using temperature-dependent NIR spectroscopy. A useful method was proposed based on the rotation of the loadings in PCA [61]. The spectra of ethanol and water were calculated from the spectra of the mixtures and the calculated spectra were found to reflect the structure in the mixture. Moreover, the structure of water at low temperatures and the mechanism of the cryoprotectant dimethyl sulfoxide (DMSO) in reducing the freezing point were investigated [62, 63]. From the resolution-enhanced spectra by CWT, the hydrogen bonding of DMSO and water was found and the interaction was weakened by the existence of FA (a model compound of protein) exists in the mixture. However, the variation of spectral feature with the content of FA was not found by ATLD, implying that, although FA may slightly reduce the anti-freezing effect, DMSO is still the key component to prevent water from icing. The interaction of water and protein or polymer plays a key role in the phase transition of the macromolecules in aqueous solutions or bio-organisms. Combined with two-dimensional correlation NIR spectroscopy and Gaussian fitting, the spectral variations of different water structures during the gelation of ovalbumin were observed and the denaturation mechanism was elucidated [64]. The variation of hydration water during the aggregation of the core fragment of tau, R2/wt, induced by heparin, was also investigated [65]. The spectral features of water structures around NH and CH groups were found in the loadings of PCA, and the variation of the structures during the aggregation was analyzed by 2D correlation spectroscopy. To understand the function of water on the thermal stability of the proteins in a confined environment, the water structures in reverse micelles (RMs) were studied [66]. The spectral feature of bridging water, which connects protein and the inner surface of RMs, was found by PCA. The bridging water may be the reason for enhancing the thermal stability of the protein in RMs. Temperature-sensitive polymers exhibit phase separation in aqueous solutions above the lower critical solution temperature (LCST). The interaction of water and the polymer is supposed to be the key factor in driving the aggregation. The interactions during the aggregation of poly(N,N-dimethylaminoethyl methacrylate) (PDMAEMA) and poly(N-isopropyl acrylamide) (PNIPAM) were investigated [67, 68]. NPCA was used to extract the information of spectral variation with temperature and concentration. It was found that, during the phase transition of PDMAEMA, the water molecules connecting the polymer chains in the loose hydrophobic structure by two hydrogen bonds are important for forming intermediate state. The water structure with three hydrogen bonds plays a key role in the stabilization of PNIPAM, which may connect the NH and CO groups in the polymer. When urea is added, the water species are destroyed, thus leading to a phase transition at a lower temperature.

9 Conclusion

NIR spectroscopy is a powerful tool for quantitative determination and structural analysis in complex systems. Chemometric methods must be applied in the spectral analysis due to the complexity of the spectra. The methods of modeling techniques, spectral preprocessing, variable selection, outlier detection, modeling transfer and discrimination analysis were developed for NIR spectra modeling. In these methods, PLS is the most widely applied modeling method. Besides, CWT, MC-UVE, RT and PCA are also powerful tools to perform spectral analysis. Furthermore, quantitative and structural information can be extracted from temperature-dependent NIR spectra by chemometric methods. Quantitative models between the spectra and the temperature or concentration can be established. Using the spectral variation with temperature, the changes of structure and interaction in liquid mixtures or aqueous systems can be observed. With the development of chemometrics, NIR spectroscopy can be a promising technique for understanding the properties or functions of compounds in chemical and biological systems.

Acknowledgments. This study was supported by the National Natural Science Foundation of China (No. 22174075), the Natural Science Foundation of Tianjin, China (No. 20JCYBJC01480), the Frontiers Science Center for New Organic Matter, Nankai University (No. 63181206), the Fundamental Research Funds for the Central Universities, Nankai University (No. 63211019) and the Haihe Laboratory of Sustainable Chemical Transformations.

References

1. Moros, J., Garrigues, S., de la Guardia, M.: Vibrational spectroscopy provides a green tool for multi-component analysis. Trends Anal. Chem. **29**, 578–591 (2010)
2. Blanco, M., Villarroya, I.: NIR spectroscopy: a rapid-response analytical tool. Trends Anal. Chem. **21**, 240–250 (2002)
3. Pasquini, C.: Near infrared spectroscopy: a mature analytical technique with new perspectives - a review. Anal. Chim. Acta **1026**, 8–36 (2018)
4. Wang, F., Chen, D., Shao, X.G.: Study on model of near-infrared spectroscopy and chemical components of cigarettes. Tob. Sci. Technol. **5**, 23–26 (2002)
5. Shao, X.G., Bian, X.H., Liu, J.J., Zhang, M., Cai, W.S.: Multivariate calibration methods in near infrared spectroscopic analysis. Anal. Methods **2**, 1662–1666 (2010)
6. Zhang, J., Cai, W.S., Shao, X.G.: New algorithms for calibration transfer in near infrared spectroscopy. Prog. Chem. **29**, 902–910 (2017)
7. Shao, X.G., Cai, W.S.: Wavelet analysis in analytical chemistry. Rev. Anal. Chem. **17**, 235–285 (1998)
8. Shao, X.G., Leung, A.K.M., Chau, F.T.: Wavelet: a new trend in chemistry. Acc. Chem. Res. **36**, 276–283 (2003)

9. Ma, C.X., Shao, X.G.: Continuous wavelet transform applied to removing the fluctuating background in near-infrared spectra. J. Chem. Inf. Comput. Sci. **44**, 907–911 (2004)
10. Cai, W.S., Li, Y.K., Shao, X.G.: A variable selection method based on uninformative variable elimination for multivariate calibration of near-infrared spectra. Chemom. Intell. Lab. Syst. **90**, 188–194 (2008)
11. Xu, H., Liu, Z.C., Cai, W.S., Shao, X.G.: A wavelength selection method based on randomization test for near-infrared spectral analysis. Chemom. Intell. Lab. Syst. **97**, 189–193 (2009)
12. Zhang, J., Cui, X., Cai, W., Shao, X.: A variable importance criterion for variable selection in near-infrared spectral analysis. Sci. China Chem. **62**(2), 271–279 (2018). https://doi.org/10.1007/s11426-018-9368-9
13. Shao, X.G., Kang, J., Cai, W.S.: Quantitative determination by temperature dependent near-infrared spectra. Talanta **82**, 1017–1021 (2010)
14. Kang, J., Cai, W.S., Shao, X.G.: Quantitative determination by temperature dependent near-infrared spectra: a further study. Talanta **85**, 420–424 (2011)
15. Cui, X., Sun, Y., Cai, W., Shao, X.: Chemometric methods for extracting information from temperature-dependent near-infrared spectra. Sci. China Chem. **62**(5), 583–591 (2019). https://doi.org/10.1007/s11426-018-9398-2
16. Sun, Y., Cai, W.S., Shao, X.G.: Chemometrics: an excavator in temperature-dependent near-infrared spectroscopy. Molecules **27**, 452 (2022)
17. Shao, X.G., Bian, X.H., Cai, W.S.: An improved boosting partial least squares method for near-infrared spectroscopic quantitative analysis. Anal. Chim. Acta **666**, 32–37 (2010)
18. Jing, M., Cai, W.S., Shao, X.G.: Multiblock partial least squares regression based on wavelet transform for quantitative analysis of near infrared spectra. Chemom. Intell. Lab. Syst. **100**, 22–27 (2010)
19. Liu, Z.C., Cai, W.S., Shao, X.G.: A weighted multiscale regression for multivariate calibration of near infrared spectra. Analyst **134**, 261–266 (2009)
20. Xu, H., Cai, W.S., Shao, X.G.: Weighted partial least squares regression by variable grouping strategy for multivariate calibration of near infrared spectra. Anal. Methods **2**, 289–294 (2010)
21. Shi, X., Cai, W.S., Shao, X.G.: Local regression method in wavelet domain and its application in near-infrared quantitative analysis, Chinese. J. Anal. Chem. **36**, 1093–1096 (2008)
22. Li, Y.K., Shao, X.G., Cai, W.S.: A consensus least squares support vector regression (LS-SVR) for analysis of near-infrared spectra of plant samples. Talanta **72**, 217–222 (2007)
23. Helland, I.S., Naes, T., Isaksson, T.: Related versions of the multiplicative scatter correction method for preprocessing spectroscopic data. Chemom. Intell. Lab. Syst. **29**, 233–241 (1995)
24. Barnes, R.J., Dhanoa, M.S., Lister, S.J.: Standard normal variate transformation and de-trending of near-infrared diffuse reflectance spectra. Appl. Spectrosc. **43**, 772–777 (1989)
25. Savitzky, A., Golay, M.J.E.: Smoothing and differentiation of data by simplified least squares procedures. Anal. Chem. **36**, 1627–1639 (1964)
26. Sjoblom, J., Svensson, O., Josefson, M., Kullberg, H., Wold, S.: An evaluation of orthogonal signal correction applied to calibration transfer of near infrared spectra. Chemom. Intell. Lab. Syst. **44**, 229–244 (1998)
27. Chen, D., Shao, X.G., Hu, B., Su, Q.D.: A background and noise elimination method for quantitative calibration of near infrared spectra. Anal. Chim. Acta **511**, 37–45 (2004)
28. Shan, R.F., Mao, Z.Y., Yin, L.H., Cai, W.S., Shao, X.G.: Discrimination of Chinese patent medicines using near-infrared spectroscopy and principal component accumulation method. Anal. Methods **6**, 4692–4697 (2014)
29. Shao, X.G., Cui, X.Y., Wang, M., Cai, W.S.: High order derivative to investigate the complexity of the near infrared spectra of aqueous solutions. Spectrochim. Acta Part A **213**, 83–89 (2019)
30. Centner, V., Massart, D.L.: Elimination of uninformative variables for multivariate calibration. Anal. Chem. **68**, 3851–3858 (1996)

31. Li, H.D., Liang, Y.Z., Xu, Q.S., Cao, D.S.: Key wavelengths screening using competitive adaptive reweighted sampling method for multivariate calibration. Anal. Chim. Acta **648**, 77–84 (2009)
32. Lucasius, C.B., Kateman, G.: Genetic algorithms for large-scale optimization in chemometrics: an application. Trends Anal. Chem. **10**, 254–261 (1991)
33. Du, G.R., Cai, W.S., Shao, X.G.: A variable differential consensus method for improving the quantitative near-infrared spectroscopic analysis. Sci. China: Chem. **55**, 1946–1952 (2012)
34. Wang, C.C., Wang, S.Y., Cai, W.S., Shao, X.G.: Silver mirror for enhancing the detection ability of near-infrared diffuse reflectance spectroscopy. Talanta **162**, 123–129 (2017)
35. Mao, Z.Y., Cai, W.S., Shao, X.G.: Selecting significant genes by randomization test for cancer classification using gene expression data. J. Biomed. Inf. **46**, 594–601 (2013)
36. Shao, X.G., Du, G.R., Jing, M., Cai, W.S.: Application of latent projective graph in variable selection for near infrared spectral analysis. Chemom. Intell. Lab. Syst. **114**, 44–49 (2012)
37. Shao, X.G., Zhang, M., Cai, W.S.: Multivariate calibration of near-infrared spectra by using influential variables. Anal. Methods **4**, 467–473 (2012)
38. Chen, D., Shao, X.G., Hu, B., Su, Q.D.: Simultaneous wavelength selection and outlier detection in multivariate regression of near-infrared spectra. Anal. Sci. **21**, 161–166 (2005)
39. Chen, D., Cai, W.S., Shao, X.G.: An adaptive strategy for selecting representative calibration samples in the continuous wavelet domain for near-infrared spectral analysis. Anal. Bioanal. Chem. **387**, 1041–1048 (2007)
40. Liu, Z.C., Cai, W.S., Shao, X.G.: Outlier detection in near-infrared spectroscopic analysis by using Monte Carlo cross-validation. Sci. China Ser. B Chem. **51**, 751–759 (2008)
41. Li, X.Y., Cai, W.S., Shao, X.G.: Correcting multivariate calibration model for near infrared spectral analysis without using standard samples. J. Near Infrared Spectrosc. **23**, 285–291 (2015)
42. Liu, Y., Cai, W.S., Shao, X.G.: Linear model correction: a method for transferring a near-infrared multivariate calibration model without standard samples. Spectrochim. Acta Part A **169**, 197–201 (2016)
43. Zhang, J., Cui, X.Y., Cai, W.S., Shao, X.G.: Modified linear model correction: a calibration transfer method without standard samples. NIR News **29**, 24–27 (2018)
44. Li, P., Du, G.R., Cai, W.S., Shao, X.G.: Rapid and nondestructive analysis of pharmaceutical products using near-infrared diffuse reflectance spectroscopy. J. Pharm. Biomed. Anal. **70**, 288–294 (2012)
45. Xu, Z.H., Liu, Y., Li, X.Y., Cai, W.S., Shao, X.G.: Discriminant analysis of Chinese patent medicines based on near-infrared spectroscopy and principal component discriminant transformation. Spectrochim. Acta Part A **149**, 985–990 (2015)
46. Liu, J.J., Xu, H., Cai, W.S., Shao, X.G.: Discrimination of industrial products by on-line near infrared spectroscopy with an improved dendrogram. Chin. Chem. Lett. **22**, 1241–1244 (2011)
47. Wang, Y., Ma, X., Wen, Y.D., Liu, J.J., Cai, W.S., Shao, X.G.: Discrimination of plant samples using near-infrared spectroscopy with a principal component accumulation method. Anal. Methods **4**, 2893–2899 (2012)
48. Fan, M., Liu, X., Yu, X., Cui, X., Cai, W., Shao, X.: Near-infrared spectroscopy and chemometric modelling for rapid diagnosis of kidney disease. Sci. China Chem. **60**(2), 299–304 (2016). https://doi.org/10.1007/s11426-016-0092-6
49. Cui, X.Y., Yu, X.M., Cai, W.S., Shao, X.G.: Water as a probe for serum-based diagnosis by temperature-dependent near-infrared spectroscopy. Talanta **204**, 359–366 (2019)
50. Cui, X.Y., Zhang, J., Cai, W.S., Shao, X.G.: Chemometric algorithms for analyzing high dimensional temperature dependent near infrared spectra. Chemom. Intell. Lab. Syst. **170**, 109–117 (2017)

51. Cui, X.Y., Liu, X.W., Yu, X.M., Cai, W.S., Shao, X.G.: Water can be a probe for sensing glucose in aqueous solutions by temperature dependent near infrared spectra. Anal. Chim. Acta **957**, 47–54 (2017)
52. Shan, R.F., Zhao, Y., Fan, M.L., Liu, X.W., Cai, W.S., Shao, X.G.: Multilevel analysis of temperature dependent near-infrared spectra. Talanta **131**, 170–174 (2015)
53. Han, L., Cui, X.Y., Cai, W.S., Shao, X.G.: Three-level simultaneous component analysis for analyzing the near-infrared spectra of aqueous solutions under multiple perturbations. Talanta **217**, 121036 (2020)
54. Shao, X.G., Cui, X.Y., Yu, X.M., Cai, W.S.: Mutual factor analysis for quantitative analysis by temperature dependent near infrared spectra. Talanta **183**, 142–148 (2018)
55. Wang, M.Y., Cui, X.Y., Cai, W.S., Shao, X.G.: Temperature-dependent near-infrared spectroscopy for sensitive detection of glucose. Acta Chim. Sinica **78**, 125–129 (2020)
56. Cui, X.Y., Zhang, J., Cai, W.S., Shao, X.G.: Selecting temperature-dependent variables in near-infrared spectra for aquaphotomics. Chemom. Intell. Lab. Syst. **183**, 23–28 (2018)
57. Cui, X.Y., Cai, W.S., Shao, X.G.: Glucose induced variation of water structure from temperature dependent near infrared spectra. RSC Adv. **6**, 105729–105736 (2016)
58. Tan, J.H., et al.: Knowledge-based genetic algorithm for resolving the near-infrared spectrum and understanding the water structures in aqueous solution. Chemom. Intell. Lab. Syst. **206**, 104150 (2020)
59. Sun, Y., Cui, X.Y., Cai, W.S., Shao, X.G.: Understanding the complexity of the structures in alcohol solutions by temperature-dependent near-infrared spectroscopy. Spectrochim. Acta Part A **229**, 117864 (2020)
60. Zhu, X.W., Cui, X.Y., Cai, W.S., Shao, X.G.: Temperature dependent near infrared spectroscopy for understanding the hydrogen bonding of amines. Acta Chim. Sinica **76**, 298–302 (2018)
61. Shao, X.G., Cui, X.Y., Liu, Y., Xia, Z.Z., Cai, W.S.: Understanding the molecular interaction in solutions by chemometric resolution of near-infrared spectra. ChemistrySelect **2**, 10027–10032 (2017)
62. Zhao, H.T., Sun, Y., Guo, Y.C., Cai, W.S., Shao, X.G.: Near infrared spectroscopy for low-temperature water structure analysis. Chem. J. Chin. Univ. **41**, 1968–1974 (2020)
63. Su, T., Sun, Y., Han, L., Cai, W.S., Shao, X.G.: Revealing the interactions of water with cryoprotectant and protein by near-infrared spectroscopy. Spectrochim. Acta Part A **266**, 120417 (2022)
64. Ma, L., Cui, X.Y., Cai, W.S., Shao, X.G.: Understanding the function of water during the gelation of globular proteins by temperature-dependent near infrared spectroscopy. Phys. Chem. Chem. Phys. **20**, 20132–20140 (2018)
65. Sun, Y., Ma, L., Cai, W.S., Shao, X.G.: Interaction between tau and water during the induced aggregation revealed by near-infrared spectroscopy. Spectrochim. Acta Part A **230**, 118046 (2020)
66. Wang, S.Y., Wang, M., Han, L., Sun, Y., Cai, W.S., Shao, X.G.: Insight into the stability of protein in confined environment through analyzing the structure of water by temperature-dependent near-infrared spectroscopy. Spectrochim. Acta Part A **267**, 120581 (2022)
67. Wang, L., Zhu, X.W., Cai, W.S., Shao, X.G.: Understanding the role of water in the aggregation of poly(N, N-dimethylaminoethyl methacrylate) in aqueous solution using temperature-dependent near-infrared spectroscopy. Phys. Chem. Chem. Phys. **21**, 5780–5789 (2019)
68. Ma, B., Wang, L., Han, L., Cai, W.S., Shao, X.G.: Understanding the effect of urea on the phase transition of poly(N-isopropylacrylamide) in aqueous solution by temperature-dependent near-infrared spectroscopy. Spectrochim. Acta Part A **253**, 119573 (2021)

Current Status and Future Trends in Sensor Miniaturization

Christian W. Huck(✉), Krzysztof B. Bec, and Justyna Grabska

Institute of Analytical Chemistry and Radiochemistry, University of Innsbruck, Innrain 80-82, 6020 Innsbruck, Austria
Christian.W.Huck@uibk.ac.at

Abstract. Ongoing miniaturization of spectrometers creates a perfect synergy with the common advantages of NIR spectroscopy. The combination of portability and direct on-site application with high-throughput and non-invasive way of analysis is a decisive advantage in various industries. The characteristics of miniaturized NIR sensors are noted versus benchtop spectrometers regarding the performance, applicability, and optimization of methodology. These devices remarkably increase the flexibility of analysis; however, various factors affect their performance in different analytical scenarios. Currently, it is a focused research direction to perform systematical evaluation studies of the accuracy and reliability of various miniaturized spectrometers that are based on different technologies, e.g. Fourier transform (FT)-NIR, micro-optoelectro-mechanical system (MOEMS) Hadamard mask or linear variable filter (LVF) coupled with array detector.

Progressing technology is accompanied by innovative data-analytical methods, integrated into the package of a micro-NIR analytical method to improve its accuracy, reliability and applicability. Advanced calibration methods (e.g. artificial neural networks ANN, nonlinear regression) directly improve the performance of miniaturized instruments in challenging analysis and balance the accuracy of these instruments towards laboratory spectrometers. Two-dimensional correlation spectroscopy (2D-COS) provides insight into the relative sensitivity observed between different instruments in specific NIR bands. The quantum mechanical simulation of NIR spectra reveals the wavenumber regions where the best correlated spectral information resides, and unveils the interactions of the target analyte with surrounding matrix, to ultimately enhance the information gathered from the NIR spectra. This set of methods enables the intelligent design of future micro-NIR analyzes, critically important for samples with a complex matrix.

Keywords: Sensors · Miniaturization · Handheld · Portable · Natural medicines · Environment

1 Introduction

During the last four decades, near-infrared (NIR) spectroscopy (800–2500 nm; 12,500–4000 cm^{-1}) has become one of the most attractive and used methods for analysis for the following reasons: it represents a non-destructive analytical tool allowing a fast

X. Chu et al. (Eds.): ICNIR 2021, *Sense the Real Change: Proceedings of the 20th International Conference on Near Infrared Spectroscopy*, pp. 59–72, 2022.
https://doi.org/10.1007/978-981-19-4884-8_5

and simultaneous qualitative and quantitative characterization of a wide variety of samples [1–3]. The last decade marked a rapid acceleration in the continuing trend of the miniaturization of NIR spectrometers (Fig. 1). These devices remarkably increase the flexibility of analysis; however, attention needs to be paid to the various factors affecting their performance in different scenarios [1–5]. Currently, it is a focused and very active research direction to perform systematical evaluation studies of the analytical accuracy and reliability of various miniaturized spectrometers available at the market [4].

NIR spectroscopy a particularly potent tool for analyzing natural products and their constituents [5]. The challenging characteristics of such analyses are commonly encountered in agri-food sector [6, 7], medicinal plant sector of pharmaceutical industry [8, 9], as well as environmental monitoring and ecology studies [10, 11], However, NIR spectra are intrinsically very complex with highly convoluted signals, which becomes an extremely strong feature in the case of complex, multi-constituent samples such as those of natural origin [12, 13]. As the result of chemical diversity and often also challenging physical properties, such as specific surface texture, presence of layers resulting from biological structure of plant tissue [14], e.g. husks, and even the presence of color, miniaturized NIR instruments often face serious challenges in natural product analysis. Therefore, the applicability and analytical performance of miniaturized NIR spectrometers in such applications is the topic of extensive feasibility research, a factor of critical importance for establishing practical applications with high reliability.

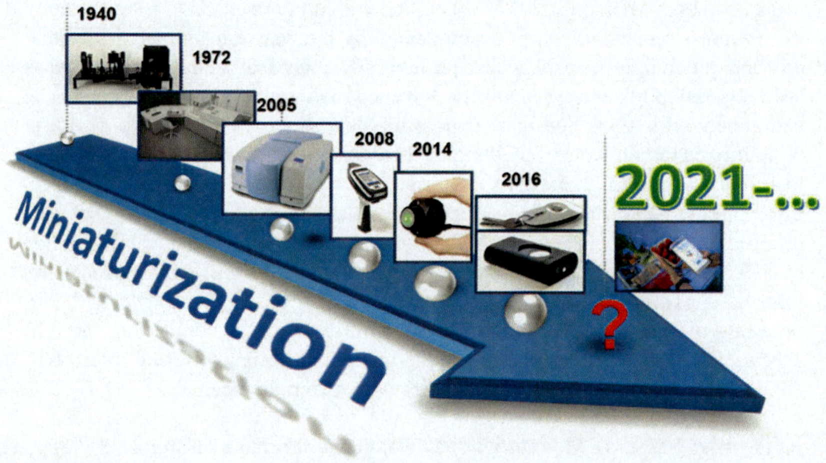

Fig. 1. The accelerating trend in the miniaturization of NIR spectrometers with visible recent breakthroughs in the flexibility and deployability factors.

This problem connects to a variety of scenarios, in which design and knowledge-based optimization of an analytical pathway to successful miniaturized NIR analysis maximizes the practical gain from using this modern technology. In this proceeding review article overviewed are the research directions at this wide-horizon of investigation, currently advanced by the Institute of Analytical Chemistry and Radiochemistry, University of Innsbruck (Austria). The importance of using combined tools, integrated into NIR analytical method, which improve its accuracy, reliability, and applicability, is discussed (Fig. 2). Advanced methods of calibration (e.g. Artificial Neural Networks) directly improve the performance of miniaturized instruments in analyzing complex and challenging samples, equalizing the accuracy of these instruments with benchtop spectrometers. Two-dimensional correlation spectroscopy (2D-COS) yields insights into the relative sensitivity observed between different instruments towards specific NIR bands [15]. Quantum mechanical simulation of NIR spectra unveils the role that the interaction of the targeted analyte with the surrounding water molecules play in enhancing the information acquired from NIR spectra [16]. The simulation also enables interpreting the instrumental difference observed between different handheld sensors in light of the chemical information on a given constituent, and to predict the performance level of a spectrometer in similar analysis. This suite of methods enables intelligent design of future NIRS analysis, which is particularly important for moist samples with complex matrix [17].

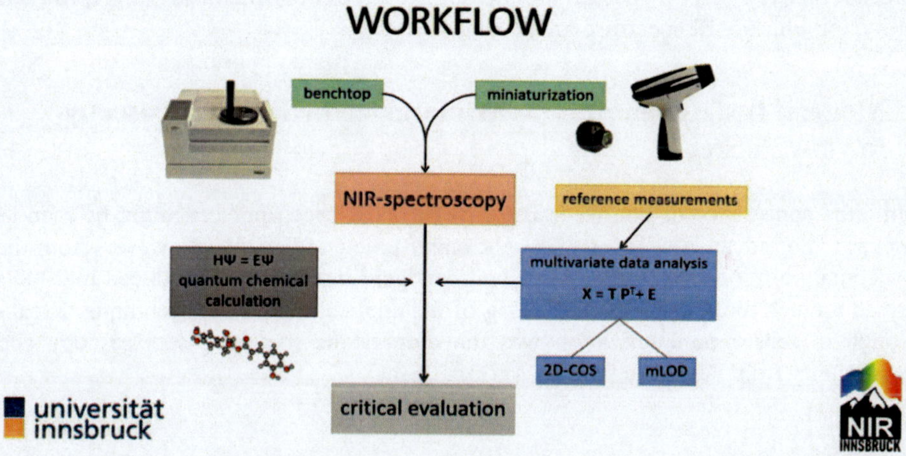

Fig. 2. The scheme of multi-planar methodology used for increasing the accuracy, flexibility and applicability of modern NIR spectroscopy.

2 New Discoveries, Novel Concepts and Innovative Applications of NIR Spectroscopy

2.1 Fundamental Properties of NIR Spectroscopy

To understand the current status of NIR spectroscopy as an analytical technique, it is important to consider decades of history of its advancement [18]. The discovery of NIR spectroscopy occurred in 19th century, when Knut Angstrom measured NIR and MIR spectra (down to 2000 cm^{-1}) of simplest chemical compounds, e.g. CO, CO2, CH4, HCl. Interestingly, he discovered that the NIR bands were much weaker in intensity than those he observed in the MIR region. This principal feature, of course, grants NIR spectroscopy the first of this major advantage: non-destructive application, as the sample does not need to be 'diluted' in non-absorbing medium to provide a well-resolved spectrum. However, this important fact remained unnoticed until Carl Norris from the United States Department of Agriculture (USDA) who needed an analytical method to assess grain quality non-destructively. Norris not only recognized the potential of NIR spectroscopy, but also applied statistical methods to create a calibration model based on NIR data and further use it to analyze grain in high-throughput manner. Thus, the next prime advantage of this technique, rapid and cost-effective character of analysis, were established. On the other hand, while superior in practical terms to its peer, i.e. MIR technique, NIR spectroscopy also suffers from inferior chemical specificity as, prior to recent breakthrough in spectra interpretability, it lacks the chemical 'fingerprinting' ability, for which MIR spectroscopy is valued [19].

3 Modern Instrumentation – Miniaturized NIR Spectroscopy in your Pocket

While the appearance of Fourier-transform (FT) NIR spectrometers might be seen as another major advance, it indeed brought major practical benefits. However, from the conceptual point-of-view, the portable and handheld instruments introduced in 2000's formed a much more decisive re-shaping of the applications of this technique. Breakthrough in sensor miniaturization was the cornerstone transiting spectroscopy and spectrometry from lab to the field [1–4] (Fig. 3).

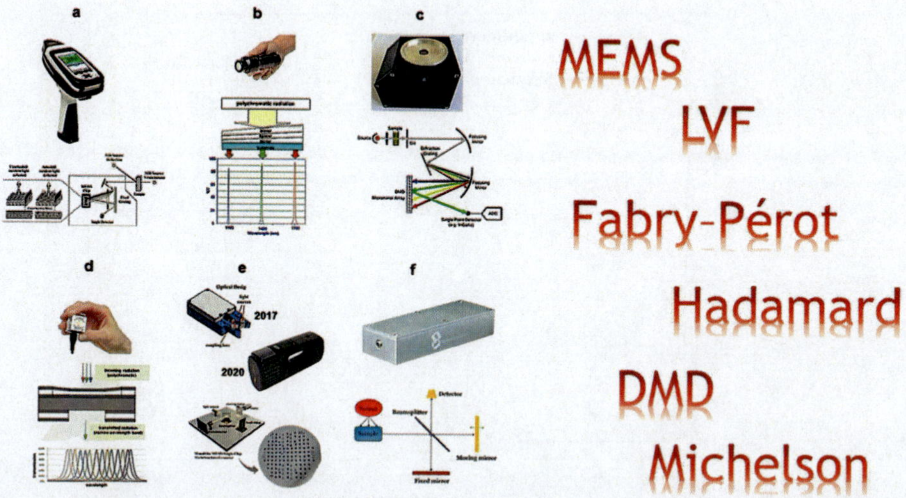

Fig. 3. Principles of wavelengths selectors built into different handheld NIR spectrometers: (a) MEMS Hadamard mask – microPHAZIR, Thermo Fisher Scientific, Waltham, USA; (b) LVF –MicroNIR Pro ES 1700, VIAVI, Santa Rosa, USA; (c) MEMS DMD – implementation of DLP NIRscan module, Texas Instruments, Dallas, USA; (d) MEMS Fabry–Perot interferometer – NIRONE Sensor S, Spectral Engines, Helsinki, Finland; (e) MEMS Michelson interferometer – NeoSpectra, Si-Ware, Cairo, Egypt; (f) MEMS Michelson interferometer with a large mirror – nanoFTIR NIR, SouthNest Technology, Hefei, China.

4 Novel Methods Provide Detailed Understanding of NIR Spectra

Regardless of the application, the intrinsic complexity of NIR spectra and the difficulty in their direct interpretation remains a limiting factor. In contrast to mid-infrared (MIR) and Raman techniques, NIR spectroscopy has been hindered in forming practically accessible synergy with computational chemistry [20]. In recent few years, however, advances in the tools of computational chemistry have created an opportunity to make a step beyond this barrier [12, 21–23].

This led to considerable advances in the applicability of the methods of theoretical chemistry to NIR spectroscopy [24]. Accurate simulation of NIR spectra of reasonably large molecules largely improves our comprehension of NIR spectra and offers an opportunity to make a step beyond this barrier. Quantum mechanical simulation of NIR spectra of a variety of compounds, which are significant from the point of view of physiochemical and analytical spectroscopy. The examples range from basic molecules (alcohols, nitriles, carboxylic acids) [25–28], to complex molecules with importance to biophysical science (fatty acids, nucleobases) [29, 30], material science and industry [31], and analytical chemistry (vitamins, natural drugs, polyphenols, alkaloids, food adulterants) [32–35].

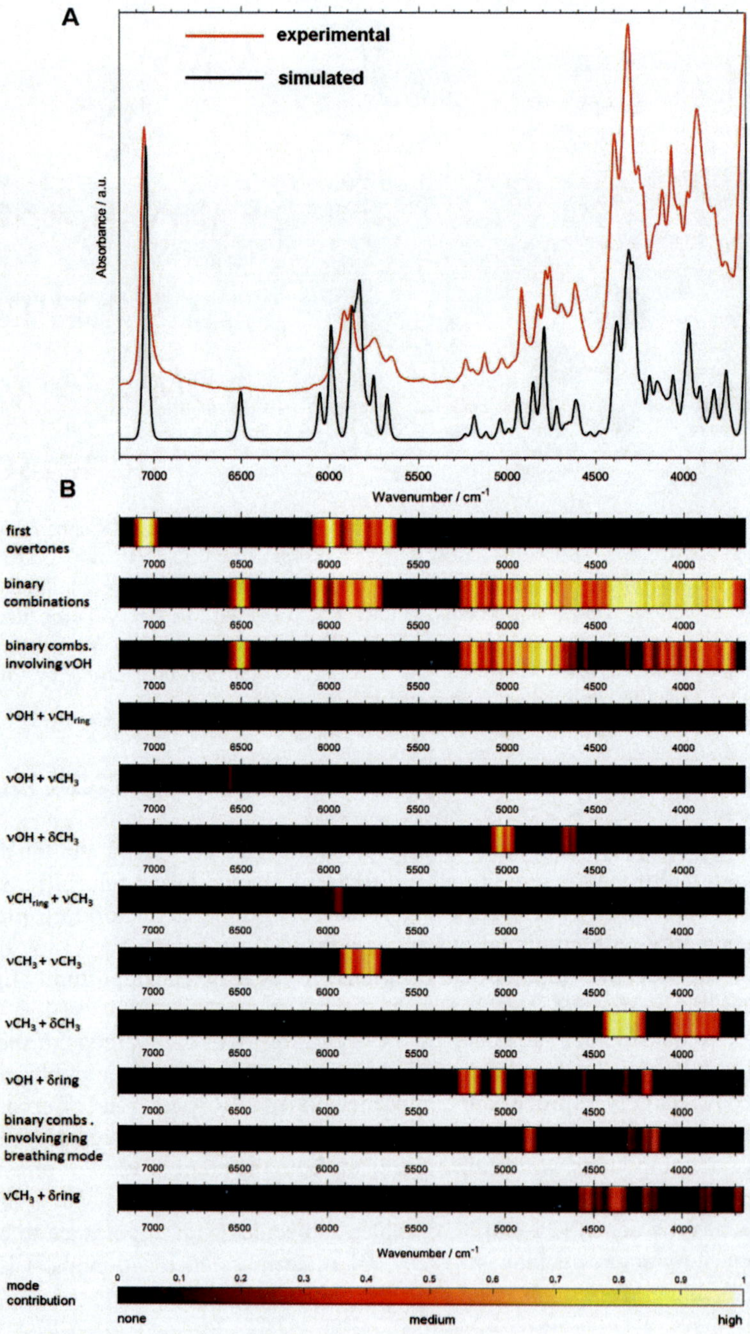

Fig. 4. Simulated NIR spectrum of thymol with depicted contributions of vibrational modes to NIR intensity. Reproduced with permission from Ref. [37]

A remarkable potential arises from the growing applicability of anharmonic computations to solving the problems, which arise in both basic and analytical NIR spectroscopy [36]. Highly convoluted, overlapping nature of NIR spectra can be successfully dissected in theoretical spectra, as presented in the example of thymol molecule – an important anti-oxidant presents in numerous medicinal plants (Fig. 4) [37]. The elucidated rich information stemming from numerous NIR bands can subsequently be used to improve basic understanding of NIR spectroscopy as well as to advance its applications.

This detailed information on chemical origin of each NIR band enables new approaches to supporting analytical applications. For example, one may select the best suited sensor for intended analysis by assessing its suitability to measure characteristic absorption regions of the targeted constituent. Miniaturized NIR spectrometers most often can only measure fragments of NIR spectrum (Fig. 5), and hence, only selective chemical information can be acquired by these devices. Accurate spectra simulation, such as the example fo thymol discussed above (Fig. 4), enable full understanding of which NIR vibrations a specific sensor can acquire, and thus, a best suited instrument for targeted analysis can be selected.

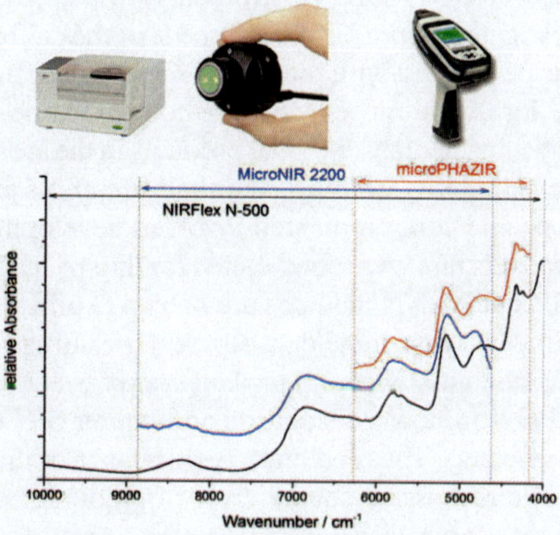

Fig. 5. In contrast to benchtop FT-NIR instruments measuring full NIR wavenumber region, the access of miniaturized NIR spectrometers to chemical information is selective resulting from only narrow, at best partially overlapping, fragments of NIR spectra measured by these sensors.

Furthermore, detailed interpretation of PLS regression factors becomes possible as well, providing deep insight into the critically important connection between the chemical information present in the sample and the analytical framework [38]. Interpretation, in chemical sense, of the meaningful variables in chemometric models becomes possible this way [39–41].

These three essential pillars, each of which takes advantage of a detailed understanding of the chemical information in the processed spectra, enable a 'smart', knowledge-based design and optimization of an analytical approach in modern miniaturized NIR spectroscopy.

5 Miniaturization vs. Advanced Calibration Methods

As explained, miniaturized technology besides revolutionary practical advantages, also implies necessary limitations in optical/spectral capabilities of these often very compact instruments. Advanced calibration algorithms, e.g. Artificial Neural Networks (ANN) and non-linear regression methods (e.g. Gaussian Process Regression, GPR) significantly improve the prediction performance of portable NIR spectrometers superior performance to PLSR no/minimal penalty of handheld vs. benchtop spectroscopy in difficult analyses (e.g. moisture analysis in plant matrices) advanced calibration enables the miniaturized spectrometers to nearly match the performance of benchtop instruments.

Mayr et al. [17] evidenced that the analytical performance of miniaturized spectrometers in a challenging scenario, i.e. that involves chemically complex plant matrices, can be significantly improved with advanced calibration methods. In the explored analytical scenario, the quantification of the moisture content in 192 samples of dried plant extract composed of five different plants from different geographical origins harvested at various times within two years. The spectra measurements involved a rich suite of spectrometers; two benchtop (NIRFlex N-500 and MPA I) and three miniaturized (microPHAZIR, MicroNIR 2200 and MicroNIR 1700ES) instruments were used to acquire the spectral datasets for all 192 investigated samples. For the needs of the calibration, reference analysis of the moisture was performed with the use of Karl Fischer titration method, which is the industry standard for this purpose. Moisture content is one of the most important quality parameters of the intermediate and final products in the area of natural medicines. It determines the market suitability, stability and shelf life of the product and requires to be constantly monitored. The calibration step involved development and evaluation of PLSR, GPR and ANN models that were constructed for the spectral sets from each instrument. The non-linear GPR and ANN methods are known to offer substantially improved performance in the case of less than ideal data-sets, e.g. resulting from difficult nature of the analyzed sample (i.e. chemically complex plant matrix), reduced quality of the spectra (i.e. narrow spectral region, lower resolution and poorer S/N typically accepted for the miniaturized spectrometers). The prediction performance of those calibration models was evaluated through the root-mean square error of prediction (RMSEP) determined for an independent test set (Table 1).

Thus, the penalty to the accuracy resulting from the hardware miniaturization can be compensated by using GPR or ANN calibration. In this case, the miniaturized spectrometers offered the prediction performance at the level of the benchtop instruments (Table 1). Moreover, the samples in native state proved to be more difficult to analyze for all evaluated instruments when using PLSR calibration [7].

Table 1. Important Bold values highlight the best prediction performance for each of the spectrometers. Reproduced (CC-BY 4.0 license) from Ref. [17].

	NIR spectrometer	Regression		ANN			
		PLSR	GPR	Number of hidden neurons			
				1	2	3	4
Drying agent	NIRFlex N-500	0.27	0.31	0.35	0.36	0.32	0.39
	MPA I	0.27	0.32	0.39	0.35	0.39	0.33
	microPHAZIR	**0.37**	**0.30**	**0.27**	**0.30**	**0.31**	**0.48**
	MicroNIR 2200	**0.32**	**0.30**	**0.33**	**0.29**	**0.33**	**0.33**
	MicroNIR 1700ES	**0.28**	**0.28**	**0.25**	**0.33**	**0.30**	**0.34**
Native	NIRFlex N-500	0.45	0.36	0.73	0.55	0.41	0.43
	MPA I	0.47	0.44	0.54	0.68	0.56	0.59
	microPHAZIR	**0.54**	**0.60**	**0.59**	**0.53**	**0.58**	**0.48**
	MicroNIR 2200	**0.43**	**0.38**	**0.32**	**0.35**	**0.44**	**0.44**
	MicroNIR 1700ES	**0.50**	**0.43**	**0.46**	**0.70**	**0.67**	**0.72**

6 NIR Sensor Fusion

A direct improvement of performance of sensors with mutually exclusive operational spectral regions may be accomplished by data fusion. Several cost-effective miniaturized NIR sensors appeared at the market, specifically intended for food analyses [4, 42–44]. These devices are potentially very interesting for small-scale manufacturing of food, which is becoming increasingly popular. Efficient quality control in such production poses considerable difficulties, and final product quality depends on the supplied ingredients. Miniaturized NIR spectrometers are particularly promising in such scenarios, but often manifest inferior performance compared to benchtop NIR spectrometers; in certain cases, no successful calibration by PLSR can be obtained for some sensors in particularly challenging analyses [17, 40]. As discussed in the previous Section, advanced calibration methods offer promising potential in such cases. Further gains can be obtained with sensor fusion, which offers convenient uplift in performance by combining spectra measured in different wavenumber regions for extended access to chemical information for more reliable calibration.

In the study of Bec et al. [45], benchtop (Buchi NIRFlex N-500) and three miniaturized (MicroNIR 1700 ES, Tellspec Enterprise Sensor and SCiO sensor) were evaluated and optimized within calibration framework based on PLSR and Gaussian Process Regression (GPR) methods for prediction of protein content in fitness bars containing edible insect material. The analyzed protein content in the calibration series was between 19.3 and 23.0% (Table 2). For non-destructive analysis of intact bars, the RMSEP values from PLSR prediction were determined to be 0.611% (benchtop) and remained in the range of 0.545 to 0.659% (miniaturized spectrometers). The prediction by GPR models was 0.506% (benchtop) and 0.482–0.580% (miniaturized). Taking into account milled samples, the corresponding RMSEP values for PLSR prediction were improved to 0.210% for the benchtop spectrometer, but remained in the inferior range of 0.525−0.571% for the miniaturized devices. However, RMSEP values for GPR prediction for miniaturized spectrometers were noticeably improved to 0.230% (MicroNIR 1700 ES), 0.326% (Tellspec) and 0.338% (SCiO). In combination with the PLSR method, portable instruments showed significantly lower predictive performance measured by RMSEP values determined for data from an independent test-set. Using a nonlinear GPR calibration method significantly improves the accuracy of prediction of miniaturized spectrometers, with MicroNIR achieving the performance equal to the stationary instrument, while only slightly worse performance was achieved by Tellspec and SCiO sensors.

A further increase in the reliability of the analysis can be achieved by low-level fusion of the spectra from the spectrometers measuring largely different, only partially overlapping fragments of NIR spectrum, such as SCiO and Tellspec instruments considered in that study. Since Tellspec and SCiO sensors cover highly complementary VIS/SW-NIR and NIR wavelength ranges, an attempt has been made to aggregate (i.e. concatenate) data from these sensors. In GPR calibration and test-set validation performed on fused data (Tellspec + SCiO), RMSEP values were improved to 0.517% (for intact samples) and 0.295% (for milled samples). Because Tellspec and SCiO sensors are consumer-oriented devices with superior affordability, it is still a cost-effective and economical option to use them together to increase accuracy and reliability of the analysis. Fused-sensor NIR spectroscopy, with combined strengths of multiple miniaturized portable instruments, can be successfully used for rapid, non-destructive analysis of total protein content with better performance than the sensors used separately [45].

Table 2. The parameters of the regression models constructed for the fused spectral data from the two cost-effective miniaturized NIR spectrometers (Tellspec Enterprise Sensor and SCiO Sensor) for the analysis of protein content in intact and milled bars. Analyzed protein concentration range: 19.3–23.0% (w/w). Reproduced (CC-BY 4.0 license) from Ref. [45].

	Intact		Milled	
	PLSR	GPR	PLSR	GPR
Pretreatment	SNV, SG2 (25 SP)	SNV, SG2 (25 SP)	SG1 (11 SP)	SG1 (11 SP)
R^2 (Cal)	0.41	0.9	0.53	0.99
R^2 (CV)	0.28	0.55	0.48	0.9
RMSEC [%]	0.654	0.272	0.580	0.0002
RMSECV [%]	0.723	0.574	0.620	0.263
R^2 (TSV)	0.38	0.64	0.51	0.89
RMSEP [%]	**0.671**	**0.517**	**0.596**	**0.295**

SG – Savitzky-Golay (1, 2–first, second derivative); SP – smoothing point; SNV – Standard Normal Variate.

7 Summary of Current Trends and Future Prospects

NIR spectroscopy is a powerful tool for qualitative and quantitative analysis involving all kinds of samples used in various industries. Unique value of NIR spectroscopy is best manifested in analysis of natural products, e.g. plant medicines, agricultural crops and also food raw material and processed products. In the same applications, availability of miniaturized, fully portable spectrometers provides revolutionary advantages, which additionally form perfect synergy with common strengths of NIR spectroscopy. With miniaturized sensors, rapid, high-throughput analysis with no/minimal sample preparation can be performed on-site.

Accompanying progress in technology, recent breakthrough in spectra-analytical methods and fundamental science decisively changed our understanding of NIR spectra and largely expanded the potential for new applications. Accurate theoretical simulation of NIR spectra, interpretation of the chemical information and how it contributes to the instrumental difference between sensors based on various optical solutions, or sensor fusion combined together bring a new perspective on NIR spectroscopy.

These qualities directly translate to thriving development of new applications of NIR spectroscopy as it becomes even more widely adopted in critically important sectors of public interest in industrial, agricultural, and environmental contexts.

Conflict of Interest. The authors declare no conflict of interest.

References

1. Ozaki, Y., Huck, C.W., Tsuchikawa, S., Engelsen, S.B.: Near-Infrared Spectroscopy. Springer, Singapore (2021). https://doi.org/10.1007/978-981-15-8648-4
2. Ozaki, Y., Huck, C.W., Beć, K.B.: Near-IR spectroscopy and its applications. In: Gupta, V.P. (ed.) Molecular and Laser Spectroscopy. Advances and Applications. Elsevier, San Diego, pp. 11–38 (2018)
3. Huck, C.W., Beć, K.B., Grabska, J.: Near infrared spectroscopy in natural product research. In: Meyers, R.A. (ed.) Encyclopedia of Analytical Chemistry: Applications, Theory and Instrumentation. John Wiley & Sons, pp. 1–29 (2020)
4. Beć, K.B., Grabska, J., Huck, C.W.: Principles and applications of miniaturized near-infrared (NIR) spectrometers. Chem. Eur. J. **27**(5), 1514–1532 (2021)
5. Beć, K.B., Grabska, J., Huck, C.W.: NIR spectral analysis of natural medicines supported by novel instrumentation, methods of data analysis and interpretation. J. Pharm. Biomed. Anal. **193**, 11368 (2020)
6. Beć, K.B., Grabska, J., Huck, C.W.: Portable spectroscopy applications in food, feed and agriculture. In: Crocombre, R.A., Leary, P.E., Kammrath, B.W. (ed.) Portable spectroscopy and spectrometry 2, Applications. John Wiley & Sons, Ltd, pp. 299–324 (2021)
7. Beć, K.B., Grabska, J., Huck, C.W.: Issues in hyperspectral traceability of foods. In: Comprehensive Foodomics, Cifuentes, vol. 3, Elsevier, pp. 258–289 (2021)
8. Beć, K.B., Grabska, J., Huck, C.W.: Infrared and near-infrared spectroscopic techniques for the quality control of herbal medicines. In: Mukherjee, P.K. (ed.) Evidence-Based Validation of Herbal Medicine, Elsevier (2022)

9. Beć, K.B., Grabska, J., Huck, C.W.: Miniaturized near-infrared spectroscopy in natural product analysis. Current and future directions. In: Gupta, V.P. (ed.) Molecular and Laser Spectroscopy- Advances and Applications, vol. 3, Elsevier (2022)
10. Beć, K.B., Grabska, J., Huck, C.W.: Current and future applications of IR and NIR spectroscopy in ecology, environmental studies, wildlife and plant investigations. In: Cozzolino, D. (ed.) Comprehensive Analytical Chemistry, vol. 95 (2021)
11. Beć, K.B., Grabska, J., Huck, C.W.: Near-infrared (NIR) sensors in environmental analysis. In: Reference Module in Biomedical Sciences. Elsevier (2021)
12. Czarnecki, M.A., Beć, K.B., Grabska, J., Hofer, T.S., Ozaki, Y.: Overview of application of NIR spectroscopy to physical chemistry. In: Ozaki, Y., Huck, C.W., Tsuchikawa, S., Engelsen, S.B., (eds.) Near-Infrared Spectroscopy. Springer, Singapore, pp. 297−330 (2021).https://doi.org/10.1007/978-981-15-8648-4_13
13. Beć, K.B., Grabska, J., Huck, C.W.: Near-infrared spectroscopy in bio-applications. Molecules, **25**(12), 2948 (2020)
14. Beć, K.B., Grabska, J., Bonn, G.K., Popp, M., Huck, C.W.: Principles and applications of vibrational spectroscopic imaging in plant science: a review. Front. Plant Sci. **11**, 1226 (2020)
15. Kirchler, C.G., et al.: Critical evaluation of spectral information of benchtop vs. portable near-infrared spectrometers: quantum chemistry and two-dimensional correlation spectroscopy for a better understanding of PLS regression models of the rosmarinic acid content in Rosmarini folium. Analyst, **142**, 455–464 (2017)
16. Beć, K.B., Grabska, J., Ozaki, Y.: Advances in anharmonic methods and their applications to vibrational spectroscopies. In: Wójcik, M.J., Nakatsuji, H., Kirtman, B., Ozaki, Y. (eds.) Frontiers of Quantum Chemistry. Springer, Singapore, pp. 438–512 (2017). https://doi.org/10.1007/978-981-10-5651-2_20
17. Mayr, S., et al.: Challenging handheld NIR spectrometers with moisture analysis in plant matrices: performance of PLSR vs. GPR vs. ANN modelling. Spectrochim. Acta. A. **249**, 119342 (2021)
18. Beć, K.B., Grabska, J., Huck, C.W.: Physical principles of infrared spectroscopy. In: Cozzolino, D. (ed.) Comprehensive Analytical Chemistry, vol. 95 (2020)
19. Beć, K.B., Grabska, J., Huck, C.W.: Biomolecular and bioanalytical applications of infrared spectroscopy – a review. Analytica. Chimica. Acta **1133**, 150–177 (2020)
20. Krzysztof, B., Beć, B., Grabska, J., Hofer, T. S.: Introduction to quantum vibrational spectroscopy. In: Ozaki, Y., Huck, C., Tsuchikawa, S., Engelsen, S.B. (eds.) Near-Infrared Spectroscopy, pp. 83–110. Springer, Singapore (2021). https://doi.org/10.1007/978-981-15-8648-4_5
21. Beć, K.B., Grabska, J., Huck, C.W., Ozaki, Y.: Quantum mechanical simulation of near-infrared spectra. Applications in physical and analytical chemistry. In Ozaki, Y., Wójcik, M.J., Popp, J. (eds.) Molecular Spectroscopy: A Quantum Chemistry Approach, vol. 2. Wiley-VCH, Weinheim, Germany, pp. 353–388 (2019)
22. Beć, K.B., Grabska, J., Huck, C.W.: Current and future research directions in computer-aided near-infrared spectroscopy a perspective. Spectrochim. Acta. A, **254**, 119625 (2021)
23. Beć, K.B., Huck, C.W. (eds.) Advances in near infrared spectroscopy and related computational methods. MDPI (2020)
24. Ozaki, Y., et al.: Advances, challenges and perspectives of quantum chemical approaches in molecular spectroscopy of the condensed phase. Chem. Soc. Rev. **50**, 10917–10954 (2021)
25. Beć, K.B., Grabska, J., Huck, C.W., Czarnecki, M.A.: Effect of conformational isomerism on NIR spectra of ethanol isotopologues spectroscopic and anharmonic DFT study. J. Mol. Liq. **310**, 113271 (2020)

26. Beć, K.B., Grabska, J., Huck, C.W., Czarnecki, M.A.: Spectra–structure correlations in isotopomers of ethanol (CX3CX2OX, X = H, D): combined near-infrared and anharmonic computational study. Molecules, **24**, 2189 (2019)
27. Grabska, J., Beć, K.B., Ozaki, Y., Huck, C.W.: Temperature drift of conformational equilibria of butyl alcohols studied by near-infrared spectroscopy and fully anharmonic DFT. J. Phys. Chem. A **121**(9), 1950–1961 (2017)
28. Beć, K.B., Karczmit, D., Kwaśniewicz, M., Ozaki, Y., Czarnecki, M.A.: Overtones of νCN vibration as a probe of structure of liquid CH3CN, CD3CN, and CCl3CN: combined infrared, near-infrared, and Raman spectroscopic studies with anharmonic density functional theory calculations. J. Phys. Chem. A, **123**, 4431–4442 (2019)
29. Grabska, J., Beć, K.B., Ishigaki, M., Huck, C.W., Ozaki, Y.: NIR spectra simulations by anharmonic DFT-saturated and unsaturated long-chain fatty acids. J. Phys. Chem. B **122**(27), 6931–6944 (2018)
30. Beć, K.B., Grabska, J., Ozaki, Y., Czarnecki, M.A., Huck, C.W.: Simulated NIR spectra as sensitive markers of the structure and interactions in nucleobases. Sci. Rep. **9**, 17398 (2019)
31. Beć, K.B., Grabska, J., Badzoka, J., Huck, C.W.: Spectra-structure correlations in NIR region of polymers from quantum chemical calculations. The cases of aromatic ring, C=O C≡N and C-Cl functionalities. Spetrochim. Acta. A, **262**, 120085 (2021)
32. Beć, K.B., Grabska, J., Huck, C.W., Mazurek, S., Czarnecki, M.A.: Anharmonicity and spectra-structure correlations in MIR and NIR spectra of crystalline menadione (vitamin K3). Molecules, **26**, 6779 (2021)
33. Beć, K.B., Huck, C.W.: Breakthrough potential in near-infrared spectroscopy: spectra simulation. A review of recent developments. Front. Chem. **7,** 48 (2019)
34. Grabska, J., Beć, K.B., Ozaki, Y., Huck, C.W.: Anharmonic DFT study of near-infrared spectra of caffeine. Vibrational analysis of the second overtones and ternary combinations. Molecules, **26**, 5212 (2021)
35. Grabska, J., Beć, K.B., Kirchler, C.G., Ozaki, Y., Huck, C.W.: Distinct difference in sensitivity of NIR vs. IR bands of melamine to inter-molecular interactions with impact on analytical spectroscopy explained by anharmonic quantum mechanical study. Molecules, **24**, 1402 (2019)
36. Grabska, J.: Current frontiers in quantum chemical simulations of NIR spectra – polymers, biomolecules, aqueous matrix, and interpretation of instrumental difference of handheld spectrometers. NIR News, **32**, 7−14 (2021)
37. Beć, K.B., Grabska, J., Kirchler, C.G., Huck, C.W.: NIR spectra simulation of thymol for better understanding of the spectra forming factors, phase and concentration effects and PLS regression features. J. Mol. Liq. **268**, 895−902 (2018)
38. Grabska, J., Beć, K.B., Mayr, S., Huck, C.W.: Theoretical simulation of near-infrared spectrum of piperine. Insight into band origins and the features of regression models. App. Spectr. **75**, 1022−1032 (2021)
39. Mayr, S., et al.: Quantification of silymarin in silybi mariani fructus: challenging the analytical performance of benchtop vs. handheld NIR spectrometers on whole seeds. Planta. Med. **87**, 1−13 (2021)
40. Mayr, S., Beć, K.B., Grabska, J., Schneckenreiter, E., Huck, C.W.: Near-infrared spectroscopy in quality control of piper nigrum: a comparison of performance of benchtop and handheld spectrometers. Talanta, **223**(2), 121809 (2021)
41. Mayr, S., et al.: Theae nigrae folium: comparing the analytical performance of benchtop and handheld near-infrared spectrometers. Talanta **221**, 121165 (2021)

42. Crocombe, R.A.: Portable spectroscopy. Appl. Spectrosc. **72**, 1701–1751 (2018)
43. Beć, K.B., Grabska, J., Siesler, H.W., Huck, C.W.: Handheld near-infrared spectrometers: where are we heading? NIR News **31**, 28–35 (2020)
44. Giussani, B., Gorla, G., Riu, J.: Analytical chemistry strategies in the use of miniaturised NIR instruments: an overview. Crit. Rev. Anal. Chem. (2022)
45. Beć, K.B., Grabska, J., Plewka, N., Huck, C.W.: Insect protein content analysis in handcrafted fitness bars by NIR spectroscopy. Gaussian process regression and data fusion for performance enhancement of miniaturized cost-effective consumer-grade sensors. Molecules, **26**, 6390 (2021)

Near Infrared Spectroscopy in China

Xiaoli Chu[1(✉)] and Hongfu Yuan[2]

[1] Research Institute of Petroleum Processing of SINOPEC, Beijing 100083, China
cxlyuli@sina.com

[2] College of Materials Science and Engineering, Beijing University of Chemical Technology, Beijing 100029, China

Abstract. The research and application of near infrared spectroscopy in China has a history of 40 years. This paper mainly introduces the research and application status of near infrared spectroscopy in China, so that the world peers can fully and systematically understand the actual situation of near infrared spectroscopy in China and sense the changes of China from NIR side.

Keywords: China · Review · Chemometrics · Applications · On-line

In the 1980s, near infrared (NIR) spectroscopy was first applied in agriculture in China. In the 1990s, NIR spectroscopy was gradually accepted as a fascinating analytical technology by some analytical scientists and factory managers. In the 2000s, NIR spectroscopy was researched and applied widely in China. In 2009, the China Council of NIR Spectroscopy (CCNIRS) was established, and the Council has played key roles in the Chinese NIR spectroscopy community since. In the past two decades, NIR has developed rapidly in China, mainly in academic research, instrument development, and practical applications in various areas.

Academically, China has published more than 20 monographs on NIR spectroscopy, imaging and their applications. Since the first Chinese Conference on NIR was held in 2006, China has successfully held eight conferences, with more than 200 participants attending each conference. The 3rd and 6th Asian NIR Conference was also held in China in 2010 and 2018. In China, there are about 40 research teams engaged in the research and application of NIR. Every year, they publish about 200 papers in academic journals at home and abroad and apply for about 100 patents (Fig. 1).

In terms of Chemometrics, Academician Yu Ruqin and Professor Liang Yizeng's team have achieved many innovative research results. For example, the variable selection method, competitive adaptive reweighting algorithm (CARS), has attracted the most extensive attention. Professor Shao Xueguang's team has also obtained abundant research results in Aquaphotomics. In addition, there are more than 10 research teams in China that specialize in the study of Chemometrics, and have put forward many innovative algorithms.

X. Chu et al. (Eds.): ICNIR 2021, *Sense the Real Change: Proceedings of the 20th International Conference on Near Infrared Spectroscopy*, pp. 73–78, 2022.
https://doi.org/10.1007/978-981-19-4884-8_6

Fig. 1. The 6th Chinese conference on near infrared spectroscopy held in Wuhan in march 2016

Chinese instrument manufacturers have developed grating array, Fourier transform and digital micromirror spectrometers with independent intellectual property rights. On this basis, a variety of commercial NIR portable and on-line analyzers applied to the measurement of various liquid and solid samples have been manufactured. Some commercial instruments have exported to Europe, Russia and some countries of Southeast Asia.

China has carried out many practical applications with its own characteristics and achieved considerable economic and social benefits. More than 100 NIR Chinese standard methods has been established, involving grains, chemicals, foods, feeds, textiles, tobaccos and so on (Fig. 2).

Fig. 2. Several Chinese NIR monographs on NIR

In the grain analysis, taking soybeans as an example, NIR has been used to screen high quality seeds rapidly and nondestructively in breeding. Many grain growers also purchase NIR analyzer to guide the planting management of soybeans. In soybean trading, the quality of soybean is predicted by NIR on the spot, and the price is determined according to the protein content. NIR has played an important role in many aspects of edible oil processing. For example, in the leaching process, the oil content of the extracted meal can be determined on-line, and the processing parameters can be adjusted in time to achieve the best leaching effect and significantly improve the yield of oil (Fig. 3).

Fig. 3. Several Chinese standard methods on NIR

In the field of fruit, the portable NIR analyzer is used for fruit tree growth management, providing scientific data for fertilization, irrigation and other cultivation management. In the harvest season, the use of NIR to determine the best harvest time of fruits has become an important means to increase the income for fruit farmers. In fruit screening, Chinese companies have independently developed online sorting equipment suitable for large size fruits such as watermelons, as well as large assembly-line sorting devices for small size fruits such as oranges.

As for tea, NIR provides technical support for geographical indication Protection of Chinese tea, and has been used for quality monitoring in the process of tea processing, such as green tea fixation, oolong tea roasting, black tea fermentation, etc., which can more objectively judge the degree of fermentation, make the fermentation process more scientific and reasonable (Fig. 4).

Fig. 4. On-line near infrared spectroscopy for detecting the quality of tea production process

Traditional Chinese medicine (TCM) is a treasure of the Chinese nation. NIR has been applied to the rapid analysis of the quality and content of specific components of TCM. Many large TCM companies in China have installed on-line NIR analyzer to monitor the extraction, purification, concentration and mixing production process, to achieve the optimal control and ensure the uniform and stable quality of TCM products (Fig. 5).

Fig. 5. On line real-time monitoring of extraction of active components of traditional Chinese medicine by near infrared spectroscopy

China has developed a NIR rapid drug detection system since 2004. For nearly 20 years of application and accumulation, the system has established a complete NIR drug database and identification models of counterfeit drug. At present, there are more than 400 mobile drug detecting vehicles in China, which are used for rapid on-site drug supervision in the countryside, town and urban areas (Fig. 6).

Fig. 6. The drug testing vehicles developed by the national institutes for food and drug control

The ancient Silk Road once made great contributions to the world's economic development and cultural prosperity, and the development of silkworm industry still exerts an important influence. The separation of male and female silkworm pupae is an important step in the silkworm seed production enterprises. China has developed a high-speed NIR sorting device for male and female silkworm pupae. The device can sort more than 10 silkworm pupae per second with an accuracy rate of 98%.

In China, many famous liquor enterprises use NIR to monitor the fermentation process of fermented grains, selecting and controlling the best fermentation conditions, so as to ensure product quality, increase output and reduce energy consumption at the same time. NIR is also used in the fermentation production process of vinegar, soy sauce and other condiments.

In large scale process industries, such as refineries and petrochemical plants, NIR has become one of the representative technologies of modern intelligent factory. China has established a complete NIR database of crude oil based on thousands of samples, which plays an important role in petroleum exploitation, trade and processing. The combination of the on-line NIR and the advanced process control and optimization system has realized the optimal operation in many Chinese refinery units (Fig. 7).

Fig. 7. On-line near infrared spectroscopy probe for monitoring chemical reaction process

NIR is used by Chinese geologists to quickly identify mineral types in the field and obtain the geographical distribution characteristics of different minerals in a specific area, which can provide a basis for further drilling work, save a lot of exploration costs and improve the efficiency of prospecting.

With the help of big data, Internet of Things and cloud computing technology, Chinese tobacco enterprises have built a NIR big data system platform to achieve the goals of center modeling, resource sharing and intelligent analysis, which can be used for quality control of all aspects of tobacco processing.

China has made a lot of scientific research work in functional near infrared spectroscopy, especially in brain functional imaging technology. Many research results are published in international journals with great influence, and some of them have been applied in psychiatry.

In the future, NIR in China will continue to make substantial progress in the width and depth of spectroscopy theory, instrument development, NIR imaging and their applications.

Acknowledgments. We would express our sincere respect to the people who has made the contributions to the progress of near infrared spectroscopy technology in China.

References

1. Chu, X.L.: China council of near infrared spectroscopy established in Beijing. NIR News **20**, 3–4 (2009)
2. Chu, X.L.: Guest editorial: near infrared spectroscopy in China. J. Near Infrared Spec. **23**(5), v–vii (2015)
3. Chu, X.L., Qian, W.: In memory of Dr Lu Wanzhen, champion and pioneer for the adoption of near infrared technologies in China. J. Near Infrared Spec. **23**(5), 269 (2015)
4. Chu, X.L., Lu, J.: Research and application status of near infrared technology in China. NIR news. **25**(6), 13–15 (2014)
5. Guo, L.H., Yuan, H.F., Chu, X.L.: Report on the sixth asian NIR symposium (ANS2018). NIR News **29**(7), 5–7 (2018)
6. Yang, Z.L., Cai, L.W., Han, L.J.: Review of standards of near infrared spectroscopy methods. J. Near Infrared Spec. **29**(6), 313–320 (2021)

Agriculture, Food and Forestry

Measurement of Gingerols and 6-shogaol in Ginger Using Near-Infrared Spectroscopy

Joel B. Johnson[1,2(✉)], Janice S. Mani[1,2], Kerry B. Walsh[1,2], and Mani Naiker[1,2]

[1] School of Health, Medical and Applied Sciences, Central Queensland University, Rockhampton, QLD, Australia
joel.johnson@cqumail.com

[2] Institute for Future Farming Systems, Central Queensland University Australia, Rockhampton, QLD, Australia

Abstract. Ginger is well known for its unique pungent flavour and numerous health benefits, both of which can be attributed to the gingerol derivatives. Quantification of these compounds is important for assessing ginger quality for processing purposes, as their levels can vary with cultivar and growing conditions. However, current analytical techniques are too expensive and time-consuming for in-house monitoring of ginger quality. Hence the objective of this work was to investigate the potential for estimating the content of gingerol derivatives in dried, ground ginger using near-infrared spectroscopy. These compounds are found at relatively low concentrations (1000–6000 mg kg^{-1}), which could make detection by NIRS challenging. In this study, reflectance NIR spectra between 10,000–4,000 cm^{-1} were collected from 100 dried, powdered samples, using a spinning cup and integrating sphere module on an Thermo Scientific Antaris II FT-NIR Analyzer. The 6-gingerol, 8-gingerol, 10-gingerol and 6-shogaol content of the samples was measured in 90% methanol extracts using high-performance liquid chromatography (HPLC). The performance of partial least squares regression (PLSR) and support vector regression (SVR) were then evaluated, with results validated against a dependent test set. The NIRS method showed promise for the prediction of 6-gingerol content, with $R^2_{cv} > 0.7$, RMSECV < 450 mg kg^{-1} (on a mean concentration of 4400 mg kg^{-1}) and RPD of approximately 2. Similarly, prediction of 6-shogaol content gave an $R^2_{cv} > 0.6$, RMSECV < 110 mg kg^{-1} (on a mean concentration of 1440 mg kg^{-1}) and RPD of 1.6. A higher accuracy was found for PLSR compared to SVR. Although the results could be further improved, the detection of these compounds at low concentrations in a matrix as complex as ginger is notable. With further refinement, NIRS may be suitable for the rapid estimation of major pungent compounds in dried ginger.

Keywords: Pungent compounds · Quality assurance · Food processing

1 Introduction

Ginger is the rhizome of the *Zingiber officinale* (Roscoe) plant, grown for its characteristic pungent flavour and medicinal uses [1–3]. Globally, over 4 million tons of ginger

X. Chu et al. (Eds.): ICNIR 2021, *Sense the Real Change: Proceedings of the 20th International Conference on Near Infrared Spectroscopy*, pp. 81–90, 2022.
https://doi.org/10.1007/978-981-19-4884-8_7

are produced per year, with a market value of approximately US $3.8 billion [4]. The largest producers of ginger include India, Nigeria and China [4].

The principal pungent compounds found in fresh ginger are the gingerols, a group of *O*-methoxyphenyl alkyl ketones with differing alkyl side chain lengths [5]. The most abundant gingerol is 6-gingerol, followed by 10-gingerol and 8-gingerol [6]. When ginger is dried at elevated temperatures, gingerols undergo an elimination dehydration reaction to form their respective shogaols [7, 8]. Given the organoleptic importance of the gingerols and shogaols [9] and the fact that their levels can vary depending on the ginger cultivar, growing conditions and storage conditions, methods for the routine assessment of these compounds would be greatly beneficial for the quality assurance of ginger used for processing purposes [10]. Although payment by industry is currently solely on a weight basis, in-house methods for the assessment of gingerol and shogaol contents could be exploited to pay farmers based on the quality of their produce. A range of analytical methods have been used for the assessment of gingerols and shogaols in ginger, including high-performance liquid chromatography (HPLC) [11], liquid chromatography-mass spectrometry (LC-MS) [12], gas chromatography-mass spectrometry (GC–MS) [13] and high performance thin layer chromatography (HPTLC) [14, 15]. However, all of these analytical methods are relatively expensive, potentially destructive to the samples and time-consuming, making them unsuitable for use in a real-time quality assessment setting.

Given the prospect of near-infrared spectroscopy (NIRS) for the assessment of a wide range of other analytes in food matrices [16, 17], this work investigated the potential of NIRS for the rapid analysis of gingerol derivatives in dried ginger. Previous authors have used NIRS and hyperspectral imaging for the assessment of moisture content in ginger (as a surrogate measure of quality) [18, 19] and there is also one report of using NIRS for the analysis of gingerols and 6-shogaol in 80 samples of sulfur-fumigated and nonfumigated ginger [20].

2 Materials and Methods

One hundred samples of dried, powdered processing-grade ginger were sourced from a commercial ginger processor in Queensland, Australia [21]. The samples were obtained from six different growers and two harvest years (2018 and 2019), thus encompassing a wide range of environmental variation. All samples (each comprising multiple rhizomes) were dried using a continuous belt drying system under consistent conditions (approx. 35–40 °C) and ground to a powder (approx. <1 mm particle size) [21].

Near-infrared (NIR) spectra were collected in triplicate from the powdered ginger samples using a FT-NIR instrument (Nicolet Antaris Near IR Analyzer, USA) over the spectral range 10,000–4,000 cm^{-1}. The instrument was used in reflectance mode, using a spinning cup attachment in conjunction with an integrating sphere. Approximately 10 g of powder was packed in the sample cup for each replicate scan. The resolution was set at 8.0 cm^{-1} with 64 scans per sample.

The pungent compounds were extracted from the ginger samples in duplicate, using a maceration-based extraction protocol previously used for the analysis of phenolic compounds [22]. Briefly, approximately 0.5 g of ginger powder was extracted twice with

90% methanol, made to a total volume of 15 mL and syringe filtered (LabServ 0.2 μm nylon membrane). The four principal pungent compounds shown in Fig. 1 (6-gingerol, 8-gingerol, 10-gingerol and 6-shogaol) were analyzed using previously reported HPLC methods [21]. The method used a reversed phase C18 column (Agilent Eclipse XDB-C18; 150 × 4.6 mm; 5 μm pore size), injection volume of 5 μL, column temperature of 27 ± 0.8 °C, detection wavelength of 230 nm and flow rate of 1 mL min^{-1}. The mobile phase gradient began at 30% methanol and 70% water, ramping to reach 60% methanol at 2 min, 63% at 10 min, 65% at 16 min and 100% at 28 min. The analytes were quantified using external calibrations of authentic standards between 10–100 mg L^{-1} (Sigma-Aldrich Australia). The limit of detection of the HPLC method was <1 mg L^{-1} (corresponding to < 30 mg kg^{-1} in the powdered samples), with a repeatability of 0.24–0.63% relative error (corresponding to 2–16 mg kg^{-1}).

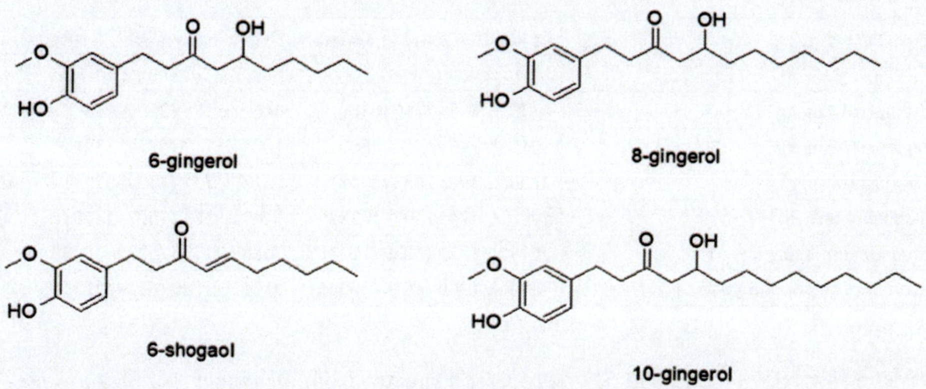

Fig. 1. The structures of the four pungent compounds quantified in the dried ginger samples in this work.

The NIR spectra were pre-processed using a variety of methods, including standard normal variate (SNV) transformation, and the first and second derivatives using the Savitzky-Golay algorithm – either alone or in combination with SNV transformation.

The dataset was divided into a calibration set (75%, or 225 spectra) and dependent test set (25%), using a Kennard-Stone algorithm. Partial least squares regression (PLSR) analysis was conducted using custom scripts in R Studio running R 4.0.2 [23]. The maximum number of model components was limited to 10 to avoid model overfitting. Cross-validation of the PLSR models was conducted using the leave-one-out method, with the optimum number of components selected from the RMSECV scree plot. To compare the performance of non-linear modelling techniques, support vector regression (SVR) was also conducted in R, with the model bootstrapped 25 times to determine the optimum "sigma" and "C" values to use. The developed models were then applied to the dependent test set to assess their prediction accuracy.

3 Results and Discussion

The descriptive statistics (mean, standard deviation, minimum and maximum) for each of the parameters measured in this study are summarized in Table 1. The average error of the HPLC method (measured as the relative standard deviation between duplicate extracts of the same ginger sample) was 1.93% for 6-gingerol, 1.54% for 6-shogaol, 2.85% for 8-gingerol and 1.23% for 10-gingerol (n = 100 samples analysed in duplicate). The highest level of variation was observed in the 6-gingerol content (coefficient of variation of 19.2% across all samples), whilst the lowest level of variation was observed in 6-shogaol content (CV of 12.2% across all samples).

Table 1. Descriptive statistics for the parameters measured in this study, obtained using HPLC analysis. Results are provided separately for the calibration and validation sets.

Parameter	Calibration set (225 spectra)	Dependent test set (75 spectra)
	Mean ± SD (range)	Mean ± SD (range)
6-gingerol (mg kg^{-1})	4375 ± 868 (2756–6488)	4469 ± 775 (2926–6488)
8-gingerol (mg kg^{-1})	1015 ± 196 (645–1480)	1007 ± 177 (687–1375)
10-gingerol (mg kg^{-1})	2230 ± 396 (1453–3258)	2212 ± 391 (1509–3258)
6-shogaol (mg kg^{-1})	1440 ± 174 (1007–1878)	1447 ± 183 (1007–1878)
Sum gingerols (mg kg^{-1})	7619 ± 1329 (5239–10488)	7688 ± 1196 (5239–10488)
Sum of pungent compounds (mg kg^{-1})	**9060 ± 1389 (6514–12366)**	**9136 ± 1260 (6514–12366)**

The raw NIR spectra and SNV-processed spectra from all ginger samples are presented in Fig. 2. The spectra were similar to that previously reported for dried ginger [20]. The major peaks were located at 8320 cm^{-1} (attributed to CH second overtone from carbohydrates), 6820 cm^{-1} (second overtone of OH from moisture in the sample or other R-OH bonds present), 5170 cm^{-1} (first overtone of OH bond) and 4740 cm^{-1} (amide bond from protein in the sample). Smaller peaks were observed at 6340 cm^{-1} (tentatively assigned to the first overtone of CH_3/ArCH), 5900–5500 cm^{-1} (first overtone of CH and CH_2 stretch) and 4310 cm^{-1} (CH deformation stretch of CH_2 and CH_3 bonds).

It was noted that for the PLSR method, the use of the raw spectra or SNV-transformed spectra gave the best results for all analytes. As shown in Table 2, the PLS regressions showed promise for the estimation of 6-gingerol content in the ginger samples (R^2_{cv} = 0.72; residual prediction deviation (RPD) = 1.89). The actual vs. predicted values of 6-gingerol (in mg kg^{-1}) for the 300 spectra are shown in Fig. 3. Examination of the loadings plot (Fig. 4) revealed the major contributing wavelengths for the first two model components to be 5173 and 4963 cm^{-1}. The first wavelength may be related to the first overtone of water or amide bonds, while no common functional groups are known to occur at the second wavelength. Hence it may be more specific to the characteristic structure of gingerol compounds (Fig. 1), perhaps as the shoulder of a peak overlapping with other spectral features (e.g. the R-OH group absorbing at ~4855 cm^{-1}). Negative loadings were also found at 5038 and 4490 cm^{-1}. Again, the first of these wavelengths does not correspond well with any common functional groups, while the second wavelength may be related to the CH_3 combination band.

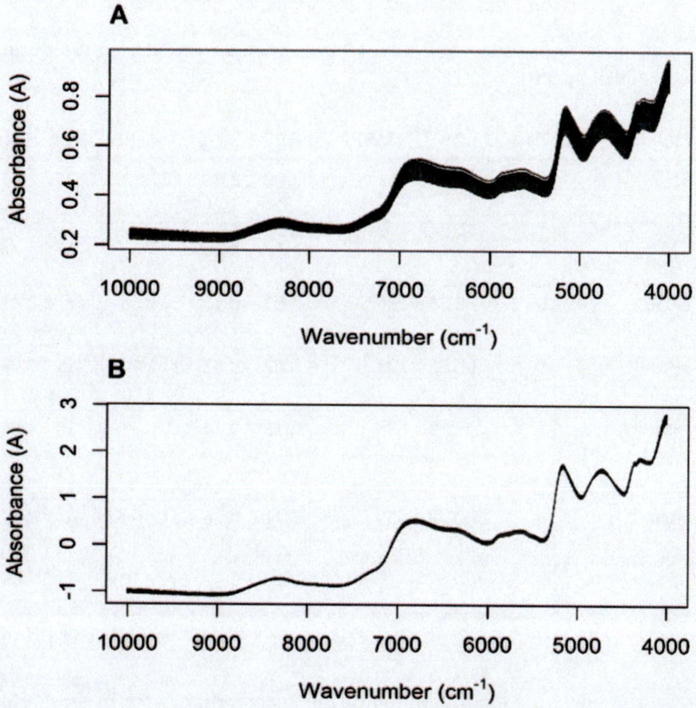

Fig. 2. (a) The raw NIR log 1/R spectra of the 100 ginger samples (n = 300 spectra). (b) The SNV-processed spectra of the ginger samples.

To confirm the loadings of the PLSR model were related to the functional bonds of 6-gingerol, the difference spectra was calculated between the samples containing the highest 6-gingerol content (6488 mg kg^{-1}) and the sample containing the lowest 6-gingerol content (2756 mg kg^{-1}) (Fig. 5). This difference spectra appeared similar to the combined loadings plots for the first two principal components, with negative peaks at 8400 cm^{-1} (2nd overtone of CH), 6800 cm^{-1} (2nd overtone of OH) and 5100 cm^{-1} (overtone of water or amide bands), and positive peaks at 5400 cm^{-1} and 4750 cm^{-1} (CH_3 combination band). The similar regions to the loadings plot (Fig. 4) confirmed that the PLSR model was targeting the correct NIR regions to detect the presence of 6-gingerol in the ginger samples.

The PLSR models also allowed the prediction of 8-gingerol content ($R^2_{cv} = 0.77$; RPD = 2.11) and 10-gingerol content ($R^2_{cv} = 0.77$; RPD = 2.10) with good prediction accuracy (RPD > 2) [24]. However, the model for the prediction of 6-shogaol content performed more poorly. This may be due to the lower range of variation found between samples for this compound, or potentially due to the challenges of discriminating 6-shogaol from the other gingerol constituents.

NIRS could also be used to predict the sum of gingerol content (R^2_{cv} = 0.77; RPD = 2.08) and sum of gingerol and shogaol content (R^2_{cv} = 0.77; RPD = 2.10) with good accuracy.

Table 2. Optimum models for the prediction of the specified analytes in dried ginger samples using PLSR on the NIR spectra.

Parameter	Pre-processing	Factors	R^2_{cv}	RMSECV	RPD	R^2_{test}	RMSEP	Bias	Slope	Intercept
6-gingerol (mg kg^{-1})	SNV	8	0.75	434	2.00	0.63	477	95	1.05	−142
8-gingerol (mg kg^{-1})	None	9	0.78	92	2.13	0.75	89	−11	1.04	−50
10-gingerol (mg kg^{-1})	None	9	0.75	196	2.02	0.81	169	−11	0.95	90
6-shogaol (mg kg^{-1})	None	9	0.62	108	1.62	0.62	114	17	0.98	48
Sum gingerols (mg kg^{-1})	None	9	0.78	628	2.12	0.74	601	18	1.00	37
Sum of pungent compounds (mg kg^{-1})	None	9	0.78	659	2.11	0.76	612	41	1.01	−57

Table 3. Optimum models for the prediction of the specified analytes in dried ginger samples using SVR on the NIR spectra.

Parameter	Pre-processing	R^2_{cv}	RMSECV	RPD	R^2_{test}	RMSEP	Bias	Slope	Intercept
6-gingerol (mg kg^{-1})	1d21	0.69	504	1.72	0.57	518	99	1.16	−597
8-gingerol (mg kg^{-1})	SNV + 1d21	0.70	112	1.75	0.70	98	−6	1.16	−170
10-gingerol (mg kg^{-1})	SNV + 1d21	0.73	212	1.87	0.80	177	5	1.14	−299
6-shogaol (mg kg^{-1})	SNV + 1d21	0.58	120	1.45	0.60	122	10	1.33	−467
Sum gingerols (mg kg^{-1})	SNV	0.63	838	1.59	0.70	676	35	1.21	−1573
Sum of pungent compounds (mg kg^{-1})	SNV	0.65	835	1.66	0.70	708	39	1.22	−1931

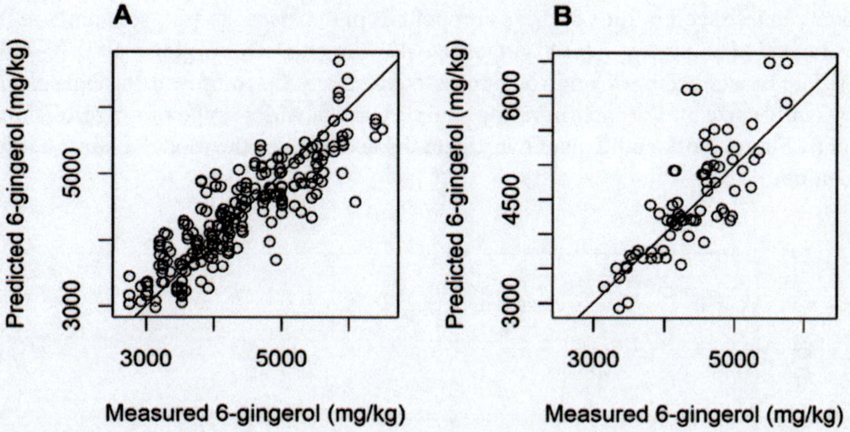

Fig. 3. Actual vs. predicted results for the 6-gingerol content of the ginger samples, using PLSR in the calibration set (A) and test set (B).

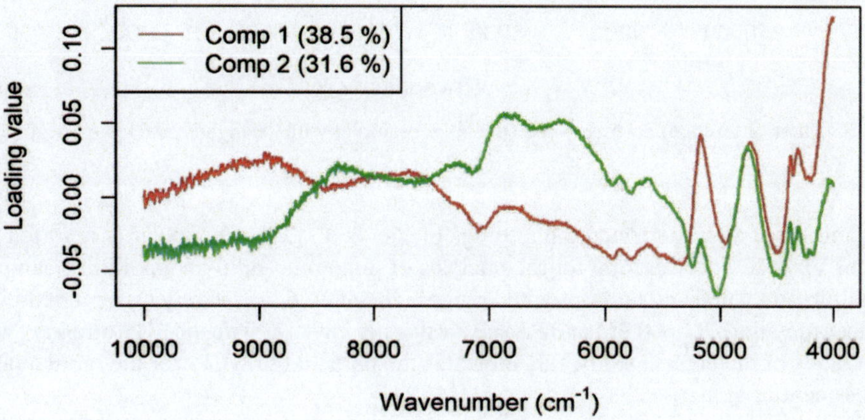

Fig. 4. Loadings plot for the first two components of the 6-gingerol PLSR prediction model.

The model results for SVR were less accurate compared to PLSR in all cases (Table 3), with RPD values < 2 and their RMSEP values being around 4–10% higher than those obtained through PLSR modelling. According to Nicolaï et al. [24], these RPD values (<2) suggest that these models would be best suited to screening purposes only. Furthermore, the results indicated that non-linear modelling techniques were unlikely to significantly improve upon the accuracy of the PLSR models. However, it is possible that emerging non-linear techniques not investigated here – such as artificial neural networks – could provide increased accuracy.

This preliminary study involving 100 samples demonstrates promise for the use of NIRS in estimation of individual gingerol compounds and the sum of gingerols or all pungent compounds. This is evident in the relatively high correlation between the results obtained by HPLC analysis and those predicted using NIRS. It is also worth noting the relatively low concentrations of these compounds in the matrices assessed, ranging from an average of 1013 mg kg^{-1} for 8-gingerol to 4398 mg kg^{-1} for 6-gingerol (0.10–0.44% w/w). The model accuracy and robustness could likely be improved through use of a larger sample size and/or incorporating samples with a wider range of gingerol/shogaol contents. Future work could also investigate the accuracy of the model against a test set population.

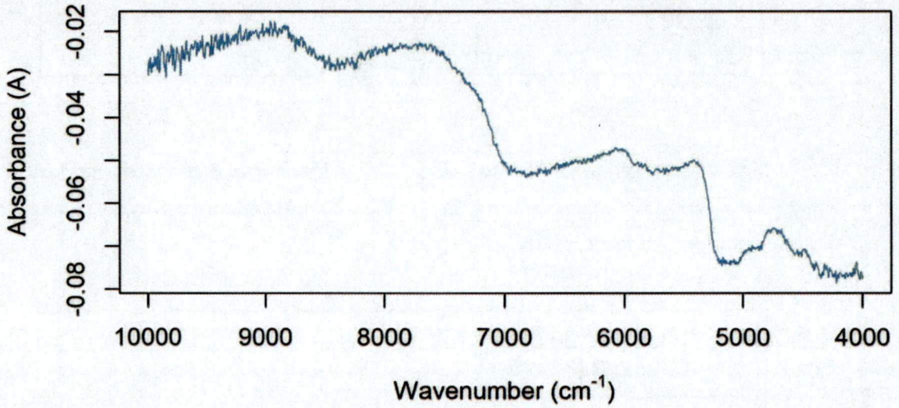

Fig. 5. Mean difference spectra between the ginger samples with the highest and lowest 6-gingerol contents.

This work also confirmed the findings of Yan et al. [20], who recently reported the use of Vis-NIR spectroscopy for the analysis of gingerols and 6-shogaol in 80 samples of sulfur-fumigated and nonfumigated ginger. They found a relatively higher coefficient of determination ($R^2 > 0.95$) and RMSEP values of 289–1190 mg kg^{-1}. Combined with the results of the present study, this indicates the potential of NIRS for the rapid quality assessment of ginger.

4 Conclusion

This study investigated the potential of NIRS for the quantification of gingerol derivatives in dried, processing-grade ginger powder. NIRS showed promise for the prediction of 6-gingerol, 8-gingerol and 10-gingerol contents, as well as for the total gingerol and total pungent content ($R^2_{cv} = 0.72$–0.77; RPD = 1.89–2.11). However, the prediction of 6-shogaol content did not perform as well ($R^2_{cv} = 0.62$; RPD of 1.63). For all analytes, PLSR showed superior results compared to SVR. With further refinement, NIRS may be suitable for the rapid estimation of major pungent compounds in dried ginger.

Acknowledgments. We wish to thank our supplier of the ginger powder. This work was supported by a New Staff Grant (RSH/5343) awarded by CQUniversity to one of the authors (MN). One of the authors (JJ) acknowledges support from the Australian Government in the form of a Research Training Program Scholarship.

References

1. Govindarajan, V.S., Connell, D.W.: Ginger-chemistry, technology, and quality evaluation: Part 2. Crit. Rev. Food Sci. **17**, 189–258 (1983)
2. Kubra, I.R., Rao, L.J.M.: An impression on current developments in the technology, chemistry, and biological activities of ginger (*Zingiber officinale* Roscoe). Crit. Rev. Food Sci. **52**, 651–688 (2012)
3. Semwal, R.B., Semwal, D.K., Combrinck, S., Viljoen, A.M.: Gingerols and shogaols: Important nutraceutical principles from ginger. Phytochemistry **117**, 554–568 (2015)
4. FAO. FAOSTAT (2022). http://www.fao.org/faostat/en/#data/QC. Accessed 12 Jan 2022
5. Raghuveer, K.G., Govindarajan, V.S.: Evaluation of spices and oleoresins VII. Gas chromatographic examination of gingerol, shogaol and related compounds in ginger. J. Food Qual. **2**, 41–54 (1979)
6. Yudthavorasit, S., Wongravee, K., Leepipatpiboon, N.: Characteristic fingerprint based on gingerol derivative analysis for discrimination of ginger (*Zingiber officinale*) according to geographical origin using HPLC-DAD combined with chemometrics. Food Chem. **158**, 101–111 (2014)
7. Connell, D., Sutherland, M.: A re-examination of gingerol, shogaol, and zingerone, the pungent principles of ginger (*Zingiber officinale* Roscoe). Aust. J. Chem. **22**, 1033–1043 (1969)
8. Johnson, J.B., Mani, J.S., Naiker, M..: Gingerol, shogaol and paradol: the chemistry of pungent ginger constituents. In: Queensland Annual Chemistry Symposium, pp. 89–90 (2020)
9. Narasimhan, S., Govindarajan, V.S.: Evaluation of spices and oleoresin-VI-pungency of ginger components, gingerols and shogoals and quality. Int. J. Food Sci. Tech. **13**, 31–36 (1978)
10. Johnson, J.B., Mani, J.S., White, S., Brown, P., Naiker, M.: Pungent and volatile constituents of dried Australian ginger. Curr. Res. Food Sci. **4**, 612–618 (2021)
11. Schwertner, H.A., Rios, D.C.: High-performance liquid chromatographic analysis of 6-gingerol, 8-gingerol, 10-gingerol, and 6-shogaol in ginger-containing dietary supplements, spices, teas, and beverages. J. Chromatogr. B **856**, 41–47 (2007)
12. Ashraf, K., Mujeeb, M., Ahmad, A., Ahmad, S., Ahmad, N., Amir, M.: Determination of gingerols in ginger by ultra-high performance liquid chromatography-tandem mass spectrometry. Anal. Lett. **47**, 2120–2128 (2014)
13. Nishidono, Y., Saifudin, A., Deevanhxay, P., Tanaka, K.: Metabolite profiling of ginger (*Zingiber officinale* Roscoe) using GC-MS and multivariate statistical analysis. J. Asia-Jpn. Res. Inst. Ritsumeikan Univ. **2**, 1–14 (2020)
14. Salmon, C.N.A., Bailey-Shaw, Y.A., Hibbert, S., Green, C., Smith, A.M., Williams, L.A.D.: Characterisation of cultivars of Jamaican ginger (*Zingiber officinale* Roscoe) by HPTLC and HPLC. Food Chem. **131**, 1517–1522 (2012)
15. Rai, S., Mukherjee, K., Mal, M., Wahile, A., Saha, B.P., Mukherjee, P.K.: Determination of 6-gingerol in ginger (*Zingiber officinale*) using high-performance thin-layer chromatography. J. Sep. Sci. **29**, 2292–2295 (2006)
16. Johnson, J.B., Walsh, K., Naiker, M.: Application of infrared spectroscopy for the prediction of nutritional content and quality assessment of faba bean (*Vicia faba* L.). Legume Sci. **2**, e40 (2020)
17. Cortés, V., Blasco, J., Aleixos, N., Cubero, S., Talens, P.: Monitoring strategies for quality control of agricultural products using visible and near-infrared spectroscopy: a review. Trends Food Sci. Tech. **85**, 138–148 (2019)
18. Lin, X., Sun, D.W.: Investigation of moisture distribution of ginger slices and splits during hot-air drying and rehydration procedures by NIR hyperspectral imaging. In: SPIE Defense + Commercial Sensing, SPIE, p. 11421 (2020)

19. Xue, L., Cai, J., Liu, M.H., Li, J., Yan, L.Y.: Determination of moisture content in ginger using three variable selection methods combined with Vis/NIR. Chin. Agric. Mech. **02**, 2 (2012)
20. Yan, H., et al.: Rapid and practical qualitative and quantitative evaluation of non-fumigated ginger and sulfur-fumigated ginger via Fourier-transform infrared spectroscopy and chemometric methods. Food Chem. **341**, 128241 (2021)
21. Johnson, J.B., Mani, J.S., White, S., Brown, P., Naiker, M.: Quantitative profiling of gingerol and its derivatives in Australian ginger. J. Food Compos. Anal. **104**, 104190 (2021)
22. Johnson, J., Collins, T., Skylas, D., Quail, K., Blanchard, C., Naiker, M.: Profiling the varietal antioxidative content and macrochemical composition in Australian faba beans (*Vicia faba* L.). Legume Sci. **2**, e28 (2020)
23. R Core Team: A language and environment for statistical computing. R Foundation for Statistical Computing, Vienna, Austria (2020)
24. Nicolaï, B.M., et al.: Nondestructive measurement of fruit and vegetable quality by means of NIR spectroscopy: a review. Postharvest Biol. Tec. **46**, 99–118 (2007)

Spectral Separation Degree Method for Vis-NIR Spectral Discriminant Analysis of Milk Powder Adulteration

Yan Tang[1], Zeqi Chen[1], Niangen Ye[1], Haoran Lin[1], Lifang Fang[2], and Tao Pan[1](✉)

[1] Department of Optoelectronic Engineering, Jinan University, Guangzhou 510632, China
tpan@jnu.edu.cn, 466945939@qq.com

[2] Guangdong Langtao Information Technology Co., Ltd., Dongguan 523808, China

Abstract. Quick and easy spectral discrimination technology of milk powder adulterated has important application prospects. The separation degree spectrum between the two spectral populations and a wavelength selection method based on the separation degree priority combination (SDPC) were proposed here. The wavelength step by step phase-out (WSP) method was used for secondary wavelength optimization. Using Vis-NIR spectroscopy combined with SDPC-WSP-PLS-DA method, the discriminant analysis models for milk powder adulterated were established. Among them, standard normal transformation (SNV) and Norris derivative filtering (NDF) were used in turn for the spectral preprocessing with the NDF parameters of $d = 2$, $s = 11$, $g = 5$. The selected optimal SDPC-WSP-PLS-DA model used only 8 wavelengths to achieve high-precision discrimination effect in modeling (100%) and independent validation (96.0%). The results shower the feasibility of applying Vis-NIR spectroscopy to high-precision discriminant analysis of milk powder adulteration. The proposed spectral separation degree method can enhance the spectral difference of different spectral populations, extract information wavelengths, and improve the discrimination effect.

Keywords: Milk powder adulterated · Discriminant analysis · Separation degree spectrum · Separation degree priority combination · Wavelength step by step phase-out

1 Introduction

Milk powder adulteration has always been a food safety issue that has attracted much attention, and related rapid detection methods are of great significance. However, for the phenomenon of using cheap milk powder to dope high-end milk powder, there is no simple and effective detection method due to the high similarity of the ingredients. In this paper, Vis-NIR spectral discriminant analysis models of multi-element doping in milk powder were established. The separation degree spectrum between the two spectral populations was proposed, and combined with partial least squares-discrimination analysis (PLS-DA), a novel wavelength selection method based on the separation degree

X. Chu et al. (Eds.): ICNIR 2021, *Sense the Real Change: Proceedings of the 20th International Conference on Near Infrared Spectroscopy*, pp. 91–96, 2022.
https://doi.org/10.1007/978-981-19-4884-8_8

priority combination (SDPC) was also proposed. Further, the wavelength step by step phase-out (WSP) [1, 2] method was used for secondary optimization to improve the discrimination accuracy of milk powder adulteration.

2 Materials and Methods

2.1 Experimental Materials, Spectrometer and Measurement

Three brands of milk powder (Beiyinmei, Junlebao, Yili, not in order as I, II, III) were collected through regular commercial channels. Among them, I-brand (20 bags) was used as the unadulterated pure samples (negative), 3 samples per bag, 60 samples in total. II-brand and III-brand were added to I-brand and mixed into adulterated samples (positive), 91 samples in total, including one-component doping (II doping into I, adulteration rate ranged 3%, 6%,..., 93%, 96%, 100%, 33 samples in total), one-component doping (III doping into I, adulteration rate ranged 3%, 6%,..., 93%, 96%, 100%, 33 samples in total), and binary-component doping (II and III doping into I, adulteration rate of II and III both ranged 2%, 4%,..., 48%, 50%, 25 samples in total).

XDS Rapid Content™ NIR grating spectrum analyzer (FOSS, Denmark) and circular diffuse reflection sample attachment. The wavelength range was 400–2498 nm, and the wavelength interval was 2 nm. In order to maintain the balances of the number of spectra, the negative samples were measured six times (60 samples, 360 spectra) and the positive samples were measured four times (91 samples, 364 spectra).

2.2 Experimental Design of Calibration-Prediction-Validation and Evaluation Indicators

Full dataset (151 samples, 724 spectra) was divided into calibration (negative: 20 samples, 120 spectra; positive: 31 samples, 124 spectra), prediction (negative: 20 samples, 120 spectra; positive: 30 samples, 120 spectra) and validation (negative: 20 samples, 120 spectra; positive: 30 samples, 120 spectra) sets. Among them, the negative (60 samples, 360 spectra) set was randomly divided into calibration (20 samples, 120 spectra), prediction (20 samples, 120 spectra) and validation (20 samples, 120 spectra) sets; according to the uniform distribution of the adulteration rate from small to large, the positive (91 samples, 364 spectra) set was divided into calibration (31 samples, 124 spectra), prediction (30 samples, 120 spectra) and validation (30 samples, 120 spectra) sets.

According to the true type and prediction type of the samples, the eight recognition accuracy rates (RAR_C^+, RAR_C^-, RAR_C, RAR_P^+,RAR_P^-, RAR_P, RAR^+, RAR^-; %) of the positive and negative samples in calibration and prediction sets, the total recognition accuracy rate ($RAR_{M,Total}$) and their standard deviations ($RAR_{M,SD}$) were calculated respectively for modeling effect evaluation. In validation, the three recognition accuracy rates (RAR_V^+, RAR_V^-, RAR_V; %) of the positive and negative samples and their standard deviations ($RAR_{V,SD}$) were also calculated respectively for validation effect evaluation.

2.3 Separation Degree Spectrum

As shown in *Fig. 1*, the solid line and the dashed line respectively represent the negative and positive spectral populations. At any wavelength λ, the minimum, average, and maximum absorbance of negative and positive samples were $A^-_{Min}(\lambda)$, $A^-_{Ave}(\lambda)$, $A^-_{Max}(\lambda)$,$A^+_{Min}(\lambda)$, $A^+_{Ave}(\lambda)$, $A^+_{Max}(\lambda)$, respectively. The absorbance distance of the negative and positive spectral populations is called the separation degree, which is recorded as S(λ), as the follows:

$$S(\lambda) = \begin{cases} A^-_{Min}(\lambda) - A^+_{Max}(\lambda), A^-_{Ave}(\lambda) > A^+_{Ave}(\lambda) \\ A^+_{Min}(\lambda) - A^-_{Max}(\lambda), A^-_{Ave}(\lambda) < A^+_{Ave}(\lambda) \end{cases} \tag{1}$$

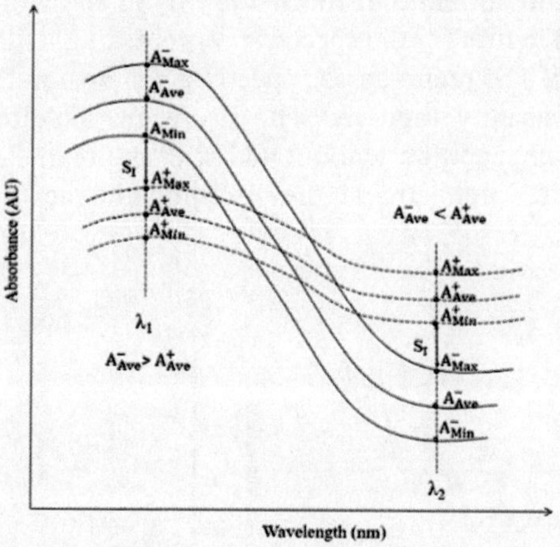

Fig. 1. Schematic diagram of the spectral separation degree of two types of spectral populations.

2.4 Wavelength Selection Based on Separation Degree Priority Combination (SDPC)

The wavelengths were sorted according to the value of separation degree from largest to smallest (total number of wavelengths: n), and *n* wavelength combinations were constructed as follows:

$$\Omega_i = \{\lambda_1, \lambda_2, \ldots, \lambda_i\}, i = 1, 2, \ldots, n \tag{2}$$

The PLS-DA model was established based on each wavelength combination. The optimal wavelength combination was determined according to total recognition accuracy rate ($RAR_{M,Total}$).

2.5 Wavelength Step by Step Phase-Out (WSP)

First, each time eliminated the wavelength, whose removing resulted in the best recognition accuracy, until only one wavelength remained. Then, the optimal model was selected from the above-mentioned process of wavelengths elimination by step-by-step phase-out mode.

3 Results and Discussion

3.1 Direct PLS-DA Models Without and with Pretreatment

The Vis-NIR spectra of milk powder adulterated are shown in Fig. 2. It can be observed that there was baseline drift in the spectra of the two types of samples, especially for the positive spectra. Standard normal transformation (SNV) and Norris derivative filtering (NDF) [3, 4] were used in turn for the spectral preprocessing, with the NDF parameters of $d = 2$, $s = 11$, $g = 5$. The preprocessed spectra are shown in Fig. 3, and the spectral baseline drift was significantly improved. The discrimination effects of direct PLS-DA models based on the full spectrum without and with pretreatment are shown in Table 1. It can be seen that after pretreatment, the discrimination accuracy and balance were both improved.

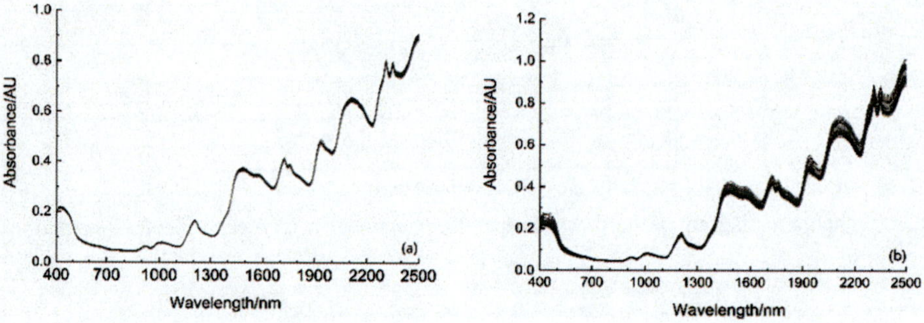

Fig. 2. Vis-NIR spectra of milk powder adulterated: (a) negative; (b) positive

3.2 SDPC-WSP-PLS-DA Models

Using the separation degree spectrum method above, the two separation degree spectra based on raw spectra and preprocessed spectra are shown in Fig. 4. Using the SDPC method, the SDPC-PLS-DA models based on raw spectra and preprocessed spectra were established, and the $RAR_{M,Total}$ both reached 100%, and the number of wavelengths were reduced to 684, 694 respectively. Further, WSP method was used for the quadratic wavelength optimization of the SDPC models above. The SDPC-WSP-PLS-DA models were established. The corresponding two wavelength combinations were 1088, 1810, 1900, 1906, 1984, 1992, 2012, 2014 nm and 488, 494, 1370, 1784, 2014, 2016, 2274, 2276 nm, respectively. The $RAR_{M,Total}$ both reached 100%.

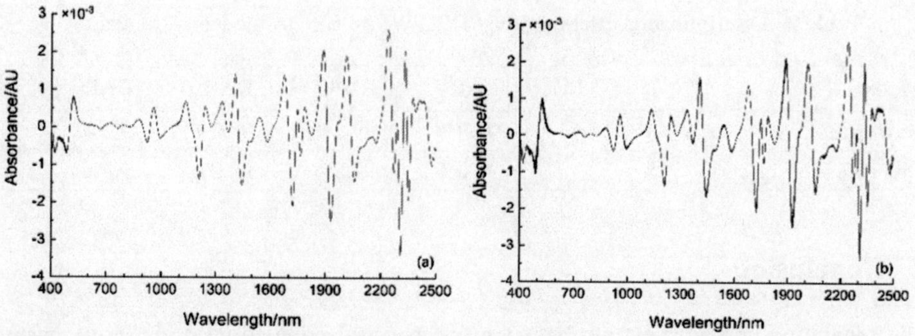

Fig. 3. The preprocessing spectra of negative and positive samples: (a) negative; (b) positive

Table 1. Discrimination effects of direct PLS-DA models without and with pretreatment in modeling.

Mode	*N*	LV	RAR_C	RAR_P	$RAR_{M, Total}$	$RAR_{M,SD}$
Raw spectra	1050	6	100%	98.2%	99.2%	0.9%
Preprocessed spectra	1050	18	100%	99.6%	99.8%	0.2%

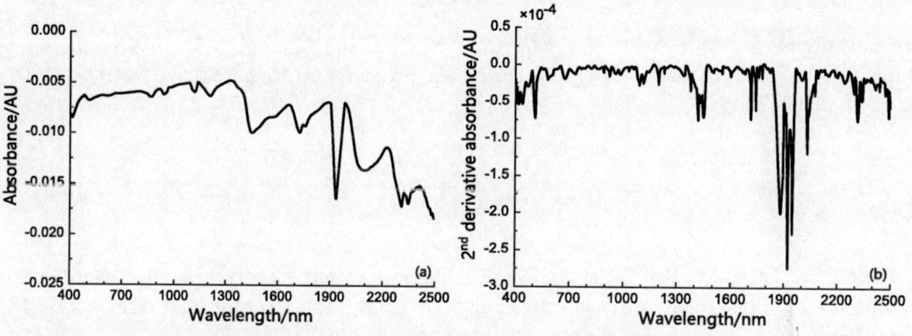

Fig. 4. Separation degree spectra based on: (a) raw spectra; (b) preprocessed spectra.

3.3 Independent Validation

The validation set (50 samples, 240 spectra) not involved in modelling were used to validate the optimal SDPC-WSP-PLS-DA models, the recognition accuracy rates of independent validation are summarized in Table 2. The results indicated that two models achieved good validation effects and the index balance of the preprocessed model was better.

Table 2. Discrimination effects of the SDPC-WSP-PLS-DA models in validation.

Mode	N	LV	RAR_V^-	RAR_V^+	RAR_V	$RAR_{V,SD}$
Raw spectra	8	6	92.9%	99.2%	96.0%	3.2%
Preprocessing spectra	8	3	94.0%	97.5%	95.8%	1.8%

4 Conclusion

The separation degree spectrum between two spectral populations and a novel wavelength selection method based on SDPC was proposed. The WSP method was used for the quadratic wavelength optimization of the SDPC models. Using Vis-NIR spectroscopy combined with SDPC-WSP-PLS-DA method, the high-precision discriminant analysis models for milk powder adulterated were established. The selected optimal SDPC-WSP-PLS-DA model used only 8 wavelengths to achieve high-precision discrimination effect in modeling ($RAR_{M,Total} = 100\%$) and independent validation ($RAR_{V,Total} = 96.0\%$). The results shower the feasibility of applying Vis-NIR spectroscopy to high-precision discriminant analysis of milk powder adulteration. The proposed spectral separation degree method can enhance the spectral difference of different spectral populations, extract information wavelengths, and improve the discrimination effect. It is also expected to be used to spectral discriminant analysis in other fields.

Acknowledgments. This work was supported by National Natural Science Foundation of China (No. 61078040) and Guangdong Province Project of China (No. 2014A020213016, No.2014A020212445).

References

1. Chen, J.M., Li, M.M., Pan, T., et al.: Rapid and non-destructive analysis for the identification of multi-grain rice seeds with near-infrared spectroscopy. Spectrochim Acta A. **219**, 179–185 (2019)
2. Yang, Y.H., Lei, F.F., Zhang, J., et al.: Equidistant combination wavelength screening and step-by-step phase-out method for the near-infrared spectroscopic analysis of serum urea nitrogen. J. Innov. Opt. Health Sci. **12**, 1950018 (2019)
3. Norris, K.H.: Applying Norris derivatives understanding and correcting the factors which affect diffuse transmittance spectra. NIR News **12**, 6–9 (2001)
4. Pan, T., Zhang, J., Shi, X.W.: Flexible vitality of near-infrared spectroscopy –Talking about Norris derivative filter. NIR News **31**, 24–27 (2020)

An Exploration into the Optimization of Feature Wavelength Screening Methods in the Processing of Frozen Fish Classification Data in Near Infrared Spectroscopy

G. Cheng[1], S. Meng[1], S. Liu[1], Y. Jiao[1], X. Chen[2], W. Zhang[1], H. Wen[1], W. Zhang[3], B. Wang[2], and X. Xu[2(✉)]

[1] The Key Laboratory of Weak-Light Nonlinear Photonics, Ministry of Education, School of Physics, Nankai University, Tianjin 300071, China

[2] College of Artificial Intelligence, Nankai University, Tianjin 300350, China
xuxx@nankai.edu.cn

[3] Lianyungang Customs P.R.C, Lianyungang 222042, China

Abstract. To effectively classify imported frozen fish, we propose a characteristic wavelength selection method based on two-dimensional correlation spectroscopy (2DCOS), which reduces spectral variables required for analysis and improves the accuracy and efficiency of classification, among the data obtained by near-infrared spectroscopy (NIRS). In the experiments, near-infrared spectral were collected from *Pollachius, Theragra chalcogramma, Gadous macrocephaius,* and *Melanogrammus aeglefinus* of the family Gadidae, comparing different preprocessing algorithms and selecting multiple scattering corrections. The 2DCOS between the four Cod samples were then constructed. Based on the autocorrelation spectrum of the synchronous 2DCOS, the relative intensities at wavelengths 1580 nm, 1744 nm, and 1900 nm were obtained to be almost zero, and the two highest peaks in the autocorrelation spectrum were at 1550–1580 nm and 1744–1900 nm, as well as the spectra in these two bands, were highly correlated, so the two bands 1550–1580 nm and 1744–1900 nm were filtered out from the complete spectrum. The results are the accuracy of the training set of the waveband SVM filtered based on the 2DCOS is 94.58%, and the accuracy of the validation set can reach up to 97.30%. The study shows that the proposed spectral data compression method based on the 2DCOS technique has a high compression rate and high classification accuracy.

Keywords: Two-dimensional correlation spectroscopy · Near infrared spectroscopy · God · Data compression

1 Introduction

Cod is the common name for the demersal fish genus Gadus, belonging to the family Gadidae. Cod Fish is mostly cold-water fish that live in the lower and middle depths of

X. Chu et al. (Eds.): ICNIR 2021, *Sense the Real Change: Proceedings of the 20th International Conference on Near Infrared Spectroscopy*, pp. 97–107, 2022.
https://doi.org/10.1007/978-981-19-4884-8_9

the ocean and are widely distributed throughout the world's oceans [1]. The Gadidae family and the hake family contain many economically important, world-class fish of high economic value.

Traditional methods of classifying cod species mainly use such as chemical, instrumental and sensory analyses [2], which are labor-, material- and financially-intensive. However, for the near-infrared spectroscopy (NIRS), which uses sophisticated information processing and extraction techniques, NIRS is an analytical instrument with advanced information measurement and information processing techniques for specific material targets [3]. It has high analytical efficiency and stable analytical results. In particular, it also has the advantage of non-destructive measurement and easy operation. In recent years it has begun to gain widespread interest in the seafood industry. NIRS is not only suitable for evaluating the quality of samples but has also been extensively studied for sample classification and species identification. Many scholars have now used NIRS in combination with a variety of chemometric methods, for example, Application of NIRS for rapid quantification of acid and peroxide in crude peanut oil coupled multivariate analysis [4]; Fast discrimination and quantification analysis of Curcumae Radix from four botanical origins using NIRS coupled with chemometrics tools [5]; NIRS coupled with chemometric algorithms for the prediction of cadmium content in rice samples [6]; Quality assessment of instant green tea using portable NIR spectrometer [7]; Synthesis, vibrational spectroscopy and X-ray structural characterization of novel NIR emitter squaramides [8].

We also observe that digital fisheries are one of the most cutting-edge and promising areas of modern fisheries, enabling rapid, non-destructive, and localized access to digital information on fisheries objects, and real-time, dynamic analysis, control, and management [9]. The bridge from the product object to the digital is built by NIR, not only giving efficient solutions to the specific problems of sorting but also extracting the characteristics of the product object into digital information. This advantage has given this technology a high degree of popularity, in line with current trends, and has given a broader platform for the joint advancement of NIRS and aquatic product classification.

In this research, samples of multiple species of similar-looking codfish were difficult to classify accurately as species when imported for inspection, as they were all white flesh samples. Near-infrared (NIR) spectra combined with two-dimensional correlation spectroscopy (2DCOS) is used to compress the spectral variables and classify the compressed spectra by support vector machine (SVM) to investigate the achievement of high-quality classification of the raw spectra with good results of compressed sensing.

2 Materials and Methods

2.1 Sample and Spectral Acquisition

The Cod Fish samples and their NIR spectra were provided by Lianyungang Customs, People's Republic of China, and included a total of 376 samples of *Pollachius virens*, *Theragra chalcogramma*, *Gadous macrocephaius*, and *Melanogrammus aeglefinus*. The NIR spectra were recorded (2nm resolution; 3 scans both for background and samples) at room temperature, by using a handheld NIR spectrometer (N210, Hanon, Shandong, China) with spectral range: 1550–1950 nm. The training sets and validation sets were divided according to the 10-fold cross-validation principle [10], and the results are shown in Table 1.

Table 1. Samples of Cod Fish analyzed by NIRS.

Sample	Number	Training set	Validation set
Pollachius virens	100	90	10
Theragra chalcogramma	100	90	10
Gadous macrocephaius	81	73	8
Melanogrammus aeglefinus	88	79	9

2.2 Data Preprocessing

Because the sample spectra may be affected by noise, sample background, and measuring instruments, the raw spectra need to be preprocessed. To reduce the influence on the model, the first derivative (1st Der), the second derivative (2nd Der), centralization, standardization, Savitzky-Golay smoothing (SG) [11], multiplicative scatter correction (MSC) [12], standard normal variate (SNV) [13] are used to preprocess the spectral data, respectively. According to the performance of SVM models, the most appropriate preprocessed method in this experiment is selected.

2.3 Establishment and Evaluation Index of SVM Model

SVM is a binary classification model and there are five widely used multi-classification SVMs, namely: One-Against-All (OAA), One-Against-One (OAO), Binary Tree (BT), Error-Correcting Output Code (ECOC), and Directed Acyclic Graph (DAG) [14]. In our research, an OAA approach to classification is used.

SVM is originally defined as a linear classifier with maximum interval on the Eigenspace [15]. However, it is difficult to satisfy such conditions in practical data classification, so the concept of optimal separation hyperplane is generalized to minimize the solution of the cost function when dealing with linearly indistinguishable data, and the cost function is defined as:

$$Y(w, \xi) = \frac{1}{2}||w||2 + C\sum_{i=1}^{N} \xi_i \tag{1}$$

where a slack variable $\xi_i \geq 0$ and penalty factor regularization parameter C are introduced, the higher the value of C, the higher the penalty for misclassified samples.

Comparison with two non-parametric classifiers (radial basis function neural network and k-nearest neighbor algorithm) in the context of spectral analysis, researchers have experimentally demonstrated that support vector machine models can directly analyze hyperspectral data in the hyperfine feature space with the highest performance and are an effective alternative to traditional pattern recognition methods [16].

In this experiment, the classification accuracy rate of the training set and test set are used as the model evaluation index. The accuracy rate is the proportion of the number of correctly classified samples N_T to the total number of samples N_R, and the formula is

$$Accuracy = \frac{N_T}{N_R} \times 100\% \tag{2}$$

In the exploration of compression sensing, the compression rate of the wavelength variable is taken as the evaluation index. The compression rate is the ratio of the wavelength variable N_C after compression to the total number of variables N_A, and its formula is

$$Compression\ rate = \frac{N_C}{N_A} \times 100\% \tag{3}$$

2.4 Two-Dimensional Correlation Spectroscopy

2DCOS is a spectroscopic research method that has been widely used in a variety of spectra. It was first applied in the field of nuclear magnetic resonance, and Noda [17] proposed in 1993 a generalized 2DCOS that includes not only sinusoidal signals such as light, electricity, and sound but also temperature, concentration, and Ph values as forms of external perturbation [18]. Because it extends the spectral signal to the second dimension, weak peaks and overlapping peaks that were originally in the one-dimensional spectrum and peaks that were disguised by noise or backbone are more clearly defined, significantly enhancing the resolution of the spectrum, and since then 2DCOS have been well used in various spectroscopic studies.

During the measurement of a spectrum, a specific external perturbation is applied to the study system, which can be any reasonable physical or chemical quantity such as electric field, magnetic field, light, heat, pressure, concentration, and pH value. It induces changes in the state, structure, or background environment of the system components, resulting in a change in the measured spectrum, which is called a dynamic spectrum. The 2DCOS is obtained by performing a two-dimensional correlation calculation on a series of obtained dynamic spectra.

Assuming that the spectral intensity of the external disturbance variable t caused by the external disturbance varies between the Tmin and Tmax is $y(v, t)$, the dynamic spectrum $\tilde{y}(v, t)$ of the system induced by the external disturbance can be defined as:

$$\tilde{y}(v, t) = \begin{cases} y(v, t) - \overline{y}(v) & T_{min} < t < T_{max} \\ 0 & otherwise \end{cases} \tag{4}$$

where $\overline{y}(v)$ is the reference spectrum of the system, the choice of reference spectrum is not unique, $\overline{y}(v)$ is often set to the average spectrum. The 2DCOS intensity $X(v_1, v_2)$ is a quantitative comparison of the change in spectral intensity at different optical variables v_1 and v_2 as the external disturbance variable t varies between Tmin and Tmax. For ease of calculation, $X(v_1, v_2)$ will be expressed as a plural form:

$$X(v_1, v_2) = \Phi(v_1, v_2) + i\Psi(v_1, v_2) \tag{5}$$

The imaginary and real parts that make up the intensity of the 2D correlation spectrum are the 2D correlation synchronous and asynchronous spectra. The synchronous and asynchronous spectral intensities can be expressed as [19]:

$$\Phi(v_1, v_2) = \frac{1}{m-1}\tilde{y}(v_1)^T\tilde{y}(v_2) \quad (6)$$

$$\Psi(v_1, v_2) = \frac{1}{m-1}\tilde{y}(v_1)^T N\tilde{y}(v_2) \quad (7)$$

where the Hilbert–Noda transformation matrix N is:

$$N_{ij} = \begin{cases} 0 & i = j \\ \frac{1}{\pi(j-i)} & i \neq j \end{cases} \quad (8)$$

3 Results and Discussion

3.1 Preprocessed Method

For the different pre-processing methods, the processed spectra were classified using SVM according to the divided training and validation sets, and the average accuracy of the results after 10 classifications were performed is shown in Table 2.

Table 2. Comparison of accuracy based on different preprocessing methods

Pretreatment methods	Training accuracy (%)	Validation accuracy (%)
Raw NIR spectra	100	78.92
MSC	99.01	94.32
Centralization	100	79.73
Standardization	100	57.57
SNV	100	22.97
1st Der	99.67	93.78
2nd Der	99.88	90.27
SG	99.88	74.59

As shown in Table 2, comparing the raw spectra with other preprocessing methods, the spectra processed by MSC have the highest classification accuracy. By observing the MSC spectra in Fig. 1b, it can be found that there are obvious characteristic absorption peaks near 1550 nm, 1670 nm, 1730 nm, and 1930 nm, which mainly come from the vibration of some specific chemical functional groups in fish. The peak at 1550nm is related to protein. The peaks at 1670nm and 1730nm related to fatty acids are largely attributed to the C–H and CH_2 vibrations. The peak at 1930nm is derived from the second-order frequency double absorption of the O-H bond in water molecules [20]. Compared with the raw spectra in Fig. 1a, the MSC spectra can obtain more useful information [21]. So, the MSC spectra were used to compress for building SVM models.

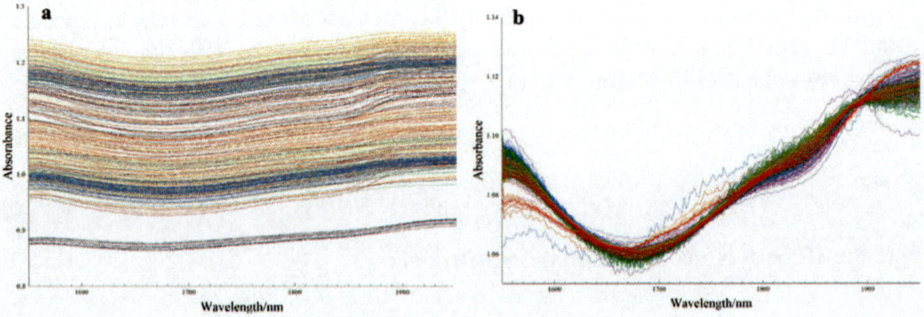

Fig. 1. NIR spectra: **a.** Raw Spectra; **b.** MSC Spectra

3.2 Selection of Feature Wavelengths

Since each spectrum contains 201 variables, modeling not only requires a large amount of calculation but also affects the accuracy of classification. To determine the optimal subset of spectral variables in classification, spectral data are discarded by selecting feature wavelengths, and then classification by SVM. In this paper, competitive adaptive reweighted sampling (CARS) algorithm and 2DCOS are selected to achieve spectral feature wavelength compression screening, and the classification results are compared.

3.2.1 Selection of Spectral Variables by CARS

The CARS [22] combined with partial least squares regression can remove the variables with small weight based on the principle of imitating the survival of the fittest through exponential decay function (EDF) and adaptive reweighted sampling (ARS). Then, the subset with the lowest Root-Mean-Square Error of Cross-Validation (RMSECV) is selected through interactive verification, which can effectively find the optimal combination of characteristic wavelengths [23]. In this study, the Monte Carlo sampling value is set to 50, the number of cross verification groups $K = 10$, and the maximum principal component number $A = 20$. As shown in Fig. 2a, with the operation of EDF in the sampling process, the number of screened variables decreases sharply and then slowly with the increase of sampling times, and finally approaches to a stable state. It indicates that CARS sample variables can be roughly selected first and then accurately selected, and improve the efficiency of screening characteristic variables of NIRS. Figure 2b shows the variation trend of RMSECV in the screening process of wavelength variables. The RMSECV value decreases, indicating the number of irrelevant variables decreases, and the RMSECV value increases, indicating that effective variables have been eliminated [24]. In Fig. 2b, the RMSECV drops to a minimum value at the 10th sampling and then fluctuates upwards. This indicates that a large number of spectral variables unrelated to the sample fish category was filtered out in the first 10 sampling operations, while some important wavelength variables related to the fish sample category were removed in the sampling operations after the 10th, resulting in a decrease in the predictive power of the model. Therefore, when RMSECV reaches the minimum value, it is taken as the optimal result. Figure 2c shows that when RMSECV reaches the minimum value, the regression coefficient of each variable is located at the position of the red vertical dot and line in

Fig. 2c. According to Fig. 2a, when the number of runs is 10, 77 wavelength variables are screened, and the compression rate of variables is 38.31%.

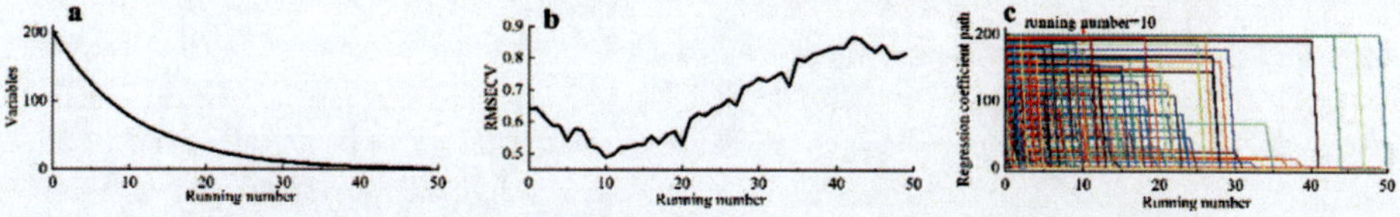

Fig. 2. Plot of variables selection by CARS method. **a.** Tendency of the number of the involved variables; **b.** the change of RMSECV value during the optimization of wavelength variables; **c.** tendency of the regression coefficients of spectral variables during the sampling runs, red signal represents the optimal number of 10 sampling runs.

3.2.2 Selection of Spectral Variables by 2DCOS

In our experiments, the 2DCOS technique was used to explore feature information between the spectra of different species, using differences between the four Cod Fish species as perturbation conditions. One MSC spectrum for each of the four cod species was randomly selected and the 2D correlation synoptic spectra and their 3D plots were plotted using inter-species differences as a perturbation as shown in Fig. 3. As shown in Fig. 3a, 2 types of autocorrelation peaks and cross peaks are present in the Synchronous 2DCOS.

Peaks located at bottom diagonal positions in Fig. 4a are autocorrelation peaks, and cross peaks located at the off-diagonal positions of a synchronous 2DCOS represent simultaneous or coincidental changes of spectral intensities observed at 2 different spectral variables [25]. The intensity of the autocorrelation peak reflects the extent to which the spectral signal changes with external perturbations at different wavelengths, i.e. the spectral band that is sensitive as the fish species changes. From the autocorrelation spectrum, Fig. 3b shows that the relative intensities at wavelengths of 1580, 1744, and 1900 nm are almost zero, and the 2 highest peaks in the autocorrelation spectrum are at 1550–1580 nm and 1744–1900 nm, where the correlation of the spectra in these 2 bands is very high. The theoretical support is that an auto peak represents the overall susceptibility of the corresponding spectral region to change in spectral intensity as an external perturbation is applied to the system[25]. The zero points mean the lowest susceptibility between four different fish species and there are no accurate chemical bond vibrations belonging to these points. In 1550–1580 nm and 1744–1900 nm, there are peaks around 1550nm which is related to protein, and peaks around 1765nm which is related to the first-order multiplier frequency of absorption of C-H bond in -CH_2-. We can explain the select bands' variables are the useful chemical bond vibration in our study, bonds that do not work in the spectral analysis are removed. The physics significance of the selected bands is the effective information of the whole spectra. It improved the signal-to-noise ratio of analyzed data. Based on the findings obtained in the autocorrelation spectrum,

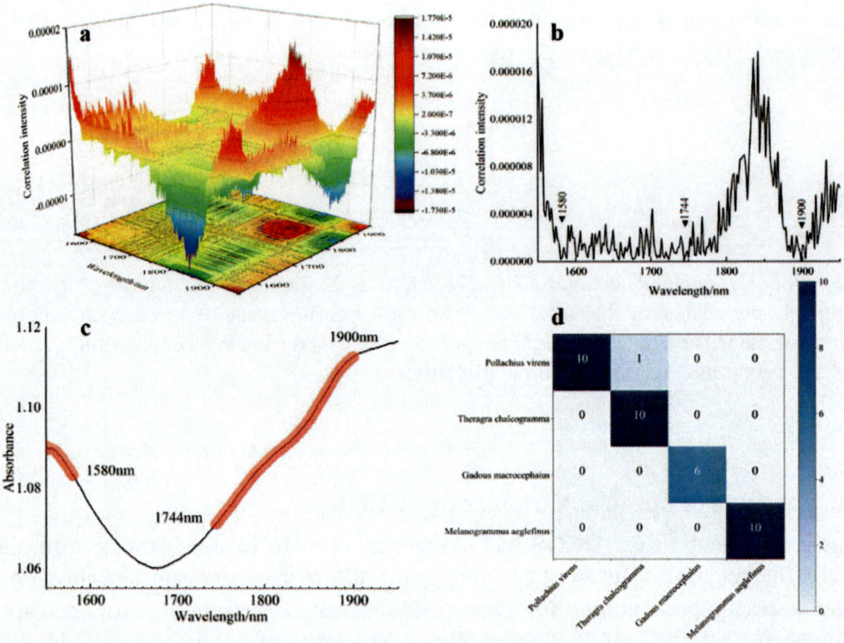

Fig. 3. Two-dimensional correlation spectroscopy: **a.** 3D stereo plots; **b.** Autocorrelation spectrum; **c.** The feature wavelength of MSC spectra; **d.** Confusion matrix of validation based on 2DCOS model

we filtered a complete MSC spectrum into two spectral band rows modeled at 1550–1580 nm, and 1744–1900 nm, as shown in Fig. 3c. Meanwhile, a variable compression rate of 47.26% was achieved by selecting 95 variables from a spectrum of 201 variables. The feature wavelengths of the Cod Fish selected by 2DCOS were classified using an SVM model and the confusion matrix of the validation set is shown in Fig. 3d.

3.2.3 SVM Model Based on Characteristic Wavelength

Our experiments compare two methods achieving an optimal combination of feature wavelengths to build a SVM model for classification. As can be seen from Table 3, comparing the modeling results of the spectra without compression, the results of variable compression using 2DCOS can achieve 97.30% of the validation accuracy in the optimal case. The compression rate of CARS is 38.31% and 2DCOS is 47.26%. But the training and validation accuracy of 2DCOS which are 94.58% and 97.30% are much better than CARS which are 93.98% and 75.68%. It can be indicated that our feature wavelengths selection algorithm based on 2DCOS can select a more useful variable so the compression is higher than CARS but accuracy is much better than CARS. In all three cases, the use of 2DCOS in compression is more efficient and more effective.

Table 3. Results in the prediction of accuracy (%) based on CARS and 2DCOS compression methods

Methods	Training accuracy	Validation accuracy	Compression rate
None	99.01	94.32	100
CARS	93.98	75.68	38.31
2DCOS	94.58	97.30	47.26

4 Conclusion

Because of the difficulty in distinguishing fish products in smart fisheries and relevant industrial chains, by comparing various preprocessed methods, this study adopts the NIRS measurement device for four different types of Cod Fish samples of NIR absorption spectra. CARS and 2DCOS were used to select the feature wavelengths, and the compressed spectra were used to build an SVM classification model to investigate the optimization of feature wavelength screening in the classification data processing. In comparison, MSC works best in preprocessing methods, with the training and validation sets achieving 99.01% and 94.32% accuracy. And the optimization of the feature wavelength screening by 2DCOS can achieve an accuracy of 94.58% for the training set and up to 97.30% for the validation set of the 2 methods. We also achieve 47.26% data compression while achieving high accuracy.

The innovations of our study can be summarised as follows: Based on the use of MSC spectra to plot 2DCOS, the wavelengths in the higher peak range of the autocorrelation spectrum are used as feature wavelengths, using a segmentation of the 2DCOS with almost zero relative intensity, compared to CARS, our method of optimizing feature wavelength screening can retain a greater degree of original spectral information and maintain a higher level of accuracy. Therefore, the use of NIRS combined with the MSC preprocessing method and 2DCOS in this study is fully feasible and effective for the optimization of characteristic wavelength screening methods in the processing of frozen fish classification data. Compared to traditional methods, NIRS is simple, portable, low cost, and does not require complex sample handling, allowing for rapid non-destructive detection. The method can also be used in studies on the detection of different qualities of cod, as well as providing research ideas for the classification and detection of other species of samples.

Acknowledgments. This work was supported by the Innovation and Entrepreneurship Training Program for College Students of Tianjin (No. 202110055320), Scientific Research Project of Nanjing Customs P.R.C (No. 2020KJ22), Jiangsu Province: Program for High-Level Entrepreneurial and Innovative Talents Introduction.

References

1. Chen, Y., Mello, L.G.S.: Growth and maturation of cod (Gadus morhua) of different year classes in the Northwest Atlantic, NAFO subdivision 3Ps. Fish Res. **42**, 87–101 (1999)
2. Comi, G., Iacumin, L., Rantsiou, K., Cantoni, C., Cocolin, L.: Molecular methods for the differentiation of species used in production of cod-fish can detect commercial frauds. Food Control **16**, 37–42 (2005)
3. Lindon, J.C., Tranter, G.E., Koppenaal, D.W.: Encyclopedia of Spectroscopy and Spectrometry, 3rd edn. Academic Press, New York (2017)
4. Haruna, S.A., et al.: Application of NIR spectroscopy for rapid quantification of acid and peroxide in crude peanut oil coupled multivariate analysis. Spectrochim. Acta A **267**, 120624 (2022)
5. Wang, L., et al.: Fast discrimination and quantification analysis of Curcumae Radix from four botanical origins using NIR spectroscopy coupled with chemometrics tools. Spectrochim. Acta A. **254**, 119626 (2021)
6. Miao, X., et al.: NIR spectroscopy coupled with chemometric algorithms for the prediction of cadmium content in rice samples. Spectrochim. Acta A **257**, 119700 (2021)
7. Sun, Y., et al.: Quality assessment of instant green tea using portable NIR spectrometer. Spectrochim. Acta A **240**, 118576 (2020)
8. Ávila-Costa, M., et al.: Synthesis, vibrational spectroscopy and X-ray structural characterization of novel NIR emitter squaramides. Spectrochim. Acta A **223**, 117354 (2019)
9. Merrifield, M., et al.: eCatch: enabling collaborative fisheries management with technology. Ecol. Inf. **52**, 82–93 (2019)
10. Zhang, G., et al.: Optimized adaptive Savitzky-Golay filtering algorithm based on deep learning network for absorption spectroscopy. Spectrochim. Acta A **263**, 120187 (2021)
11. Wu, Y., Peng, S., Xie, Q., Han, Q., Zhang, G., Sun, H.: An improved weighted multiplicative scatter correction algorithm with the use of variable selection: application to near-infrared spectra. Chemometr. Intell. Lab. Syst. **185**, 114–121 (2019)
12. Bi, Y., et al.: A local pre-processing method for near-infrared spectra, combined with spectral segmentation and standard normal variate transformation. Anal. Chim. Acta **909**, 30–40 (2016)
13. Zhang, H., Yang, S., Guo, L., Zhao, Y., Shao, F., Chen, F.: Comparisons of isomiR patterns and classification performance using the rank-based MANOVA and 10-fold cross-validation. Gene **569**, 21–26 (2015)
14. Wu, G., He, Y.: Identification of varieties of cashmere by Vis/NIR spectroscopy technology based on PCA-SVM. In: 2008 7th World Congress on Intelligent Control and Automation, pp. 1548–1552 (2008)
15. Chen, Q., Zhao, J., Fang, C.H., Wang, D.: Feasibility study on identification of green, black and Oolong teas using near-infrared reflectance spectroscopy based on support vector machine (SVM). Spectrochim. Acta A **66**, 568–574 (2007)
16. Melgani, F., Bruzzone, L.: Classification of hyperspectral remote sensing images with support vector machines. IEEE Trans. Geosci. Remote Sens. **42**, 1778–1790 (2004)
17. Noda, I.: Generalized two-dimensional correlation method applicable to infrared, raman, and other types of spectroscopy. Appl. Spectrosc. **47**, 1329–1336 (1993)
18. Guo, R., et al.: A novel systematic absence of cross peaks-based 2D-COS approach for bilinear data. Spectrochim. Acta A **220**, 117103 (2019)
19. Cheng, W., Sun, D.W., Pu, H., Wei, Q.: Heterospectral two-dimensional correlation analysis with near-infrared hyperspectral imaging for monitoring oxidative damage of pork myofibrils during frozen storage. Food Chem. **248**, 119–127 (2018)

20. Khodabux, K., L'Omelette, M.S.S., Jhaumeer-Laulloo, S., Ramasami, P., Rondeau, P.: Chemical and near-infrared determination of moisture, fat and protein in tuna fishes. Food Chem. **102**, 669–675 (2007)
21. Teye, E., Amuah, C.L.Y., McGrath, T., Elliott, C.: Innovative and rapid analysis for rice authenticity using hand-held NIR spectrometry and chemometrics. Spectrochim. Acta A **217**, 147–154 (2019)
22. Li, H., Liang, Y., Xu, Q., Cao, D.: Key wavelengths screening using competitive adaptive reweighted sampling method for multivariate calibration. Anal. Chim. Acta **648**, 77–84 (2009)
23. Liu, J., et al.: Estimation of soil organic matter content based on CARS algorithm coupled with random forest. Spectrochim. Acta A **258**, 119823 (2021)
24. Guo, H., et al.: Application of Fourier transform near-infrared spectroscopy combined with GC in rapid and simultaneous determination of essential components in Amomum villosum. Spectrochim. Acta A **251**, 119426 (2021)
25. Noda, I., Ozaki, Y.: Two-Dimensional Correlation Spectroscopy-Applications in Vibrational and Optical Spectroscopy. Johns Wiley & Sons, Chichester (2004)

Handheld NIR and PLS-DA Models for Onsite Detection of Injected Water and Discrimination of Different Injected Solutions in Tuna

S. Nieto-Ortega[1(✉)], Á. Melado-Herreros[1], I. Olabarrieta[1], G. Foti[1], G. Ramilo-Fernández[2], C. G. Sotelo[2], B. Teixeira[3,4], A. Velasco[2], and R. Mendes[3,4]

[1] AZTI, Food Research, Basque Research and Technology Alliance (BRTA), Parque Tecnológico de Bizkaia, Astondo Bidea, Edificio 609, 48160 Derio, Bizkaia, Spain
snieto@azti.es

[2] Instituto de Investigaciones Marinas (CSIC), Eduardo Cabello 6, 36208 Vigo, Spain

[3] Department for the Sea and Marine Resources, Portuguese Institute for the Sea and Atmosphere (IPMA), Avenida Doutor Alfredo Magalhães Ramalho, 6, 1495-165 Algés, Portugal

[4] Interdisciplinary Center of Marine and Environmental Research (CIIMAR), University of Porto, Rua das Bragas 289, 4050-123 Porto, Portugal

Abstract. A handheld near infrared (NIR) spectroscopy device, with a wavelength range from 900 nm to 1650 nm and coupled with two Partial Least-Squares Discriminant Analysis (PLS-DA) models, has been used to demonstrate its applicability as a proof of concept for quality monitoring of bigeye tuna (*Thunnus obesus*). First, a classification model was created to discriminate between injected and non-injected tuna samples. Then, a second classification model was developed to discriminate between non-injected and each water and additives treatment used. The results were promising, showing both models good results in the validation dataset. The first model, with 8 latent variables (LV), had an error-rate of 0.08 and an accuracy value of 0.93. It showed a good discrimination between injected and non-injected samples. The second model, with 10 LV, presented an error rate of 0.15 and an accuracy of 0.88. The discrimination between treatments was good even when protein hydrolysate solutions were used (sensitivity = 0.81; specificity = 0.99; precision = 0.87), a case which is typically hard to detect with accurate destructive analysis. This work opens new possibilities for onsite inspection in the fish industry, where NIR could be used as a complementary tool for the detection of water solutions in tuna.

Keywords: Quality control · Water addition · Additives · Seafood · Chemometrics

1 Introduction

Seafood products, due to their high commercial value, are susceptible to fraudulent techniques such as incorrect labelling. One typical example is the incorporation of water

X. Chu et al. (Eds.): ICNIR 2021, *Sense the Real Change: Proceedings of the 20th International Conference on Near Infrared Spectroscopy*, pp. 108–117, 2022.
https://doi.org/10.1007/978-981-19-4884-8_10

and additives. This is a practice usually performed to compensate moisture losses during harvest, processing and storage of fish [1]. It may be justified on some technological grounds, such as to reduce the drip-loss, with a possible antioxidant effect or to retain nutrients [2]. However, the line between adding water and additives to make up these losses and an excessive water addition in order to obtain economic illicit gain is very thin [1]. Even though this procedure is allowed, if the amount of added water represents more than the 5% of the weight of the product, it must be indicated in the label [3], and sometimes this declaration is not present.

Regarding this practice, the most common procedure involves the muscle injection of brines, usually composed of water, salt phosphates and functional proteins and peptides through a multi-needle injection [4–6]. Phosphates are a group of food additives (E450, E451, E452) used for several technological processes [7]. Functional proteins and peptides have also been used for increasing the water binding properties of the products, leading to an improved stability, texture and a higher water holding capacity (WHC) [8, 9].

Nowadays, the official control methods to detect the water addition in seafood are based on analytical methods, which involve the analysis of moisture and protein of the samples [10], due to an existent physiological correlation between moisture and protein levels [11]. However, these methods are time-consuming, they destroy part of the sample and generate toxic residues. Furthermore, there are some problems related with the detection of water addition using protein hydrolysates, as not only do they increase the water content, but also the protein level, so that the moisture/protein ratio is not affected as much as expected in order to detect the injection. The use of non-destructive technologies in food quality control is gaining attention due to its advantages over traditional methods. They are rapid, allow the inspection of most of the production and are non-invasive. During the last years some techniques have been used to perform the quality control of fish. Some examples are the use of Raman spectroscopy [12] or hyperspectral imaging [13], to avoid the mislabelling of frozen-thawed fish fillets. Other techniques, such as electrical impedance, have been used to detect the freshness of meat and fish [14]. Among all, NIR has emerged as a valid solution for detecting adulterants in seafood. For example, Khodabux *et al.* [15] determine moisture, protein and fat content in tuna using NIR. Reis *et al.* [16] used this technique to discriminate between fresh and frozen/thawed tuna fillets. Ghidini *et al.* [17] developed a tool based on NIR to control the histamine in tuna. However, the NIR devices used for fish quality control are usually laboratory equipment, which are not handheld and do not allow onsite determinations. Moreover, although some works have demonstrated the usefulness of NIR to detect water addition in tuna [18], up to authors' knowledge, this is the first time that this technology is used to discriminate between non-injected and injected tuna and to ensure that all the solutions that can be used for the injection are well classified.

The objective of this work was to demonstrate that a Near Infrared Spectroscopy (NIRs) handheld device, in combination with chemometrics classification methods, is a valid complementary technique to traditional approaches for the onsite detection of added water and the discrimination of different injected solutions in tuna samples.

2 Materials and Methods

For the experiment, 7 loins of bigeye tuna (*Thunnus obesus*) captured in the FAO 34 fishing area (Portugal) were purchased in 2018. They were cut in 60 portions of approximately 500 g and divided in 6 groups. 10 remained without any injection, and 50 (10 portions per treatment) were injected with a 10% weight injection of 5 different solutions:

- A. 3% salt
- B. 3% salt + 3% polyphosphates
- C. 3% salt + 5% polyphosphates
- D. 3% salt + 5% hydrolysate prepared in house from four-spot megrim *Lepidorhombus boscii*
- E. a polyphosphate commercial solution (Pescamine 150, Vaessen-Schoemaker, Deventer, The Netherlands).

A handheld NIR (MicroNIR OnSite spectrometer from Viavi, Italy), working in the range from 900 nm to 1650 nm was used to measure the NIR spectra of the tuna samples. It has two integrated vacuum tungsten lamps and works with a 128-pixel InGaAs photodiode array with a spectral resolution of < 1.225% of centre wavelength (1% typical). All the tuna portions were scanned before and after the injection process and, for each sample, 8 scans were acquired in different parts of the portion. The number of NIR scans collected in each treatment is shown in Table 1.

Table 1. Number of scans acquired per sample.

Treatment	N° Scans/sample
Control	480
A	80
B	80
C	80
D	80
E	80

A scheme of the methodology followed during the experiment is shown in Fig. 1.

After the spectral measurements, and to evaluate if the injection procedure was efficient, the water/protein ratio was calculated in all the samples and an ANOVA analysis was performed between the control and the samples injected with the different treatments (using the software Statgraphics centurion XVI from Statgraphics Technologies, Inc., The Plains, VA, USA). For that purpose, moisture and protein content were determined in all the tuna portions. Moisture was analysed by standard gravimetric analysis [19], and protein content was determined by the Dumas combustion method, according to the methodology developed by Saint-Denis and Goupy [20].

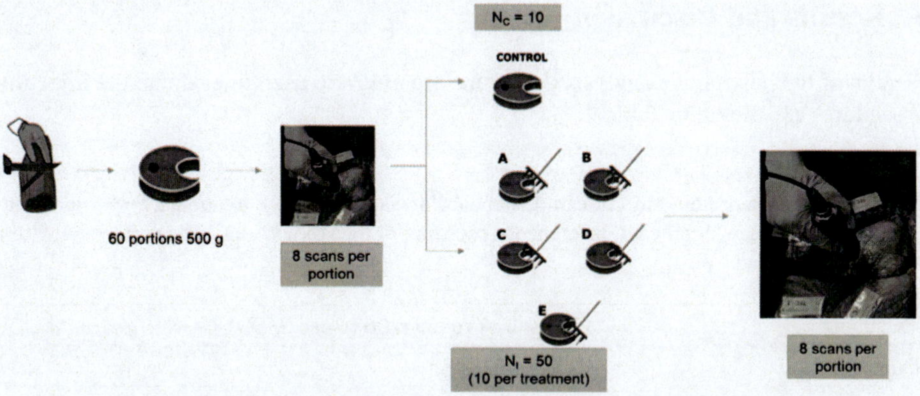

Fig. 1. Methodology followed during the experimentation

Once all the spectral data was collected, they were analysed using Matlab 2020b (The Mathworks, Natick, MA, USA). Before performing the classification analysis, data were pre-processed. Several techniques were tested, such as Savitzky-Golay 1st and 2nd derivatives, Standard Normal Variate (SNV) with and without detrend, and different combinations of them. After selecting the optimal pre-processing method data were autoscaled. Then, two different datasets were defined: one for calibration purposes, with 80% of the data, and the other for validation, with the remaining 20%. This was done by applying the Duplex algorithm, which divides the data based on the Euclidean distance between all the points. It is an iterative process where, in the first loop, the two farthest points are assigned to the calibration set. After that, the second loop assigns the next two points with the greatest distance to the validation set. The process continues until all the points are assigned to one of the sets [21].

Using the Classification toolbox (version 5.4), developed by Milano Chemometrics and QSAR Research group [22], two different Partial Least Squares Discriminant Analysis (PLS-DA) models were developed, based on the Bayes theorem. The first one (model 1) had the purpose of discriminating between injected and non-injected samples. The second model (model 2) was aimed to differentiate between the control samples and the 5 different injection solutions used, to find if all the groups were correctly classified by the technology.

In both models NIR data represented the independent variable X while Y, the dependent categorical variable, was codified as a dummy matrix. In the first PLS-DA model developed, model 1, it was expressed in binary code (as 1 s and 0 s). Model 2 expressed Y as values between 1 and 6 (being 1 s the non-injected samples and numbers between 2 – 6 the samples injected with each treatment).

Calibration dataset was used to train the models. Then, a Venetian Blinds cross-validation (CV) with 5 CV groups was performed to select the optimal number of latent variables (LV). Finally, the models were validated using the validation dataset. The performance was evaluated using several figures of merit, such as the accuracy, the error rate, the sensitivity and the specificity for both the cross validation and the validation dataset.

3 Results and Discussion

Results of the destructive analysis (moisture/protein ratio used to evaluate the injection procedure) are shown in Table 2.

Table 2. Moisture/protein ratio (mean ± standard deviation) in each treatment. ^{a,b,c} and ^{d} consecutive letters mean significant differences according to the ANOVA analysis performed with a 95% of confidence level.

Treatment	Moisture/Protein ratio (mean ± SD)
Control	2.76 ± 0.09^{a}
A	2.93 ± 0.10^{bc}
B	2.94 ± 0.09^{c}
C	3.02 ± 0.10^{d}
D	2.88 ± 0.08^{b}
E	2.97 ± 0.06^{cd}

As it can be seen in Table 2, control samples present the lowest moisture/protein ratio (2.76 ± 0.09), and it is statistically different from all the injected samples, demonstrating that the injection procedure was efficient in all the cases. The ratio in the treatments also show some statistically significant differences, showing that some treatments retain more added water than others. Treatment C (3% salt + 5% polyphosphates) retains the highest amount of water (ratio of 3.02), being the most efficient, while treatment D (3% salt + 5% protein hydrolysate) had the moisture/protein ratio closest to control samples. This may be due to the fact that, in this case, not only the moisture content was increased, but also the total protein level, as the solution has protein hydrolysates. This is a clear example of the problems encountered when trying to detect added water using the moisture/protein ratio in samples injected with protein-based solutions.

Results of both PLS-DA models are shown in Tables 3 and 4.

Table 3. Results of the PLS-DA model developed to discriminate between injected and non-injected samples (model 1).

		Pre-processing	LV	Error-rate	Accuracy	Sensitivity	Specificity	Precision
CV	Non-injected	2° derivative Savitzky-Golay (window 5, order 2) + autoscaling	8	0.05	0.95	0.96	0.95	0.96
	Injected					0.95	0.96	0.95
Validation	Non-injected			0.07	0.93	0.92	0.94	0.94
	Injected					0.94	0.92	0.92

As it can be seen in Tables 3 and 4, the two PLS-DA models developed show a good discrimination between groups of samples. The results are slightly better in the case of

Table 4. Results of the PLS-DA model developed to discriminate between non-injected and each treatment (model 2).

		Pre-processing	LV	Error-rate	Accuracy	Sensitivity	Specificity	Precision
CV	Non-injected	2° derivative Savitzky-Golay (window 5, order 2) + autoscaling	10	0.10	0.91	0.89	0.96	0.97
	A					0.92	0.99	0.91
	B					0.79	0.97	0.75
	C					0.80	0.97	0.70
	D					0.94	0.98	0.84
	E					0.90	0.97	0.78
Validation	Non-injected			0.15	0.88	0.92	0.94	0.94
	A					0.88	0.99	0.88
	B					0.82	0.96	0.70
	C					0.76	0.97	0.76
	D					0.81	0.99	0.87
	E					0.92	0.99	0.92

model 1, created to differentiate injected and non-injected samples. It presents, for the validation set, an error-rate of 0.07, an accuracy of 0.93 and, for each class, values of sensitivity, specificity and precision greater than 0.90 in all the cases (0.92, 0.94 and 0.94 for non-injected samples and 0.94, 0.92 and 0.92 for injected samples, respectively). It also has the lower complexity, with 8 LV. However, results are also good in model 2. It performs an acceptable separation between all the categories (non-injected and the five different treatments used to inject samples) with an error-rate of 0.15 and an accuracy of 0.88 for the validation set. In this second case the model is a little more complex, with 10 LV. Thus, these results show the viability of NIR to discriminate onsite injected tuna samples from non-injected fish portions, ensuring that all the treatments are well separated.

It is worth mentioning the good classification performed for the treatment D (3% salt + 5% hydrolysate prepared in house from four-spot megrim *Lepidorhombus boscii*) in model 2, with sensitivity, specificity and precision values of 0.81, 0.99 and 0.87, respectively. The detection with destructive methods of these solutions usually presents problems, as it has been seen when calculating the ratio, because the injection of hydrolysate-based treatments results in an increase of both water and protein content. Therefore, the moisture/protein ratio in injected samples remains substantially unchanged with respect to the non-injected case. Considering that, this proof of concept opens new possibilities to complement the water/protein ratio analysis in this kind of injection treatments.

Loadings for latent variable 1 (LV1) and latent variable 2 (LV2) of both PLS-DA models are shown in Figs. 2 and 3. Although a lot of peaks are observed, it is still possible to identify the regions related to the different chemical compounds.

As it can be seen in Figs. 2 and 3, LV1 of model 1 is quite similar to LV1 of model 2, and the same happens to LV2. Both models show loadings with almost the same spectral shape and peaks at similar wavelengths. This suggests that the two models selected similar vibrational features for the discrimination between injected vs non-injected and

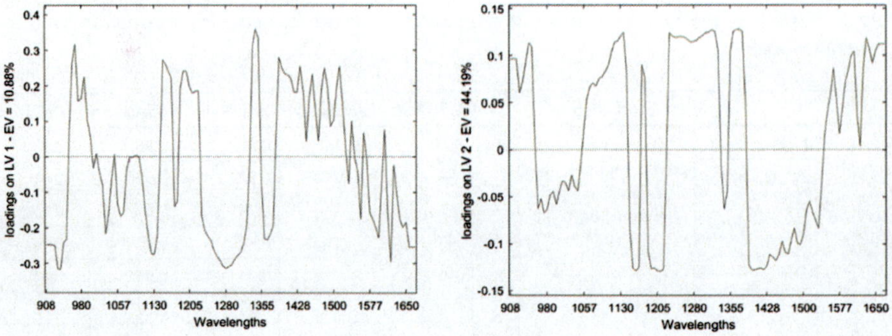

Fig. 2. Loadings of LV1 (left) and LV2 (right) of the PLS-DA model (model 1) developed to discriminate between non-injected and injected samples.

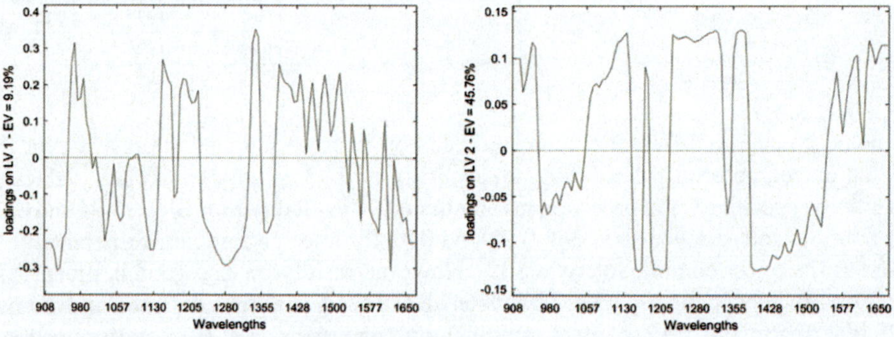

Fig. 3. Loadings of LV1 (left) and LV2 (right) of the PLS-DA model (model 2) developed to discriminate between non-injected and each treatment.

injected vs type of treatment. LV2 retains the greatest amount of variance in both cases (44.19% in the first model and 45,76% in the second). However, the first LV does not necessarily retain the highest explained variance [23] if the covariance between X and Y is the biggest, due to LV being constructed oriented along the directions of the maximal covariance [24].

Water is a strong absorber in the NIR region. Therefore, it is expected that LV of both models are strongly dominated by the water signal, since food systems with a high-water content usually have absorption bands at wavelengths close to those of pure water [25] (specially in this case, where additional water solutions are intentionally added). Observing the loadings (Figs. 2 and 3), it can be seen that LV1 and LV2 of model 1 and 2 present regions with strong peaks situated close to the main water absorption bands. They are located around 970 nm, 1200 nm and 1450 nm, and they are caused by the second overtone of the OH stretching band, the combination of the first overtone of the O-H stretching and bending band and the first overtone of the OH-stretching band, respectively [25, 26].

In the central part of the spectra, it can be seen (in LV1 and LV2 of both models) a region with a strong influence, which has a big peak around 1280 nm. According to

Laub-Ekgreen *et al.* [27], the region between 1100 nm and 1300 nm is closely related to the sodium chloride, salt that has been added to all the samples except the control ones. This region overlaps the peaks due to water absorption at 1200 nm. Although electrolytes cannot absorb in the NIR spectra, salt solutions can be detected by this technology, due to the effect they cause on the hydrogen bonds of water, explaining why the water absorption band at 1200 nm has influence in this region.

Some authors have related the spectral bands around 1510 nm and 1690 nm with the differences among samples in the protein content [28], related with the N-H overtones. In both models the LVs present peaks around those wavelengths which suggests that the algorithm is detecting the differences in the crude protein content of the samples. This could explain the good classification result for the samples treated with the D solution where added proteins are present. However, it is usually difficult to identify the peaks, due to their closeness to absorption bands of other compounds such as water or fat.

4 Conclusion

In this work, the potential of NIR sensing technology in combination with the PLS-DA classification algorithm for the detection of added water and the discrimination of several additives in fresh tuna was explored. The models were robust enough to differentiate not only the injected from the control samples, but also to discriminate between the different treatments used in the injection. The detection was good even when a protein hydrolysates-based solution was added, for which identification through accurate analytical methods used to be difficult. This work could be of great interest as a way of performing a non-destructive rapid inspection to take real time decisions at different points of the fish value chain.

Acknowledgments. The research leading to these results has received funding from the Interreg Atlantic Area Project EAPA_87/2016, SEATRACES: Smart Traceability and Labeling ToolBox for a Sustainable Seafood Production. The authors acknowledge the Basque Government - Department of Economic Development, Sustainability and Environment – Directorate of Quality and Food Industries for the scholarship of S. Nieto-Ortega. B. Teixeira acknowledges the Portuguese Foundation for Science and Technology (FCT), the European Social Fund (FSE) and the Ministry of Education and Science for supporting a grant (Ref. SFRH/BPD/92929/2013). A. Velasco acknowledges the Spanish Ministry of Science, innovation and technology for the contract of technical staff (PTA2016–12254-I). This paper is contribution n° 1110 from AZTI, Food Research, Basque Research and Technology Alliance (BRTA).

References

1. van Ruth, S.M., Brouwer, E., Koot, A., Wijtten, M.: Seafood and water management. Foods **3**, 622–631 (2014)
2. Kent, M., Knöchel, R., Daschner, F., Berger, U.K.: Composition of foods using microwave dielectric spectra. Eur. Food Res. Technol. **210**, 359–366 (2000)
3. EU, Regulation (EU) No 1169/2011 of the European Parliament and of the Council of 25 October 2011 on the provision of food information to consumers, Off. J. Eur. Union 304, 18–63 (2011)

4. Åsli, M., Mørkøre, T.: Brines added sodium bicarbonate improve liquid retention and sensory attributes of lightly salted Atlantic cod. LWT Food Sci. Technol. **46**, 196–202 (2012)
5. Kin, S., Schilling, M.W., Smith, B.S., Silva, J.L., Jackson, V., Kim, T.J.: Phosphate type affects the quality of injected catfish fillets. J. Food Sci. **75**, S74–S80 (2010)
6. Zhao, Q., Klonowski, I., Karlsdottir, M.G., Arason, S., Thorarinsdottir, K.A.: Effects of injection of protein solutions prepared from fish by-products on yield and chemical properties of chilled and frozen saithe (*pollachius virens*) fillets. J. Aquat. Food Prod. Technol. **22**, 258–269 (2013)
7. Kulaev, I.S., Vagabov, V.M., Kulakovskaya, T.V.: The Biochemistry of Inorganic Polyphosphates. John Wiley & Sons, West Sussex (2004)
8. Thorarinsdottir, K.A., Gudmundsdottir, G., Arason, S., Thorkelsson, G., Kristbergsson, K.: Effects of added salt, phosphates, and proteins on the chemical and physicochemical characteristics of frozen cod (*gadus morhua*) fillets. J. Food Sci. **69**, SNQ144–SNQ152 (2004)
9. Morrisey, M., DeWitt, C.A.: Value-Added Seafood. In: Seafood processing: Technology, Quality and Safety. John Wiley & Sons, Chichester (2014)
10. Mendes, R., Schimmer, O., Vieira, H., Pereira, J., Teixeira, B.: Control of abusive water addition to *Octopus vulgaris* with non-destructive methods. J. Sci. Food Agric. **98**, 369–376 (2018)
11. Yeannes, M.I., Almandos, M.E.: Estimation of fish proximate composition starting from water content. J. Food Compos. Anal. **16**, 81–92 (2003)
12. Velioğlu, H.M., Temiz, H.T., Boyaci, I.H.: Differentiation of fresh and frozen-thawed fish samples using Raman spectroscopy coupled with chemometric analysis. Food Chem. **172**, 283–290 (2015)
13. Qin, J., et al.: Detection of fish fillet substitution and mislabeling using multimode hyperspectral imaging techniques. Food Control **114**, 107234 (2020)
14. Zhao, X., Zhuang, H., Yoon, S.C., Dong, Y., Wang, W., Zhao, W.: Electrical impedance spectroscopy for quality assessment of meat and fish: a review on basic principles, measurement methods, and recent advances. J. Food Qual. **2017**, 6370739 (2017)
15. Khodabux, K., L'Omelette, M.S.S., Jhaumeer-Laulloo, S., Ramasami, P., Rondeau, P.: Chemical and near-infrared determination of moisture, fat and protein in tuna fishes. Food Chem. **102**, 669–675 (2007)
16. Reis, M.M., Martínez, E., Saitua, E., Rodríguez, R., Pérez, I., Olabarrieta, I.: Non-invasive differentiation between fresh and frozen/thawed tuna fillets using near infrared spectroscopy (Vis-NIRS). LWT Food Sci. Technol. **78**, 129–137 (2017)
17. Ghidini, S., et al.: Histamine control in raw and processed tuna: a rapid tool based on NIR spectroscopy. Foods **10**, 885 (2021)
18. Melado-Herreros, Á., et al.: Comparison of three rapid non-destructive techniques coupled with a classifier to increase transparency in the seafood value chain: Bioelectrical impedance analysis (BIA), near-infrared spectroscopy (NIR) and time domain reflectometry (TDR). J. Food Eng. **322**, 110979 (2022)
19. AOAC. Official Methods of Analysis of the Association of Official Analytical Chemists International (18th edn.), AOAC international, Washington DC. (2005)
20. Saint-Denis, T., Goupy, J.: Optimization of a nitrogen analyser based on the Dumas method. Anal. Chim. Acta **515**, 191–198 (2004)
21. Snee, R.D.: Validation of regression models: methods and examples. Technometrics **19**, 415–428 (1977)
22. Ballabio, D., Consonni, V.: Classification tools in chemistry. Part 1: linear models. PLS-DA Anal. Methods **5**, 3790–3798 (2013)
23. Kramer, R.: Chemometric Techniques for Quantitative Analysis. Marcel Dekker, New York (1998)

24. Nicolaï, B.M., Theron, K.I., Lammertyn, J.: Kernel PLS regression on wavelet transformed NIR spectra for prediction of sugar content of apple. Chemom. Intell. Lab. Syst. **85**, 243–252 (2007)
25. Buning-Pfaue, H.: Analysis of water in food by near infrared spectroscopy. Food Chem. **82**, 107–115 (2003)
26. Clevers, J.G.P.W., Kooistra, L., Schaepman, M.E.: Using spectral information from the NIR water absorption features for the retrieval of canopy water content. Int. J. Appl. Earth Obs. Geoinf. **10**, 388–397 (2008)
27. Laub-Ekgreen, M.H., Martínez-López, B., Jessen, F., Skov, T.: Non-destructive measurement of salt using NIR spectroscopy in the herring marinating process. LWT Food Sci. Technol. **97**, 610–616 (2018)
28. Mosry, N., Sun, D.-W.: Robust linear and non-linear models of NIR spectroscopy for detection and quantification of adulterants in fresh and frozen-thawed minced beef. Meat Sci. **93**, 292–302 (2013)

Identification of Variety and Age of Abalones Based on Near-Infrared Spectroscopy

Huang Yangming, Gao Jingxian, Tang Guo, Xiong Yanmei(✉), and Min Shungeng(✉)

College of Science, China Agricultural University, Beijing 100193, People's Republic of China
{xiongym,minsg}@cau.edu.cn

Abstract. In this research, NIR combined with chemometrics is applied to identification of abalone variety and age in order to decrease losses in sales and farming. Identification of Green disc abalone age can be realized using principal component analysis (PCA). Partial least square discrimination analysis (PLS-DA) can be divided into two categories: PLS2-DA and PLS1-DA. When a dataset has only two classes, performances of PLS2-DA and the novel approach Euclidean distance coupled to PLS1-DA (EuD-PLS1-DA) are highly similar with accuracy being over 98% for identification of Nan-Ri abalone age and identification of Green disc abalone age. EuD-PLS1-DA is superior to PLS2-DA when confronting multi-class problems, such as variety classification of abalones. Accuracy of EuD-PLS1-DA is 86.52% and 93.38% separately for calibration set and validation set, which is satisfactory; accuracy of PLS2-DA is smaller than 80%. The classification results show the usefulness of NIR linked to PLS-DA for identification of variety and age of abalones.

Keywords: Near-infrared spectroscopy · Abalones · Partial least square discrimination analysis · Variety and age identification

1 Introduction

As one of important marine edible shellfishes, quality of abalones has been paid attention owing to high nutrition. Researchers commonly employ time-consuming technologies and rely on laborious chemical analysis to understand chemical and nutritional compositions of abalones. Abalones not only have high edible value, but also have therapeutic potential and bioactive molecules [1]. Hence there are many researches about extracts of abalones and their applications [2–5]. Every part of the abalone is useful and the researches about it should be comprehensively developed for quality control. There are various kinds of abalones in the market. Without a large number of long-term experience accumulation, it is difficult to identify abalones of different varieties when their traits are very similar. Meanwhile, abalones of different varieties have great differences in sensory quality, price, living habit and culturing process, even within different ages of the same variety. Hence, it's crucial to solve the problems of identification of abalone variety and age in order to decrease losses in sales and farming.

X. Chu et al. (Eds.): ICNIR 2021, *Sense the Real Change: Proceedings of the 20th International Conference on Near Infrared Spectroscopy*, pp. 118–123, 2022.
https://doi.org/10.1007/978-981-19-4884-8_11

Near-infrared spectroscopy (NIR) has the advantages of no or little sample pre-treatment, high efficiency, low cost and non-destructive analysis and has been applied to various fields. However, there are a few studies about abalones with infrared spectroscopy [6–8]. And their focus is the processes of abalone shell formation. Although there is a work where deep learning applied to classification of life phase of abalones and the traits of respective abalones were collected, measuring data of the traits demands lots of time and human labor [9]. NIR which is time-saving is used for identification of variety and age of abalones, which is basically a blank.

NIR is often combined with chemometrics and partial least square discrimination analysis (PLS-DA) is used for problems of classification/discrimination. PLS-DA can be split into two categories [10]: PLS1-DA and over-used PLS2-DA. According to the reference [11], Euclidean distance linked to PLS1-DA (EuD-PLS1-DA) which is a novel approach is better than PLS2-DA in dealing with multi-class problems. Unfortunately, the true potential of PLS-DA basically is not developed on the basis of the reference [11]. Here, both EuD-PLS1-DA and PLS2-DA are used to validate the feasibility of identification of variety and age of abalones and a comparison between both approaches is made.

2 Materials and Methods

2.1 Collection of Samples

There are four different types of abalones: Red shell, Green disc, Nan-Ri abalones and Haliotis discus hannai Ino. Meanwhile, Green disc abalones of 9 months and Nan-Ri abalones of nursery stage were scanned using NIR for realizing identification of age. The number of collected samples is present in Table 1. All samples were obtained from Fujian province in China.

2.2 Instrument and Parameters Settings

MicroNIR^{TM}1700 portable near-infrared spectrometer (JDSU, USA) was used in the experiment. The samples were recorded within the wavelength range of 900–1700 nm. The integration time is 8000 μs and the background and dark current were corrected every 30 min. Sapphire window was installed on the optical platform of NIR. The window was attached to the foot muscle of living abalones for obtaining spectra after the abalones were cleaned with water. Owing to heterogeneity and big surface area of the foot muscle, spectra of every abalone were obtained from different points of foot muscle in different days for building robust classifiers. Hence the number of spectra is obviously greater than that of samples under the same class. In order to validate the superiority of EuD-PLS1-DA compared with PLS2-DA, every spectrum is treated as one sample. But it should be more appropriate that different measurements of one abalone are averaged as the final measurement. There are three datasets: one dataset for variety classification of abalones, the other two datasets separately for the identification between different ages of Green disc abalones and the identification between different ages of Nan-Ri abalones. The last three columns in Table 1 are labels for the three datasets: Label-1 is for variety classification; Label-2 is for identification of Green disc abalone age; Label-3 is for identification of Nan-Ri abalone age.

Table 1. Descriptions of different types of abalones

Variety	Age/months	No. sample	No. spectra	Label-1	Label-2	Label-3
Red shell abalone	18	31	94	1		
Green disc abalone	9	32	115	2	1	
	18	25	127		2	
Nan-Ri abalone	16	73	638	3		1
	Nursery stage	65	348			2
Haliotis discus hannai Ino	18	35	115	4		

2.3 Algorithms

Spectral data were processed on MATLAB platform (Version 2014a, the Math Works, Inc. USA).

2.4 Kennard-Stone Algorithm (K-S) [12]

Samples of each class in every dataset are divided into calibration set and validation set using K-S. The ratio of the number of samples in calibration set to the one in validation set is 4:1.

2.5 Principal Component Analysis (PCA)

PCA is the most important linear algorithm in decreasing the space dimension of raw data. It projects raw data into a new space where the information of raw data is saved as much as possible and a few scores can stand for raw data (principal components (PCs) form a new coordinate system). Meanwhile, it's an unsupervised pattern recognition algorithm.

2.6 PLS-DA Algorithm

PLS2-DA: In fact, current used PLS-DA mainly is PLS2-DA. A dependent variable matrix is constructed involving all classes in a dataset. The number of columns in this matrix is equal to the number of classes. Finally, only one discriminant model is produced and applies to identification of unknown samples whose other samples already were involved in the modeling.

EuD-PLS1-DA: The work [11] provides a detailed introduction about EuD-PLS1-DA. In brief, EuD is to evaluate differences between different classes and the class has maximal distance compared with current remaining classes is used for modeling (the class is assigned to 1 and remaining classes are assigned to 0); the class will be deleted after its classifier is built; using EuD-PLS1-DA will produce (K-1) classifiers (K is the number of classes in a dataset). Unlike EuD-PLS1-DA, traditional PLS1-DA employs one-versus-all strategy to build K classifiers in which one class is compared with the remaining classes and any class will not be deleted during building a series of classifiers.

2.7 Indexes to Evaluate Model Performances

The number of latent variables (LVs) is decided based on the variance explanation that exceeds 1% for both x-information and y-information. Accuracy is for calibration and validation set (Acc-cal and Acc-val, respectively) and the formulas of PLS2-DA and EuD-PLS1-DA for accuracy are the same with the work [12]. It's worth mentioning that the formula of EuD-PLS1-DA for accuracy is based on the reference [13]. Determinant coefficient of calibration and validation set (Q2cv and Q2val, respectively), root mean square error of cross validation and validation (RMSECV and RMSEV, respectively) were also shown.

3 Results and Discussion

Figure 1 presents all spectra from three datasets. From Fig. 1, Savitzky-Golay 1st derivative (9 points, the left and right sides are symmetrical) was applied to all spectra in order to avoid the influence of baseline drift on classifiers.

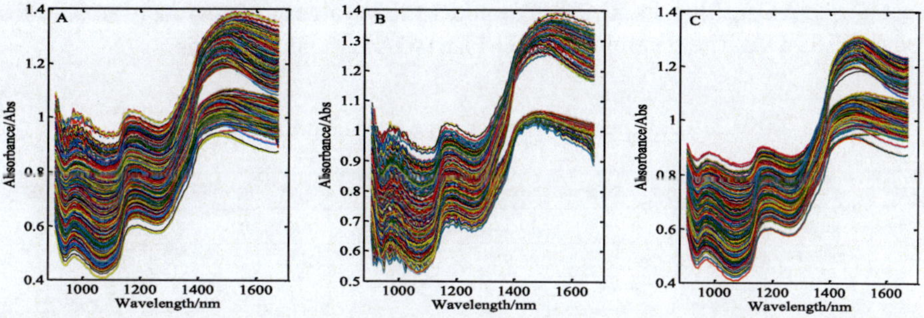

Fig. 1. Spectra from three datasets (A) Variety classification; (B) Green disc abalones; (C) Nan-Ri abalones

3.1 Datasets Processed Using PCA

Figure 2 shows new space distributions of samples from three datasets processed by PCA. The variance contribute rate (%) of the first and second principal component is also present in Fig. 2. Cumulative variance contribute rate based on the first two principal components is greater than 85% for the three datasets. Except for the dataset about Green disc abalone age, some samples of the other two datasets have an overlapping behavior, especially in terms of variety classification. PCA is unable to address variety classification and identification of Nan-Ri abalone age.

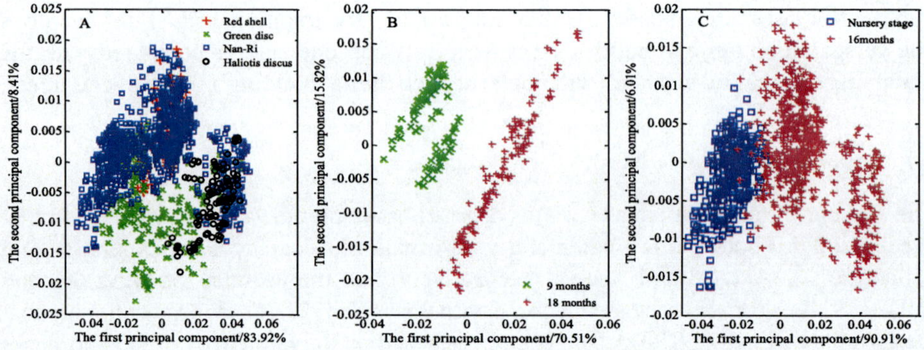

Fig. 2. Three datasets processed by PCA (A) Variety classification; (B) Green disc abalones; (C) Nan-Ri abalones

3.2 Identification of Variety and Age of Abalones by PLS-DA Approaches

LVs were decided in the range of 3–5. From Table 2, performances of classifiers are very similar between PLS2-DA and EuD-PLS1-DA when there are only two classes in a dataset. Every column in y-information in PLS2-DA has predicted values and only one column in EuD-PLS1-DA has predicted values. Hence RMSECV and RMSEV values of PLS2-DA are slightly bigger than those of EuD-PLS1-DA. When confronting multi-class problems, for example variety classification, performance of EuD-PLS1-DA is superior to that of PLS2-DA. Plus, its RMSECV and RMSEV values are obviously smaller than those of PLS2-DA. The potential of EuD-PLS1-DA is validated again.

Table 2. Classification results of PLS2-DA and EuD-PLS1-DA

Dataset	Method	Class	Acc-cal/%	Acc-val/%	Q2cv	RMSECV	Q2val	RMSEV
Variety classification	PLS2-DA		79.74	78.40	0.40	0.54	0.38	0.55
	EuD-PLS1-DA	4	86.52	93.38	0.49	0.19	0.56	0.18
		2			0.66	0.23	0.65	0.23
		1&3			0.15	0.26	0.15	0.26
Green disc abalones	PLS2-DA		100	100	0.98	0.10	0.99	0.08
	EuD-PLS1-DA	1&2	100	100	0.98	0.07	0.99	0.06
Nan-Ri abalones	PLS2-DA		98.60	98.48	0.82	0.29	0.82	0.29
	EuD-PLS1-DA	1&2	98.60	98.48	0.82	0.20	0.82	0.20

4 Conclusion

This study proves the usefulness of NIR coupled with multivariate statistical analysis for abalone varietal classification and age prediction. The classification results of EuD-PLS1-DA are satisfactory, which can meet actual demands. Again, performance of EuD-PLS1-DA is superior to that of PLS2-DA when confronting multi-class problems and they are very similar when a dataset has only two classes. It's worth noting that non-linear algorithms can be used to raise accuracy for varietal classification. Hopefully, this work can be as reference to enlarge the application range of NIR on abalones owing to its advantages.

Acknowledgments. This research has received financial support from the National Natural Science Foundation of China (Grant number: 31301685) in the collection and analysis of data.

References

1. Suleria, H.A.R., Masci, P.P., Gobe, G.C., Osborne, S.A.: Therapeutic potential of abalone and status of bioactive molecules: a comprehensive review. Crit. Rev. Food Sci. **57**, 1742–1748 (2017)
2. Liu, B., Jia, Z., Li, C., Chen, J., Fang, T.: Hypolipidemic and anti-atherogenic activities of crude polysaccharides from abalone viscera. Food Sci. Nutr. **8**, 2524–2534 (2020)
3. Jian, W., et al.: Fabrication of highly stable silver nanoparticles using polysaccharide-protein complexes from abalone viscera and antibacterial activity evaluation. Int. J. Biol. Macromol. **128**, 839–847 (2019)
4. Zanjani, N.T., et al.: Abalone hemocyanin blocks the entry of herpes simplex virus 1 into cells: a potential new antiviral strategy. Antimicrob. Agents Ch. **60**, 1003–1012 (2016)
5. Ren, L., Wu, Z., Ma, Y., Jian, W., Xiong, H., Zhou, L.: Preparation and growth-promoting effect of selenium nanoparticles capped by polysaccharide-protein complexes on tilapia. J. Sci. Food Agr. **101**, 476–485 (2021)
6. Auzoux-Bordenave, S., et al.: Ultrastructure, chemistry and mineralogy of the growing shell of the European abalone Haliotis tuberculata. J. Struct. Biol. **171**, 277–290 (2010)
7. Gaume, B., Fouchereau-Peron, M., Badou, A., Helleouet, M., Huchette, S., Auzoux-Bordenave, S.: Biomineralization markers during early shell formation in the European abalone Haliotis tuberculata, Linnaeus. Mar. Biol. **158**, 341–353 (2011)
8. Auzoux-Bordenave, S., Brahmi, C., Badou, A., de Rafelis, M., Huchette, S.: Shell growth, microstructure and composition over the development cycle of the European abalone Haliotis tuberculata. Mar. Biol. **162**, 687–697 (2015)
9. Sahin, E., Saul, C.J., Ozsarfati, E., Yilmaz, A.: Abalone life phase classification with deep learning. In: 5th International Conference on Soft Computing and Machine Intelligence (ISCMI), pp. 163–167 (2018)
10. Lee, L.C., Liong, C., Jemain, A.A.: Partial least squares-discriminant analysis (PLS-DA) for classification of high-dimensional (HD) data: a review of contemporary practice strategies and knowledge gaps. Analyst **143**, 3526–3539 (2018)
11. Huang, Y., Huang, Y., Song, X., Gao, J., Xiong, Y., Min, S.: Comparison of a novel PLS1-DA, traditional PLS2-DA and assigned PLS1-DA for classification by molecular spectroscopy. Chemometr. Intell. Lab. 209, 104225 (2021)
12. Kennard, R.W., Stone, L.A.: Computer aided design of experiments. Technometrics **11**, 137–148 (1969)
13. Lee, L.C., Jemain, A.A.: Predictive modelling of colossal ATR-FTIR spectral data using PLS-DA: empirical differences between PLS1-DA and PLS2-DA algorithms. Analyst **144**, 2670–2678 (2019)

Discrimination of Adulterated Milk Using Temperature-Dependent Two-Dimensional Near-Infrared Correlation Spectroscopy

Ming Y. Huang, Jia Long, Ren J. Yang(✉), Hai Y. Wu, Hao Jin, and Yan R. Yang

College of Engineering and Technology, Tianjin Agricultural University, Tianjin 300384, China
rjyang1978@163.com

Abstract. A discriminant method for adulterated milk was proposed using temperature-dependent two-dimensional (2D) near-infrared (NIR) correlation spectroscopy combined with N-way partial least squares discriminate analysis (NPLS-DA). Two brands of Mengniu (MN) and Sanyuan (SY) pure milk and adulterated milk with urea (0.2–20 mg.mL^{-1}) were prepared. One-dimensional (1D) NIR spectra of all samples were collected at room temperature and 30 °C−55 °C (5 °C interval). Synchronous 2D NIR correlation spectrum of each sample was calculated under the perturbation of temperature. For comparing, the discriminant models of MN brand, SY brand, and two-brands of adultetated milk were built based on 1D NIR spectra (room temperature), temperature-related three-dimensional (3D) NIR spectra, and sychronous 2D correlation spectra, respectively. For 1D NIR spectra, the discrimination accuracies of three models of MN, SY, and two-brands for unknown samples were 81.5%, 88.9%, and 85.2%, respectively. For temperature-related 3D spectra, the discrimination accuracies of three models for unknown samples were 96.3%, 96.3%, and 90.7%, respectively. For 2D correlation spectra, the discrimination accuracies of three models for unknown samples were 100%, 100%, and 98.1%, respectively. The results show that the proposed method can provide better discrimination results than 1D spectra and temperature-related 3D spectra.

Keywords: Temperature-dependent 2D NIR correlation spectroscopy · Adulterated milk · Urea · Discriminant analysis

1 Introduction

As we know, milk is a natural drink rich in nutritional value, which meet some nutritional needs of the human body. Protein in dairy products is one of the best nutrients to enhance physique and human immunity. Generally, the price of milk increases with protein content.Therefore, in order to obtain high profits, some businesses add urea to milk for improving the content of protein. It is urgent to develop a rapid, widely available, and cost-effective method to control and detect milk quality.

Near infrared (NIR) spectroscopy is as a rapid, nondestructive and multi-component analysis technology. It has important theoretical value and broad application prospects

X. Chu et al. (Eds.): ICNIR 2021, *Sense the Real Change: Proceedings of the 20th International Conference on Near Infrared Spectroscopy*, pp. 124–131, 2022.
https://doi.org/10.1007/978-981-19-4884-8_12

for food quality control based on NIR spectroscopy and multivariate method. Ni et al. [1] identified six different adulterated milk based on NIR spectroscopy and non-line pattern recognition methods, and pointed that NIR spectroscopy could distinguish the high concentration levels of adulterants in milk. Musa and Yang [2] applied NIR spectroscopy for determining the contents of water, urea, starch and goat milk in fresh cow milk. The results showed that NIR spectroscopy was feasible and reliable for detecting adulterants in cow milk. Karunathilaka et al. [3] developed a non-targeted detection method of adulterated milk powder using NIR spectroscopy and soft independent modeling of class analogy (SIMCA), and pointed that NIR spectroscopy was a promising technology for detecting adulterated milk powder. Xu [4], Yuan [5], Dong [6], Liu et al. [7] used NIR spectroscopy for detecting adulterated milk with melamine, and good results were obtained.

In order to more effectively extract the characteristic information of trace adulterants in food, two-trace two-dimensional (2T2D) correlation spectral technology [8, 9] has also been applied to the detection of adulterated food [10]. Yang et al. [11–13] firstly proposed and investigated the feasibility of synchronous (Sy) 2T2D NIR correlation spectroscopy combined with multivariate methods for detecting adulterated milk, and pointed that 2T2D NIR correlation spectra can provide better results than one-dimensional (1D) NIR spectra. Sohng et al. [14] used asynchronous (Asy) 2T2D correlation NIR spectra for identifying adulterated olive oil, and demonstrated 2T2D correlation spectral as an effective method for improving the accuracy of discrimination. Wu et al. [15] classified adulterated milk based on 2T2D NIR correlation characteristic slice spectra, and obtained good results of different brands milk. Yang [16] and Yu et al. [17] discriminated adulterated food using Sy-Asy 2T2D NIR correlation spectra. Meanwhile, the parameterization of 2T2D NIR correlation spectroscopy has also been used to calssfy adulterated milk [18, 19].

In present study, a strategy of combining temperature-dependent 2D correlation spectroscopy and N-way partial least squares discriminate analysis (NPLS-DA) for classifying pure milk and adulterated milk. Discrimination results were assessed and compared to those for using 1D NIR spectra, and temperature- related three-dimensional (3D) NIR spectra (sample × temperature × wavenumber variable).

2 Theory

Under the external perturbation of temperature, 1D NIR spectra of each sample were collected and formed into the dynamic spectral matrix $A_{n \times s}$ (n is the number of spectra, s is the number of wavenumber variables). In terms of generalized 2D correlation spectral theory [20, 21], synchronous 2D correlation spectrum $\phi(v_1, v_2)$ of each sample can be calculated

$$\phi(v_1, v_2) = \frac{1}{n-1} A^T A \tag{1}$$

where v_1 and v_2 express the wavenumber variable. T represents matrix transpose.

Synchronous 2D correlation spectrum $\phi(v_1, v_2)$ represents the similarity of spectral intensity variation at different spectral vatiables v_1 and v_2 under the perturbations of temperature.

3 Materials and Methods

Two brands of Mengniu (MN) and Sanyuan (SY) milk were purchased from local supermarkets. For each brand milk, 40 pure and 40 adulterated milk with urea (0.2–20 mg ml^{-1}) were prepared respectively. Samples temperature (30 °C–55 °C, interval 5 °C) was controlled using a Figurewater bath.

Under different temperature (including room temperature), NIR spectra of pure and adulterated milk were acquired in the range of 4000–12000 cm^{-1} using a Spectrum GX FTIR Spectrometer (Perkin-Elmer, USA) coupled with integrating sphere accessory. Each spectrum was an average of 32 scans collected at 8 cm^{-1} resolution.

Temperature-dependent synchronous 2D NIR correlation spectra of all samples were calculated in the range of 5000–7000 cm^{-1} by 2D Shige software. The discrimination models of adulterated milk were established using synchronous 2D NIR correlation spectra and NPLS-DA. All computations were performed in Matlab (The Math-works Inc., Natick, MA).

4 Results and Discussion

Figure 1a and b were NIR spectra of the MN brand of pure and adulterated milk in the range of 4000–12000 cm^{-1} at different temperatures (30 °C–55 °C, interval 5 °C). As can be seen that NIR spectral absorption intensity of pure and adulterated milk do not change significantly in the whole spectral range at the temperature range of 30 °C–55 °C. The NIR spectra of the SY brand of pure and adulterated milk are silimar with the MN brand of milk, which are not shown here. It was impossible to determine whether milk was adulterated in terms of spectral shape, position and intensity of the absorbance peaks. It can also be observed from Fig. 1 that there are two strong absorption peaks at 5186, 6886 cm^{-1} and two weak absorption peaks at 8300 and 10222 cm^{-1} for pure and adulterated milk. Therefore, spectral range of 5000–7000 cm^{-1} was used to analyze in present research.

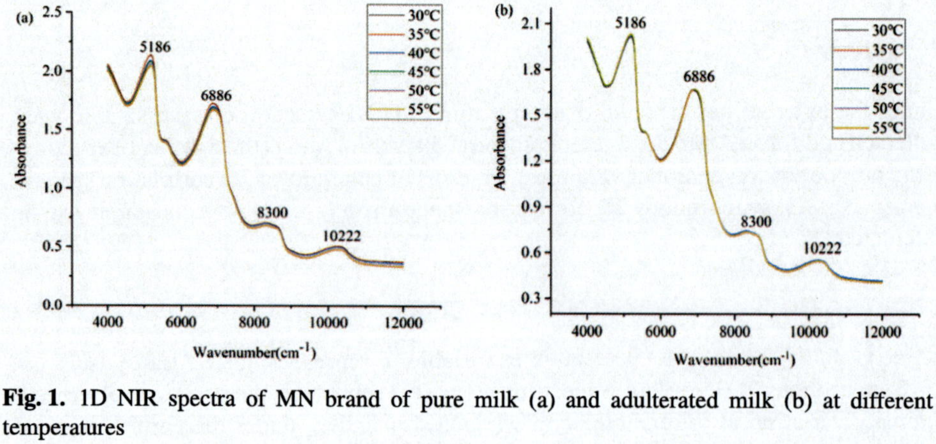

Fig. 1. 1D NIR spectra of MN brand of pure milk (a) and adulterated milk (b) at different temperatures

For 1D dynamic spectra varying with temperature, synchronous 2D NIR correlation spectra of two brands of milk samples were calculated in spectral range of 5000–7000 cm^{-1} according to Eq. 1. Figure 2a and 2b showed synchronous 2D NIR correlation spectra of MN brands of pure and adulterated milk. Comparing Fig. 2a and Fig. 2b, due to the trace of urea adulterated in milk, although the 2D correlation spectra have high spectral resolution, it is impossible to identify whether the milk is adulterated by the naked eye. Synchronous 2D NIR spectra of SY brand of pure and adulterated milk are similar to that in Fig. 2, with only subtle difference. Therefore, The method of pattern recognition is needed to determine whether milk is adulterated.

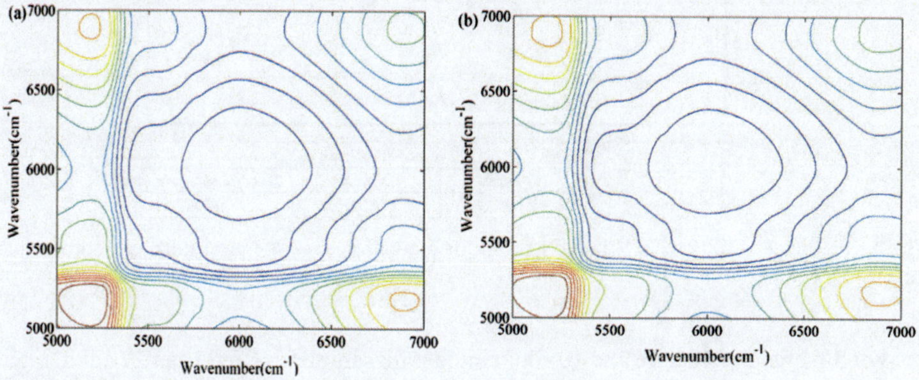

Fig. 2. Synchronous 2D NIR correlation spectra of MN brand of pure milk (a) and adulterated milk (b)

For comparison, 1D spectra (room temperature), temperature-related 3D spectra, and synchronous 2D correlation spectra were used to establish the discrimination models of pure and adulterated milk. In the discrimination models, pure milk samples were labeled as "0", and adulterated milk samples were labeled as "1". Predictive values greater than 0.5 were classified as adulterated milk, and those less than 0.5 were classified as pure milk.

For MN brand of 40 pure and 40 adulterated milk samples, two-thirds of 80 samples (including 26 pure and 27 adulterated milk samples) were selected as the calibration set. The remaining 27 samples were as the prediction set for validating models. Partial least squares discriminate analysis (PLS-DA) was used to build discrimination model based on 1D NIR spectra (80 × 1001). For 27 unknown samples in the prediction set, 2 pure and 3 adulterated milk samples were misjudged, and the discrimination accuracy was 81.5% (Fig. 3). The NPLS-DA models were built using temperature-related 3D spectra (80 × 6 × 1001) and synchronous 2D correlation spectra (80 × 1001 × 1001), respectively. The prediction results of the two NPLS-DA models for unknown samples were also shown in Fig. 3. As can be seen from Fig. 3, the correct classification rate was 96.3% based on temperature-related 3D spectra and 100% based on 2D correlation spectra.

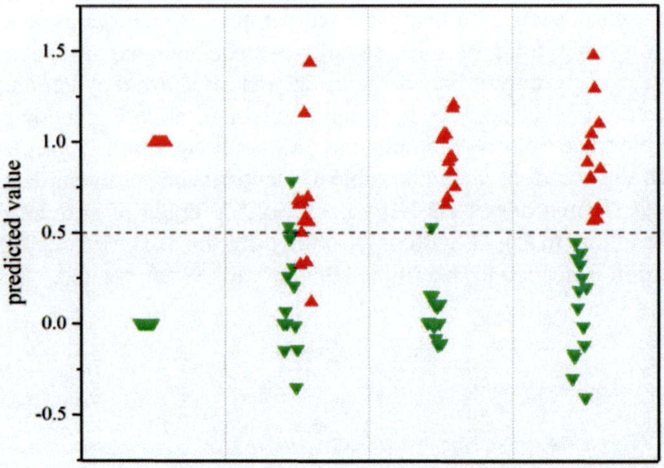

Model	Predicted		
	1D spectra	3D spectra	2D correlation spectra
Prediction set	81.5%	96.3%	100%

Fig. 3. Predicted results of MN brand of milk in prediction set (▼: pure milk, ▲: adulterated milk)

For SY brand of 40 pure and 40 adulterated milk samples, 26 pure and 27 adulterated milk samples were as calibration set, and the remaining 27 samples were as prediction set. The three discrimination models were built using 1D spectra, temperature-related 3D spectra and 2D correlation spectra, respectively. For 1D NIR spectral PLS-DA model, 3 adulterated milk samples were misjudged in the prediction set, and the discrimination accuracy of unknown samples was 88.9% (Fig. 4). For 3D spectral NPLS-DA model, only 1 adulterated milk sample was misjudged, and the discrimination accuracy of unknown samples was 96.3% (Fig. 4). For 2D correlation spectral NPLS-DA model, all samples were correctly identified, and the discrimination accuracy of unknown samples was 100% (Fig. 4).

For two brands of 80 pure and 80 adulterated milk, 52 pure and 54 adulterated milk samples were selected as the calibration set for building discrimination models of two brands of adulterated milk. Figure 5a showed the discrimination results of MN brand of pure and adulterated milk in prediction set using 1D spectra, 3D spectra, and 2D correlation spectra, respectively. There were 5, 1 and 0 samples which were misjudged respectively in prediction set, and the correct rates of the three models for unknown samples of MN milk were 81.5%, 96.3%, 100%. The prediction results of the three models for SY milk were shown in Fig. 5b. There are 3, 4 and 1 samples which were misjudged respectively in prediction set, and the correct rates of the three models for unknown samples of SY milk were 88.9%, 92.6%, 96.3%. For two-brands of 54 unknown samples in prediction set, there are 8, 5 and 1 samples which were misjudged respectively, and the correct rates were 85.2%, 90.7%, and 98.1%.

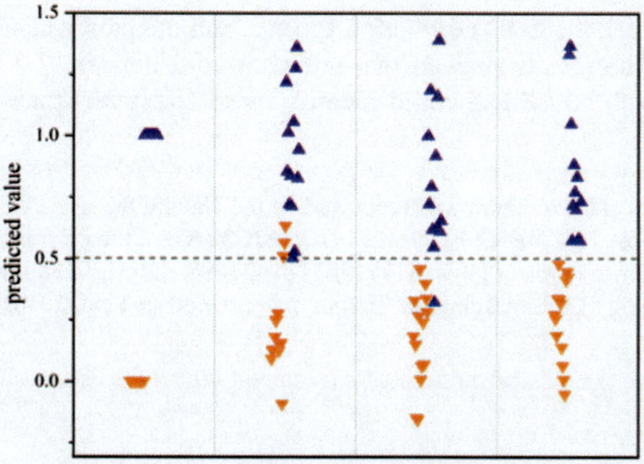

Model	Predicted		
	1D spectra	3D spectra	2D correlation spectra
Prediction set	88.9%	96.3%	100%

Fig. 4. Predicted results of SY brand of milk in prediction set (▼: pure milk, ▲: adulterated milk)

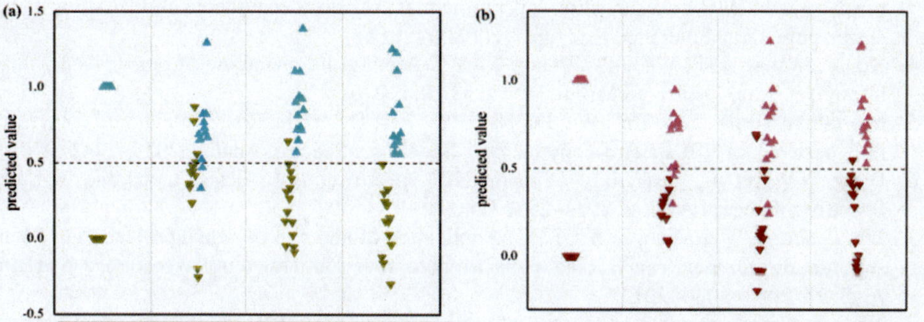

	Predicted		
	1D spectra	3D spectra	2D correlation spectra
Prediction set	81.5%	96.3%	100%

Model	Predicted		
	1D spectra	3D spectra	2D correlation spectra
Prediction set	88.9%	85.2%	96.3%

Fig. 5. Predicted results of two brand of MN milk (a) and SY milk (b) in prediction set (▼: MN pure milk, ▲: MN adulterated milk, ▼: SY pure milk, ▲: SY adulterated milk).

From the above analysis results, it can be seen that the discrimination models using 2D NIR correlation spectra can provide higher discrimination accuracy than 1D NIR spectra and temperature-related 3D NIR spectra. The reason may be that 2D correlation spectra can extract more characteristic information.

5 Conclusion

In this study, the feasibility of identifying pure and adulterated milk using temperature-dependent synchronous 2D NIR correlation spectra was investigated. The discrimination models were established based on 1D NIR spectra, temperature-related 3D NIR spectra, and temperature-dependent 2D correlation spectra, and the performance of the models were compared. The results showed that temperature-dependent 2D NIR correlation spectra coupled with NPLS-DA could identify more effectively pure and adulterated milk.

Acknowledgements. This research was supported by the Natural Science Foundation of Tianjin under the project (Nos. 18JCYBJC96400 and 14JCYBJC30400) China Natural Science Foundation Committee under the project (Nos. 41771357, 81471698 and 31201359) and the Enterprise Science and Technology Commissioner of Tianjin under the project No. 20YDTPJC01340.

References

1. Ni, L.J., Zhong, L., Zhang, X., Zhang, L.G., Huang, S.X.: Identification of adulterants in adulterated milks by near infrared spectroscopy combined with non-linear pattern recognition methods. Spectrosc. Spectral Anal. **34**, 2673–2678 (2014)
2. Musa, M.A., Yang, S.: Detection and quantification of cow milk adulteration using portable near-infrared spectroscopy combined with chemometrics. Afr. J. Agric. Res. **170**, 198–207 (2021)
3. Karunathilaka, S.R., Yakes, B.J., He, K., Chung, J.K., Mossoba, M.: Non-targeted NIR spectroscopy and SIMCA classification for commercial milk powder authentication: a study using eleven potential adulterants. Heliyon **4**, e00806 (2018)
4. Xu, Y., Wang, Y.M., Wu, J.Z., Zhang, X.C.: Detecting the melamine of pure milk by near infrared spectra. J. Infrared Millim W. **29**, 53–56 (2010)
5. Yuan, S.L., He, Y., Ma, T.Y., Wu, D., Nie, P.C.: Fast Determination of melamine content in milk base on vis/NIR spectroscopy method. Spectrosc. Spectral Anal. **9**, 2939–2942 (2009)
6. Dong, Y.W., et al.: Feasibility of using NIR spectroscopy to detect melamine in milk. Spectrosc. Spectral Anal. **2**, 2934–2938 (2009)
7. Liu, R., Yang, R.J., Miao, J., Xu, K.X.: Application of kernel orthogonal projection to latent structure discriminant analysis in the discrimination of adulterated milk. Spectrosc. Spectral Anal. **33**, 2083–2086 (2013)
8. Noda, I.: Closer examination of two-trace two-dimensional (2T2D) correlation spectroscopy. J. Mol. Struct. **1213**, 128194 (2020)
9. Noda, I.: Two-trace two-dimensional (2T2D) correlation spectroscopy–a method for extracting useful information from a pair of spectra. J. Mol. Struct. **1160**, 471–478 (2018)
10. Yang, R.J., et al.: Two-trace two-dimensional (2T2D) correlation spectroscopy application in food safety: a review. J. Mol. Struct. **1214**, 128219 (2020)
11. Yang, R.J., Liu, R., Xu, K.X., Yang, Y.R.: Discrimination of adulterated milk using NPLS-DA combined with two-dimensional correlation near-infrared spectroscopy. Acta Photonica Sinic. **42**, 580–585 (2013)
12. Yang, R.J., Liu, R., Xu, K.X., Yang, Y.R.: Discrimination of adulterated milk based on two-dimensional correlation spectroscopy (2DCOS) combined with kernel orthogonal projection to latent structure (K-OPLS). Appl. Spectrosc. **67**, 1363–1367 (2013)
13. Yang, R.J., Liu, R., Xu, K.X., Yang, Y.R.: Quantitative analysis of melamine by multi-way partial least squares model with two-dimensional near-infrared correlation spectroscopy. Proc. SPIE **8939**, 893912 (2014)
14. Sohng, W., Eum, C., Chung, H.: Exploring two-trace two-dimensional (2T2D) correlation spectroscopy as an effective approach to improve accuracy of discriminant analysis by highlighting asynchronous features in two separate spectra of a sample. Anal. Chim. Acta. **1152**, 338255 (2021)

15. Wu, H.Y., Yang, R.J., Wei, Y., Dong, G.M., Jin, H., Zeng, Y.N.: Ai CL Influence of brands on a discrimination model for adulterated milk based on asynchronous two-dimensional correlation spectroscopy slice spectra. Spectrochim Acta A. **271**, 120958 (2022)
16. Yang, R.J., et al.: Synchronous-asynchronous two-dimensional correlation spectroscopy for the discrimination of adulterated milk. Anal. Methods **7**, 4302–4307 (2015)
17. Yu, G., Yang, R., Lu, A., Tan, E.: Detection of adulterated sesame oil based on synchronous-asynchronous two-dimensional mid-Infrared correlation spectroscopy. Spectrosc. Spectral Anal. **37**, 1105–1109 (2017)
18. Yang, R.J., Liu, R., Xu, K.X., Yang, Y.R.: Classification of adulterated milk with the parameterization of 2D correlation spectroscopy and least squares support vector machines. Anal. Methods **5**, 5949–5953 (2013)
19. Miao, J., Cao, Y.Z., Yang, R.J., Liu, R., Sun, H.L., Xu, K.X.: Identification of adulterated milk based on two-dimensional correlation near-infrared spectra parameterization and BP neural network. Spectrosc. Spectral Anal. **33**, 3032–3035 (2013)
20. Noda, I.: Two-dimensional infrared (2D IR) spectroscopy: theory and applications. Appl. Spectrosc. **44**, 550–561 (1990)
21. Noda, I., Ozaki, Y.: Two-Dimensional Correlation Spectroscopy: Applications in Vibrational and Optical Spectroscopy. John Wiley & Sons, Chichester (2004)

Development of NIRS Calibrations for Seed Content of Lipids and Proteins in Contrasting White Lupin Germplasm

B. Ferrari, S. Barzaghi(✉), and P. Annicchiarico

Research Centre for Animal Production and Aquaculture, Council for Agricultural Research and Economics, Viale Piacenza 29, 26900 Lodi, Italy

stefania.barzaghi@crea.gov.it

Abstract. White lupin (*L. albus*) has high potential interest as a high-protein food or feed crop. In addition, the oil of its seed has high quality for human nutrition. Crop improvement for these traits would profit of low-cost, NIRS-based evaluation methods that could be applied to large numbers of genotypes. The aim of this work was developing and assessing calibration models for NIRS prediction of these traits, envisaging analyses either on whole grain samples or on ground samples. Samples for the reference analyses were chosen by applying the Kennard-Stone algorithm to the whole set of spectra recorded from 2342 samples, both for seeds and flours. A group of 146 samples was selected to calculate calibration models based on chemical analyses for lipid and protein contents (Soxhlet extraction and Dumas method, respectively). After chemometric elaborations of the collected NIR spectra, with a repeated double cross-validation, the best results were obtained with lupin flours spectra for the estimation of protein content, using 4 LV and a mean centering as pretreatment, with performances which are good enough for breeding purposes (RPD = 3.30). Predictions were somewhat worse with lupin flours spectra for oil content, which attained RPD = 2.46 with 2 LV and first derivative and mean center as pretreatments. Results were less satisfying for predicting protein or oil content based on whole seed spectra, a non-destructive sample scenario of special interest for selection based on individual seeds.

Keywords: *L.albus* · Protein · Oil · Double cross-validation · Calibrations

1 Introduction

White lupin (*Lupinus albus L.*) is a high-protein grain legume native of the Mediterranean region that has increasing interest as a component of functional, nutraceutical, healthy or vegan food or as a high-protein feed, owing to seed protein content in the range 34–45 g*100 g^{-1} with good content of essential amino acids [1] and the ability of its γ-conglutin protein fraction to control glycaemia [2]. In addition, white lupin has 8–12 g*100 g^{-1} seed content of oil with excellent nutritional characteristics [3].

X. Chu et al. (Eds.): ICNIR 2021, *Sense the Real Change: Proceedings of the 20th International Conference on Near Infrared Spectroscopy*, pp. 132–136, 2022.
https://doi.org/10.1007/978-981-19-4884-8_13

White lupin breeding for improved grain quality could profit of fast and low-cost NIRS screening methods for seed protein content and oil content. Conventional methods for these traits (Dumas analysis for proteins and hot organic solvents extractions for lipids) are usually performed with chemical hazardous reagents and are time-consuming; NIRS methods instead would ease the evaluation of large genotype numbers from various evaluation environments, thereby maximizing the genetic progress and taking into account the occurrence of possible genotype-by-environment interactions for these traits [4]. NIRS-based methods have high interest in this context, but their application to lupins were reportedly limited to protein content in two studies [5, 6].

The aim of this work was to develop and to assess calibration models for NIRS-based predictions of protein and oil content, envisaging analyses either on whole grain samples or on ground samples of a large number of white lupin genotypes including either landrace accessions or recent inbred lines developed by Council for Agricultural Research and Economics (CREA)'s breeding program.

2 Materials and Methods

We considered seed samples of two contrasting germplasm sets. One included 1974 samples of RIL (Recombinant Inbred Line) genotypes derived from 16 crosses performed by CREA. The other comprised 368 seed samples of landraces belonging to CREA's world collection. RIL samples were harvested in 2018 and 2019 from both primary and secondary inflorescences of the plants. The samples were analyzed by means of a NIRFlex 500 spectrometer (Büchi Italia, Cornaredo, Italy), averaging the spectra of 10 individual seeds per sample, recorded using the tablet adapter of the measurement cell, in reflectance mode. Samples were milled using a MM400 Mixer Mill (Retsch Gmbh and Co., Germany) at 30 Hz for about 40 s and analyzed as flours, using the vial adapter of the measurement cell and averaging 3 repetitions of the same samples. The flours were also measured in reflectance mode.

A group of 146 samples, was chosen for the reference analyses applying the Kennard-Stone algorithm on the whole set of spectra - both for seeds and flours - and was used to calculate calibration models to quantify lipid and protein content. The algorithm automatically selected 74 from RIL and 72 from CREA's world collection. The wet lab analyses were performed on the samples' flours.

The lipid quantification on a dry matter basis was performed as an ether extract in a Soxhlet apparatus by 6 h distillation [7]. The protein content on a dry matter basis was measured with the Dumas method [8] that consists in the complete combustion of the flour to get the nitrogen content using a CHN elemental analyser (Carlo Erba, Milano, Italy).

PLSR calibrations were calculated using the R package "Chemometrics" to perform a repeated double cross-validation as suggested by Filzmoser et al. [9] and obtain thereby more reliable estimates of the prediction performance. After converting the spectra in absorbance, various spectra pretreatments, such as standard normal variate (SNV), mean center, Savitzky-Golay first and second derivatives (2nd order, 15 points), were considered to get the best models. The models were evaluated based on their standard error in prediction (SEP), Bias and ratio of prediction to deviation (RPD) values.

3 Results and Discussion

The whole sets of NIR scans are represented in Fig. 1, where one may notice the great dispersion of the seeds' spectra caused by the phenomenon of light scattering due to both the roughness and the different size of the seeds.

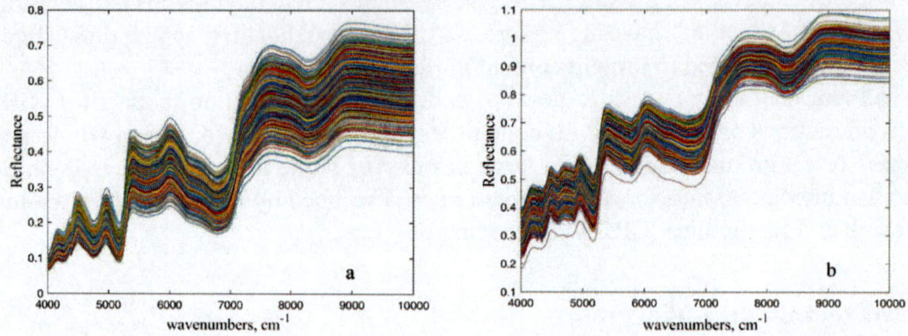

Fig. 1. Raw NIR spectra structures recorded for white lupin seeds (a) and derived flours (b).

Calibration models were calculated by associating the punctual lipids and proteins values of the samples chosen for official analyses to the relative spectra. The oil content in the samples selected for the reference analyses had mean value of 9.1 g*100 g^{-1} with a standard deviation value of 1.8 g*100 g^{-1}. The minimum and maximum values were 4.4 g*100 g^{-1} and 13.8 g*100 g^{-1} respectively. Proteins content values averaged 38.5 g*100 g^{-1} with a standard deviation value of 4.9 g*100 g^{-1}. The minimum was 19.5 g*100 $g^{-1}$and the maximum value was 50.0 g*100 g^{-1}.

The best-predictive calibration model for oil content estimated from whole grain samples was calculated by converting the spectra in absorbance and then applying a SNV and the mean center pretreatments. A standard error in prediction of 1.0 and a Bias of 0.078 were obtained using 7 latent variables (LV). This corresponded to a RPD of 1.76, which would correspond to a poorly predictive model. The best model selected for prediction of protein content from whole grain samples was calculated using the same pretreatments and 8LV: the SEP was 2.3, the Bias was −0.149 and RPD was 2.15. The scatter plots of references vs predicted values from repeated double-cross validation models are shown in Fig. 2.

NIRS-based predictions based on flour samples gave better results. The best model for predicting lipids was calculated converting the spectra in absorbance then applying the first derivative and the mean center. With 2 LV the SEP was 0.7, the Bias was −0.022 and consequently the RPD was equal to 2.46. We used only conversion to absorbance and the mean center for predicting proteins, obtaining by 4 LV a SEP of 1.5, a Bias of −0.085 and RPD = 3.3. Thus, this model can be considered sufficiently accurate for plant breeding purposes (Fig. 3).

The models with best performances on flour samples were applied to the whole set of samples both to quantify the lipid and protein content and to verify whether there was any difference between the seeds generated from the primary and secondary inflorescences.

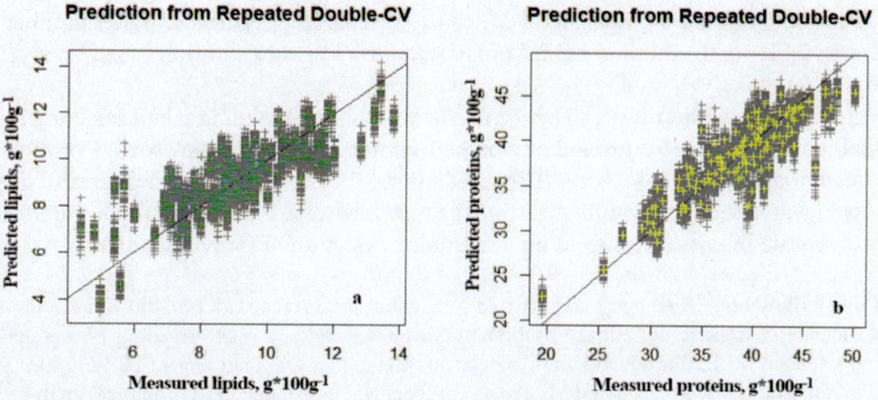

Fig. 2. Scatter plots of references vs predicted values from repeated double-cross validation for lipid content (a) and protein content (b) of lupin seeds.

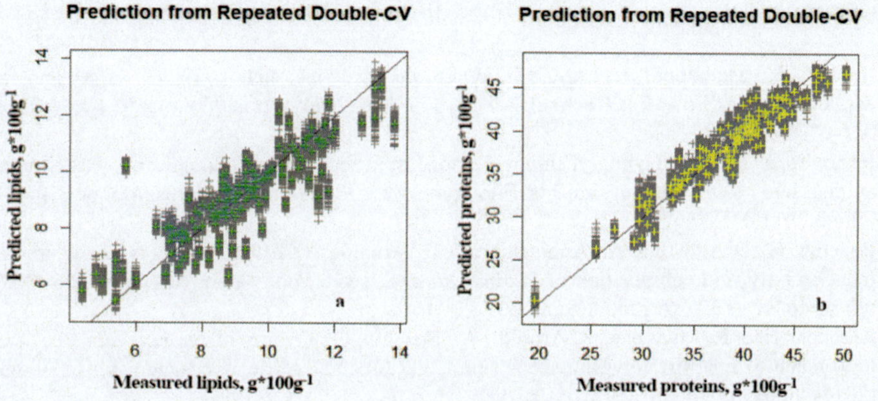

Fig. 3. Scatter plots of references vs predicted values from repeated double-cross validation for lipid content (a) and protein content (b) of lupin flours.

Results from primary and secondary inflorescences of the RIL plants or from different cropping years have been very similar (data not shown). The mean of protein and lipid contents, obtained by NIRS predictions for RIL and ecotypes, were 39.7 g*100 g^{-1} and 9.0 g*100 g^{-1} respectively. These data were largely comparable with those reported in other studies [1, 3–4].

4 Conclusion

Our results showed definitely acceptable NIRS-based predictions for protein and oil content only when using lupin flours spectra. NIRS predictions based on flour samples were highly reliable for protein quantification and moderately reliable for lipids.

Calibration based on the spectra of whole seed samples required higher number of latent variables and exhibited higher prediction errors than calibration based on ground samples.

Flour-based calibrations can be useful for white lupin breeding when selecting individual inbred lines on the ground of a bulked sample of their progeny seed. Conversely, the selection of lines based on NIRS predictions for the individual non-destructively assessed whole seed that would generate the new line were not sufficiently accurate and would require improvement to allow for a quick selection of individual plants.

Acknowledgments. This study was carried out within the Project LIVESEED: Improving the performance of organic agriculture by boosting organic seed and plant breeding efforts across Europe, funded by the European Union's Horizon 2020 under grant agreement N. 727230.

We are grateful to A. Tava and M. Crosta for technical assistance in the chemical analyses of white lupin flours.

References

1. Boukid, F., Pasqualone, A.: Lupine (*Lupinus spp.*) proteins: characteristics, safety and food applications. Eur. Food Res. Technol. **248**(2), 345–356 (2021). https://doi.org/10.1007/s00217-021-03909-5
2. Bertoglio, J.C., et al.: Hypoglycemic effect of Lupin seed gamma-Conglutin in experimental animals and healthy human subjects. Fitoterapia **82**, 933–938 (2011). https://doi.org/10.1016/j.fitote.2011.05.007
3. Boschin, G., D'Agostina, A., Annicchiarico, P., Arnoldi, A.: Effect of genotype and environment on fatty acid composition of *Lupinus albus* L. seed. Food Chem. **108**, 600–606 (2008). https://doi.org/10.1016/j.foodchem.2007.11.016
4. Annicchiarico, P., Manunza, P., Arnoldi, A., Boschin, G.: Quality of *Lupinus albus* L. (white lupin) seed: extent of genotypic and environmental effects. J. Agric. Food Chem. **62**, 6539–6545 (2014). https://doi.org/10.1021/jf405615k
5. Faluyi, M.A., Zhou, X.M., Zhang, F., Leibovitch, S., Migner, P., Smith, D.L.: Seed quality of sweet white lupin (*Lupinus albus*) and management practice in eastern Canada. Eur. J. Agron. **13**, 27–37 (2000). https://doi.org/10.1016/S1161-0301(00)00057-5
6. Annicchiarico, P., et al.: Detection and exploitation of white lupin (*Lupinus albus* L.) genetic variation for seed γ-conglutin content. J. Appl. Bot. Food Qual. **89**, 212–216 (2016). https://doi.org/10.5073/JABFQ.2016.089.027
7. AOAC Official Method of Analysis, Method 920.39, Fat (crude) or ether extract in animal feed. 18th ed., AOAC International, Gaithersburg, MD (2005)
8. Kirsten, W.J., Hesselius, G.U.: Rapid, automatic, high capacity Dumas determination of nitrogen. Microchem. J. **28**, 529–547 (1983). https://doi.org/10.1016/0026-265X(83)90011-5
9. Filzmoser, P., Liebmann, B., Varmuza, K.: Repeated double cross validation. J. Chemom. **23**, 160–171 (2009). https://doi.org/10.1002/cem.1225

Determination of Nitrogen and Phosphorus in Dairy Slurry Using Near Infrared Diffuse Reflection Spectroscopy

Mengting Li, Zengjun Yang, Shengbo Liu, Di Sun, and Run Zhao(✉)

Agro-Environmental Protection Institute, Ministry of Agriculture and Rural Affairs, Tianjin, China
zhaorun@aepi.org.cn

Abstract. Nitrogen and phosphorus are important nutrient measurement indicators for the slurry field application. Conventional wet chemical methods have been used as a routine detection way of total nitrogen (TN) and total phosphorus (TP). However, it is time-consuming, costly and destructive that cannot realize real-time and on-site detection. For the rapid and reliable determination of TN and TP in dairy farm slurry, near infrared spectroscopy (NIRS) was employed in this study. 472 samples were collected from 33 dairy farms in Tianjin. The near infrared diffuse reflectance spectra of all samples were scanned using Fourier transform near infrared spectrometer. And partial least squares models were established for quantitative analysis of TN and TP. Results were as follows: the correlation coefficient R_P were 0.92 and 0.91, the root mean square error of prediction (RMSEP) were 426.14 mg/L and 16.65 mg/L, the residual predictive deviation (RPD) were 2.73 and 2.63 for TN and TP, respectively. The prediction results of TN were better than that of TP. The results manifest that it is feasible to rapidly determine the contents of slurry TN and TP via the NIRS. This study can provide the technical support for reasonable land application of slurry.

Keywords: Dairy farm slurry · Total nitrogen (TN) · Total phosphorus (TP) · Partial least squares (PLS) · Rapid determination

1 Introduction

With acceleration of scale and intensification of dairy farming, effective management of complex slurry system is a focal point during sustainable industrial development. Returning slurry to the field is an eco-friendly addressing approach that has widely been used nowadays. The nitrogen (N) and phosphorus (P) are important measurement indicators that how to rapidly and accurately obtain their contents of the slurry is in urgent demand. A set of policies have successively been issued by the central governments since 2017, in which clearly pointed out developing quick-test methods and establishing relevant standards for the land application. Therefore, it is of great practical significance

X. Chu et al. (Eds.): ICNIR 2021, *Sense the Real Change: Proceedings of the 20th International Conference on Near Infrared Spectroscopy*, pp. 137–144, 2022.
https://doi.org/10.1007/978-981-19-4884-8_14

to establish a rapid quantitative analysis method of N and P in the slurry, ensuring scientific reuse meanwhile controlling over environmental risks.

The near-infrared spectroscopy (NIRS) [1–3], based on the correlation between chemical properties and absorption of electromagnetic radiation, was a rapid, convenient, non-destructive, cost-effective and predictive technique applied in the soil, agricultural products, food, environment and other fields [4–6]. NIRS has proved its superiority in predicting the soil properties simultaneously, such as total nitrogen (TN), soil organic matter, pH, clay content and so on [7, 8]. And it also used monitoring of several parameters (calcium, alcoholic degree, total acidity, volatile acidity, et al.) in white wine fermentations [9, 10]. In addition, NIRS has also been applied to the animal manure during the fermentation engineering for decades. Awhangboad [11], Liang [12], Yang [13] *et al.* used NIRS to successfully predict the contents of volatile fatty acids, ammonium nitrogen and TN during anaerobic digestion and composting. But to our knowledge, applied research on the valid nutrients determination in complicated slurry system of dairy farms via NIRS has rarely been reported.

Therefore, the ultimate purpose of present study was to establish a reliable and robust NIRS model that providing a fast measuring means for guiding scientific field application. It was well-off achieved by (1) quantifying TN and TP of 472 slurry samples as reference value. (2) Collecting and analyzing NIRS of slurry. (3) Forming partial least square model to predict TN and total phosphorus (TP) of slurry, respectively.

2 Materials and Methods

2.1 Sample Preparation

A total of 472 slurry samples were collected from 33 representative dairy farms of Tianjin on December 2018, March, June and September 2019, respectively. The farming scale was between 440 and 5400 heads. Sampling sites were covering throughout the whole links from the first point to the last before the field application. After collection, samples were timely delivered to the laboratory for testing.

2.2 Reference Analysis

Contents of TN and TP (reference data) in slurry samples were determined according to the national standard methods, *i.e.* Determination of Kjeldahl nitrogen and Ammonium Molybdate Spectrophotometric respectively [14, 15]. Automatic Kjeldahl nitrogen determination apparatus (Foss kjeltec 8400, Denmark) and ultraviolet-visible spectrophotometer (722E, China) were acquired to measure the slurry TN and TP, respectively.

2.3 Near Infrared Spectra Acquisition

The NIR spectra were collected by the Fourier Transform Near-infrared spectrometer (FT-NIR) from PerkinElmer Company of the United States. FT-NIR was equipped with an integrating sphere accessory and InGaAs detector. The spectral range was between

12 000–4 000 cm^{-1} with parameters of 8 cm^{-1} resolution, 2 cm^{-1} scanning interval and 64 scanning times. Place the sample cell with slurry on the rotating platform of the integrating sphere to measure the spectra.

2.4 Chemometric Analysis

2.4.1 Elimination of Abnormal Samples

Student residual and lever values were used to detect abnormal samples. The threshold value of student residual was set as 3 and −3, and the values beyond the range were considered as chemical anomalies. Lever value was an indicator for eliminating spectral anomalies. A total of 31 and 39 abnormal samples were removed from nitrogen and phosphorus samples, respectively.

2.4.2 Determination of Calibration and Prediction Set

The concentration gradient method was used to divide the samples into calibration set and validation set (Table 1). (1) Samples were arranged in ascending order of concentration. (2) In line with the method of 'taking one from two', 1/3 of modeling samples were chosen as prediction set with the rest as calibration set. Ensure that concentration range of calibration set was greater than that of prediction set.

Table 1. Sample information of calibration set and prediction set

Indicator	Classification	Sample size	Content range (mg/L)	Mean (mg/L)
Total nitrogen	Calibration set	295	14.16–5262.29	1481.70
	Prediction set	146	40.87–5218.96	1515.27
Total phosphorus	Calibration set	289	0.04–178.2	75.38
	Prediction set	144	4.08–179.02	74.52

2.4.3 Selection of Preprocessing Method

The sample system was complex that slurry was massively mixed with urine, manure, grass mustard and sludge. Thus the light scattering was strong [16]. The original spectral data not only contained chemical information of samples, but also included external interference information. Therefore, appropriate pretreatment methods need to be done before modeling. The optimal pretreatment method was baseline correction and normalized normalization for TN and TP, respectively.

2.4.4 Selection of Modeling Variables

A custom band range was used to select modeling variables. 12000–6000 cm^{-1}, 10000–6000 cm^{-1}, 10000–4000 cm^{-1} and 8000–4000 cm^{-1} were modeled in Unscrambler software. The effect of TN model in the band of 10000–4000 cm^{-1} was better than that in the whole band. The effect of TP in the whole band model was better.

2.4.5 Model Construction and Evaluation

The models were established by PLS regression. In general, model performance was evaluated by the root mean square error of prediction (RMSEP) and R_p [17, 18]. To overcome different ranges among measured samples, RPD (the radio of RMSEP to the standard deviation) was also widely proposed according to the ranges as follows: RPD > 4.0, excellent; 3.0–4.0, successful; 2.2–3.0, Useful; 1.7–2.2, moderately useful; 1.5–1.7, screening; < 1.5, poor [19, 20].

3 Results and Discussion

3.1 NIR Spectral Characteristics of Slurry Samples

Representative NIR absorbance spectra (12000−4000 cm^{-1} spectral range) collected from the slurry was shown in Fig. 1. The trend of spectra was basically same but the intensity of absorbance was different. The obvious absorption peaks near 5158 cm^{-1}, 6938 cm^{-1}, 8362 cm^{-1} and 10232 cm^{-1} were due to water absorption in the slurry. The peak attributed to combined frequency absorption of O-H antisymmetric stretching plus bending at 5158 cm^{-1}. The combined frequency absorption of O-H symmetric and antisymmetric stretching was located at 6938 cm^{-1} and 8362 cm^{-1}. The peaks near 10232 cm^{-1} caused by multiple frequency absorption of O-H symmetrical stretching plus combined absorption of O-H antisymmetric stretching [21]. Therefore, it was feasible to establish NIRS models for quantitative analyzing nitrogen and phosphorus content of slurry.

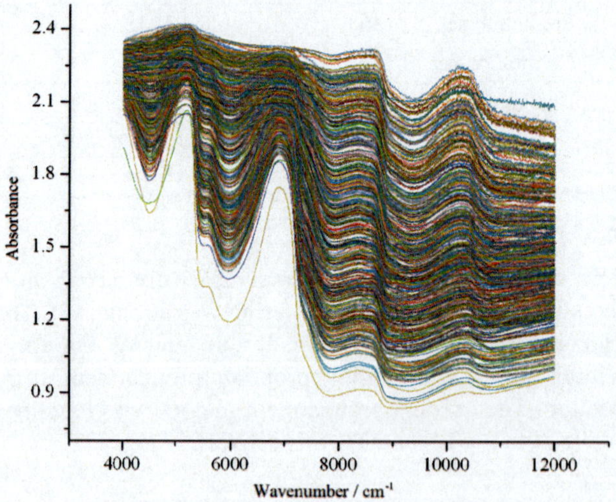

Fig. 1. Original spectra of 472 slurry samples

3.2 Selection of Principal Component Number

Root mean square error of validation (RMSECV) was often used to select the best modeling principal factor. To prevent overfitting, the study combined RMSECV with the root mean square error of calibration (RMSEC) to select the number of main factors, ensuring that RMSEP/RMSEC $\leq$ 1.2. Figure 2 showed the relationship between RMSECV and the number of main factors. For TN, when the principal factor was 11, RMSECV was the minimum meanwhile RMSEP/RMSEC = 0.92. For TP, when the main factor was 5, RMSECV was the minimum and RMSEP/RMSEC = 0.84. Therefore, 11 and 5 main factors were selected to develop NIRS model of TN and TP, respectively.

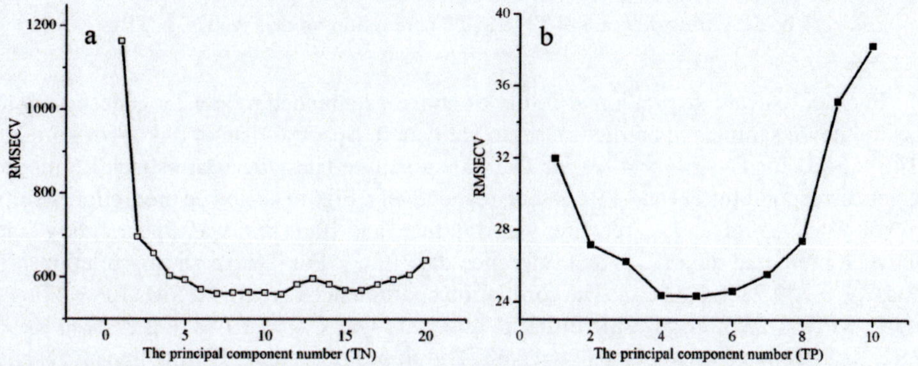

Fig. 2. Effect of principal component number on RMSECV of NIRS models (a. TN, b. TP)

3.3 Establishment, Validation, and Evaluation of NIRS Model

Based on spectral variable X (295 × 3001) of calibration set and corresponding variable Y (295 × 1), 11 main factors were selected to establish a quantitative analysis of slurry TN. Figure 3 showed prediction results. The red line is 1:1 line. Black line is linear fitting line between predicted data (C_{pc}) and reference data (C_{c0}). The more overlapped between the two lines, the smaller deviation between predicted data and reference data, and the better performance of model. The linear fitting relationship between C_{pc} and C_{c0} was $C_{pc} = 297.89 + 0.80\ C_{c0}$. The correlation coefficient (R_c) was 0.895 while RMSEC was 465.89 mg/L. The slope of fitted line was less than 1, indicating that the model's predicted data in the calibration set was smaller than the reference data. Although the fitted line deviates a bit from the ideal 1:1, all points basically fall near fitted line and 1:1 line.

On account of spectral variable X (289 × 4001) of calibration set and corresponding variable Y (289 × 1), 5 main factors were selected to form a quantitative model of slurry TP. The fitting relationship was $C_{pc} = 21.27 + 0.71\ C_{c0}$. The R_c was 0.85 while RMSEC was 19.73 mg/L. Compared to TN model, the slopes of fitted lines were all less than 1. But the difference was that although most of points also fall around fitted line and 1:1 line, the distances were far away.

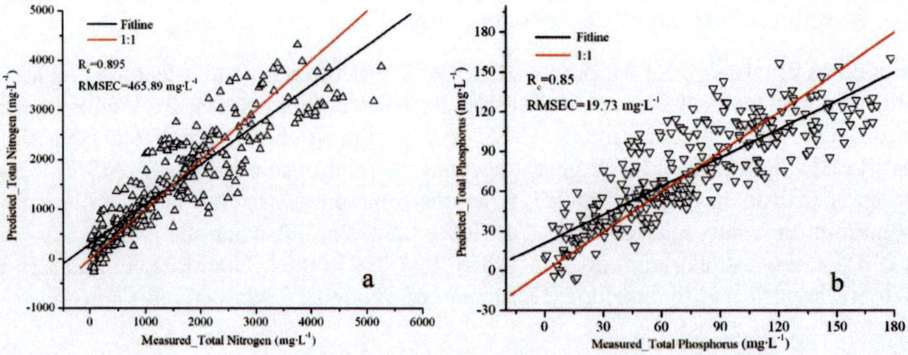

Fig. 3. Fitting diagram of TN and TP calibration models (a. TN, b. TP)

In order to evaluate predictive ability of above-established models, it is necessary to use unknown samples in prediction set to validate it. Spectral data of unknown samples (146 × 3001 for TN, 144 × 4001 for TP) was substituted into the established PLS model to get corresponding TN and TP content, respectively. Figure 4 showed prediction results of unknown samples. The red line was 1:1 line, and blue line was linear fitting line between predicted data (C_{pp}) and reference data (C_{p0}). For TN, the fitting relationship was $C_{pp} = 259.78 + 0.83\ C_{p0}$. The correlation coefficient (R_p) = 0.92, RMSEP = 426.14 mg/L, RPD = 2.73. For TP, the fitting relationship was $C_{pp} = 10.48 + 0.85\ C_{p0}$. R_p = 0.91, RMSEP = 16.65 mg/L, RPD = 2.63. The slopes of predicted fitting line of TN and TP in unknown samples were all smaller than 1:1 ideal line, indicating that predicted data was generally lower than reference data. Taking above results into consideration, it is feasible to establish NIRS model for quantitative analysis of slurry TN and TP.

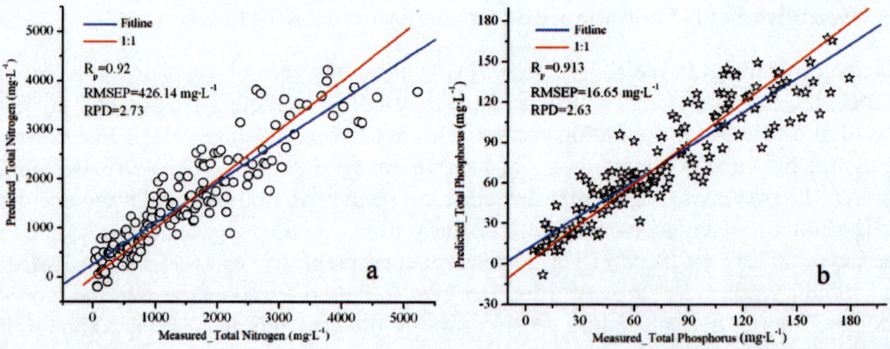

Fig. 4. Fitting diagram of predicted data and reference data of TN and TP models (a. TN, b. TP)

4 Conclusion

472 slurry samples were used to form NIRS model of TN and TP. The results showed that R_p = 0.92, RMSEP = 426.14 mg/L, RPD = 2.73 for TN, R_p = 0.91, RMSEP =

16.65 mg/L, RPD = 2.63 for TP. And it showed that both TN and TP NIRS model were successful. This study can provide the technical support for reasonable field application of scientifically recycling of slurry from dairy farms.

Acknowledgments. This work was supported by the Central Public-interest Scientific Institution Basal Research Fund (No. Y2022CG09), Innovation Team of Tianjin Dairy (Sheep) Research System (No. ITTCRS202100007) and Innovation Project of Agricultural Science and Technology by CAAS.

References

1. Crocombe, R.A.: Portable spectroscopy. Appl. Spectrosc. **72**, 1701–1751 (2018)
2. Jancewicz, L.J., et al.: Development of near-infrared spectroscopy calibrations to estimate fecal composition and nutrient digestibility in beef cattle. Can. J. Anim. Sci. **97**, 51–64 (2016)
3. Qiao, T., Ren, J., Craigie, C., Zabalza, J., Maltin, C., Marshall, S.: Quantitative prediction of beef quality using visible and NIR spectroscopy with large data samples under industry conditions. J. Appl. Spectrosc. **82**, 137–144 (2015)
4. Weeranantanaphan, J., Downey, G., Allen, P., Sun, D.W.: A review of near infrared spectroscopy in muscle food analysis: 2005–2010. J. Near Infrared Spectrosc. **19**, 61–104 (2011)
5. Liu, D., Zeng, X.A., Sun, D.-W.: NIR spectroscopy and imaging techniques for evaluation of fish quality-a review. Appl. Spectrosc. Rev. **48**, 609–628 (2013)
6. Wang, L., Sun, D.W., Pu, H.B., Cheng, J.H.: Quality analysis, classification, and authentication of liquid foods by near-infrared spectroscopy: A review of recent research developments. Crit. Rev. Food Sci. Nutr. **57**, 1524–1538 (2017)
7. Ng, W., et al.: Convolutional neural network for simultaneous prediction of several soil properties using visible/near-infrared, mid-infrared, and their combined spectra. Geoderma **352**, 251–267 (2019)
8. Yang, M.H., Xu, D.Y., Chen, S.C., Li, H.Y., Shi, Z.: Evaluation of machine learning approaches to predict soil organic matter and pH using vis-NIR spectra. Sensors **19**, 263 (2019)
9. Vestia, J., Barroso, J.M., Ferreira, H., Gaspar, L., Rato, A.E.: Predicting calcium in grape must and base wine by FT-NIR spectroscopy. Food Chem. **276**, 71–76 (2018)
10. dos Santos, C.A.T., et al.: Raman spectroscopy for wine analyses: a comparison with near and mid infrared spectroscopy. Talanta **186**, 306–314 (2018)
11. Awhangboad, L., Bendoulab, R., Roger, J.M., Béline, F.: Multi-block SO-PLS approach based on infrared spectroscopy for anaerobic digestion process monitoring. Chemom. Intell. Lab. Syst. **196**, 103905 (2020)
12. Liang, H., Huang, Y.P., Shen, G.H., Han, L.J., Yang, Z.L.: Near-infrared real-time online bypass detection of volatile fatty acids in anaerobic fermentation of manure. Trans. Chin. Soc. Agric. Eng. **36**, 220 (2020)
13. Yang, Z.L., Huang, Y.P., Shen, G.H., Mei, J.Q., Han, L.J.: Rapidly detection of key parameters in whole composting process based on online near infrared spectroscopy. Trans. Chin. Soc. Agric. Mach. **50**, 356–361 (2019)
14. Sun, D., Li, M.T., Mu, M.R., Zhao, R., Zhang, K.Q.: Rapid determination of nitrogen and phosphorus in dairy farm slurry via near-mid infrared fusion spectroscopy technology. Trans. Chin. Soc Agric. Eng. **41**, 3092–3098 (2021)
15. Li, M.T., et al.: Variation characteristics and rules of nitrogen and phosphorus contents throughout the slurry movement from scaled dairy farms in Tianjin. Trans. Chin. Soc Agric. Eng. **36**, 27–33 (2020)

16. Diwu, P.Y., Bian, X.H., Wang, Z.F., Liu, W.: Study on the selection of spectral preprocessing methods. Spectrosc. Spectral Anal. **39**, 2800–2806 (2019)
17. Liu, J.M., Jin, S., Bao, C.H., Sun, Y., Li, W.Z.: Rapid determination of lignocellulose in corn stover based on near-infrared reflectance spectroscopy and chemometrics methods. Bioresour. Technol. **321**, 124449 (2021)
18. Vaudour, E., Gilliot, J.M., Belb, L., Lefevre, J., Chehdi, K.: Regional prediction of soil organic carbon content over temperate croplands using visible near-infrared airborne hyperspectral imagery and synchronous field spectra. Int. J. Appl. Earth Obs. Geoinf. **49**, 34–48 (2016)
19. Viscarra, R.R.A., Walvoort, D.J.J., McBratney, A.B., Janik, L.J., Skjemstad, J.O.: Visible, near infrared, mid infrared or combined diffuse reflectance spectroscopy for simultaneous assessment of various soil properties. Geoderma **131**, 59–75 (2006)
20. Cabassi, G., Cavalli, D., Fuccella, R., Gallina, P.M.: Evaluation of four NIR spectrometers in the analysis of cattle slurry. Biosyst. Eng. **133**, 1–13 (2015)
21. Fan, M.L., Zhao, Y., Liu, Y., Cai, W.S., Shao, X.G.: Aquaphotomics of near infrared spectroscopy. Prog. Chem. **27**, 242–250 (2015)

Rapid Prediction of Multiple Quality Parameters in Milk Powder by Ultraviolet Spectrometry Combined with Chemometric Method

J. F. Pang[1,3], X. Huang[1], and Y. K. Li[1,2](✉)

[1] Department of Environmental Science and Engineering, North China Electric Power University, Baoding 071003, People's Republic of China
liyankun_ncepu@foxmail.com, liyankun@ncepu.edu.cn
[2] MOE Key Laboratory of Resources and Environmental Systems Optimization, Beijing 102206, People's Republic of China
[3] Department of Environmental Science and Technology, China University of Geosciences, Wuhan 430074, People's Republic of China

Abstract. The composition of milk powder (powdered milk) determines its quality and nutritional value. Currently, the standard or traditional methods that measure content of main components of milk powder have some disadvantages. In this study, ultraviolet (UV) spectroscopy combined with multivariate calibration/regression model was used to simultaneously predict the value of four main quality parameters including protein, fat, carbohydrate and energy rather than single component content in milk powder. Partial least squares (PLS) was chosen to establish regression model with the optimized number of principal factor. Without component separation/purification in the measurement with UV spectroscopy and pretreatment process in PLS modeling, good prediction results of multi-parameters were obtained with low root mean square error of prediction (RMSEP), high correlation coefficients (>0.98) and high RPD (Residual predictive deviation). By comparison, the results obtained by directly using work curve method were not satisfactory. Furthermore, PLS model acquired accurate and robust results than those of multivariate linear regression (MLR) model. It indicates that with the help of PLS, UV spectrometry is an effective, fast and simple "green" technique to simultaneously detect content of main parameters in milk powder. The proposed method could be applied to the quality control of milk powder, and be studied further to extend to quantitative analysis of milk liquid and even other food.

Keywords: Ultraviolet spectra · Chemometrics · Milk powder · Quantitative analysis

1 Introduction

In general, dairy products are important sources of nutrition for human beings, and milk powder is taken as an indispensable food in many families. The high-quality dairy

X. Chu et al. (Eds.): ICNIR 2021, *Sense the Real Change: Proceedings of the 20th International Conference on Near Infrared Spectroscopy*, pp. 145–156, 2022.
https://doi.org/10.1007/978-981-19-4884-8_15

products are in great demand, while the quality and nutritional value are closely related to the content of ingredients. Milk powder contains a variety of ingredients, furthermore, milk powder of different kinds or brands also differ in the content of ingredients. Thus, the detection of ingredients is very important for quality control of milk powder.

The main nutritional components in milk powder commonly include protein, fat, carbohydrate and mineral etc. Currently, several standard methods or traditional methods that measure content of these components have been widely used [1–5]. However, these methods usually are labor and time consuming, employ tedious operations, have low separation efficiency, and some additive solvents produce adverse effects on the operators and environment. For example, in protein detection, classical Kjeldahl method is time-consuming, and employs sophisticated operations. The sample is easily damaged and harmful gases may be released during the experiment; Dumas combustion method is carried out at high temperature, and employs specialized instrument. Moreover, both above methods are easily influenced by nonprotein nitrogen (NPN) reagents [6]. For fat detection, in alkaline hydrolysis method, the sample needs heated hydrolysis and repeated extraction with large amount of ether reagent [7]. For carbohydrate detection, high performance liquid chromatography (HPLC) is costly and cumbersome to operate.

Spectroscopy technique has been taking vital role in the analysis of sophisticated samples in food field [8–11]. Ultraviolet (UV) Spectrometer is regular laboratory equipment [12], and some researchers have made use of the UV spectra with Lambert-Beer law to measure content of protein or fat in food [7, 13]. However, many factors in the measurement process may affect the ultraviolet intensity, such as a large number of interfering substances and background effects, as well as pH value of the solution and so on. That is the reason why these methods need experimental pretreatments to extract sole ingredient for its detection. To overcome the above disadvantages, the multivariate calibration method has been adopted to help establishing the relationships between measurement data and multiple components content or categories [14–21]. Through in-depth analysis of the measured data information, "chemical separation" is replaced by "mathematical separation", therefore, simultaneous determination of multiple components is achieved, and the prediction of component content is not affected even under the coexistence of unknown interferences [22]. In this work, partial least squares (PLS) model was established to correlate main components/quality parameters with UV spectra of milk powder, and predicted content of protein, fat, carbohydrate and energy in milk powder. The results present a fast, cost-effective, accurate and robust method for simultaneous measurement of multiple parameters in milk powder based on ultraviolet spectrometry combined with PLS model.

2 Materials and Methods

2.1 Samples and Experimental

Thirteen kinds of qualified whole milk powder with Chinese famous brand (Feihe, Beingmate, Wondersun, Mengniu, Sanyuan, Yili & Gucheng) were purchased at the local supermarket. The reference value of main nutritional parameters in milk powder are shown in Table 1. Among them, protein content was determined by Kjeldahl method, fat content was determined by alkaline hydrolysis method, and HPLC was used to determine carbohydrate content [1–5]. Energy value was calculated by the formula as below [23].

$$X = A_1 \times B_1 + A_2 \times B_2 + A_3 \times B_3 \quad (1)$$

where X is amount of energy, kJ;

A_1: Protein mass, g; B_1: Energy coefficient of protein, 17 kJ g^{-1};

A_2: Fat mass, g; B_2: Energy coefficient of fat, 37 kJ g^{-1};

A_3: Carbohydrate mass, g; B_3: Energy coefficient of carbohydrate, 17 kJ g^{-1}.

Use 0.9% (mass concentration) sodium chloride (NaCl) aqueous solution to adequately dissolve 13 kinds of milk powder separately and obtain 1.5 mg mL^{-1} original solutions. Then original solutions were respectively diluted by 0.9% NaCl aqueous solution by ultrasonic machine (SB25-12; Ningbo Xinzhi Biotechnology Co. Ltd., China) for 10 min at 35.0 °C. Finally, 13 kinds of milk powder solutions were obtained with concentration of 0.06 mg mL^{-1}, 0.12 mg mL^{-1}, 0.18 mg mL^{-1}, 0.24 mg mL^{-1}, 0.30 mg mL^{-1}, 0.36 mg mL^{-1} and 0.42 mg mL^{-1}, that is 91 samples. Then these samples were measured immediately by ultraviolet-visible (UV-Vis) spectrometer.

Table 1. The reference values of main parameters in milk powder

Type	Nutrition Information per 100 g			
	Protein (g)	Fat (g)	Carbohydrate (g)	Energy (KJ)
1	20.0	23.0	48.0	2007
2	18.1	13.5	58.2	1814
3	18.5	15.0	51.1	1738
4	18.2	15.0	51.1	1733
5	18.2	16.0	52.5	1794
6	18.2	15.0	51.1	1733
7	19.0	23.0	50.0	2024
8	4.0	8.0	78.0	1600
9	19.1	24.0	48.5	2037
10	17.5	18.5	55.9	1932
11	18.2	15.0	51.1	1733
12	20.0	24.0	47.0	2050
13	27.0	18.0	45.0	1922

2.2 Ultraviolet Spectra

The Ultraviolet absorbance spectra of the samples were acquired by a Dual-beam UV-Vis spectrometer (TU-1901; Beijing General Analysis General Instrument Co. Ltd., China)

in the range of 190–400 nm against solvent as blank. The resolution was set as 0.5 nm and 1-cm path quartz cuvette was used. Each sample was repeatedly scanned 3 times, and the average spectra of three times were used.

3 Theory and Algorithm

3.1 Partial Least Squares

Partial Least Squares (PLS) method is a popular chemometric method in spectral data analysis [24, 25]. PLS reveals the linear relationship between the independent matrix (**X**) and the dependent variables (**Y**). PLS transforms **X** and **Y** into new variables at the same time, and tries to maximize the variances of the new variables and find the maximum correlation between the new variables. PLS is used to establish the regression model of multiple dependent variables to multiple independent variables, and it includes PLS1 and PLS2 algorithm. PLS1 treats multidimensional response variables as multiple single dimension variables, and PLS2 processes the response multidimensional variables simultaneously. In this study, PLS1 modeling was adopted.

3.2 Multivariate Linear Regression

Multivariate linear regression (MLR) based on least squares is a conventional regression method used to establish the linear relevance between a dependent variable and one or more independent variables [26]. The combination of multiple independent variables is used to predict or estimate the dependent variable.

In the modeling, root mean squared error of validation set (RMSEV) is used as an evaluation criterion. The results of root mean squared error of prediction set (RMSEP), the ratio of the standard deviation (SD) of the reference/true values for the prediction samples to the RMSEP (RPD) [27, 28], the correlation coefficient of determination (R^2) and p-value are compared. Lower RMSE and higher RPD values can indicate lower prediction error of the model. R^2 can quantify the relationship between the reference value and the predicted value, and p-value test can determine the reliability of the fitting equation.

$$RMSE = \left[\frac{1}{n}\sum\nolimits_{i=1}^{n}(y_i - \hat{y}_i)^2\right]^{\frac{1}{2}} \tag{2}$$

$$RPD = \left[\frac{1}{n}\sum\nolimits_{i=1}^{n}(y_i - \overline{y}_i)^2\right]^{\frac{1}{2}}/RMSEP \tag{3}$$

where $\widehat{y}_i$ is the predicted content of the ith sample, y_i is the reference content of the ith sample, n is the number of predicted samples.

4 Results and Discussion

4.1 Spectra Profiles

Figure 1(a) shows the spectra of all the milk powder solution. It seems that the spectrogram shape is similar, whereas samples with different types or concentrations have different maximal absorption values and the sharp absorption peaks locate at slightly different wavelengths (red-shift or blue-shift). To clarify, Fig. 1(b) presents a typical example of ultraviolet spectrum. Studies have shown that in dairy products, protein contains amino acid residues of tyrosine (Tyr), tryptophan (Trp) and phenylalanine (Phe), which contain conjugated double bond structure of benzene ring with maximal absorption peak around 280 nm. Based upon the property of fatty acids to absorb UV light, fat has a wide absorbance in the range of ~220 to 240 nm that corresponds mainly to conjugated dienes, and a sharp absorption peak in the range of ~202 to 215 nm that depends on total lipid concentration. While the open-chain carbohydrate has a strong absorption peak at 280 nm that corresponds to ketone group [7]. Above all, the absorption bands of multi-components in milk powder overlap and interference, hindering the resolution of the mixture through conventional spectrophotometry. For resolving mixtures, the multivariate regression has been introduced and applied to the present case.

It was also found that when the concentration of milk powder solution was too low, stable data can not be acquired, so two inverted peaks occurred. Then the two outliers were abandoned, ultimately, UV spectra data of 89 samples were remained, which were divided into a training set, a test set and a prediction set by Kennard-Stone (KS) algorithm according to a number ratio of 3:1:1.

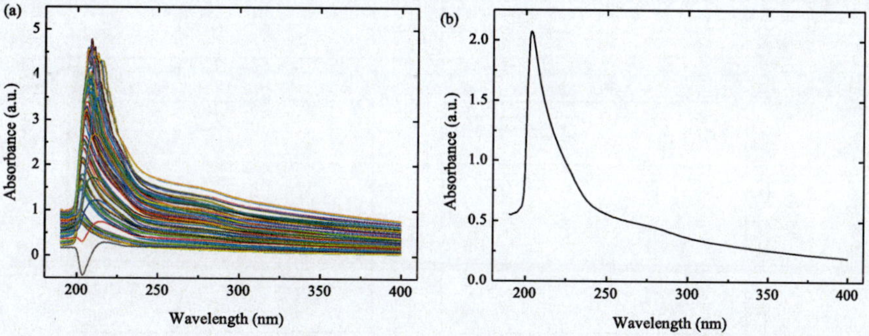

Fig. 1. Ultraviolet spectra of milk powder solution (a) UV spectra of all samples; (b) A typical spectrum example

4.2 The Results of PLS Modeling

For PLS model, the number of principal factor (n_f) has an important effect on the accuracy of prediction results. Too many factors would cause a decrease in the prediction accuracy for abundant or interfering information are included in the model. Too few factors would also lead to inaccuracy because not all necessary and characterized information are used in this performance. So, the optimal n_f was determined. Here, n_f was selected as 1–15 for investigation. The optimal n_f is determined in accordance with the smallest value of RMSEV. Figure 2 shows the relationship between RMSEVs and n_f of the four parameters including protein, fat, carbohydrate and energy value in milk powder. Overall, the RMSEV values fluctuate a lot at the beginning. As n_f increases, RMSEV values tend to decline and stabilize. After comprehensively considering the four parameters, 10 is selected as the optimal n_f.

At the optimal n_f, the four parameters were predicted. Various spectral pretreatment methods including smoothing, normalization and derivation had been tried, whereas the results had not been apparently improved. Therefore, the original spectra were directly used for modeling. Performances of PLS model in predicting the four parameters are presented in Table 2. It can be seen that low RMSEPs (0.0011–0.2247) and high RPDs (5.6213–18.2507) in PLS modeling were obtained. The values of correlation coefficient (R^2) are over 0.98 and p-values are very low, which is taken as ≈0. To clarify, the plots of the prediction values versus the reference values are also presented in Fig. 3. Obviously, the prediction results of protein with $R^2 \approx 0.998$ and RPD = 18.2507 are the best among the results of four parameters.

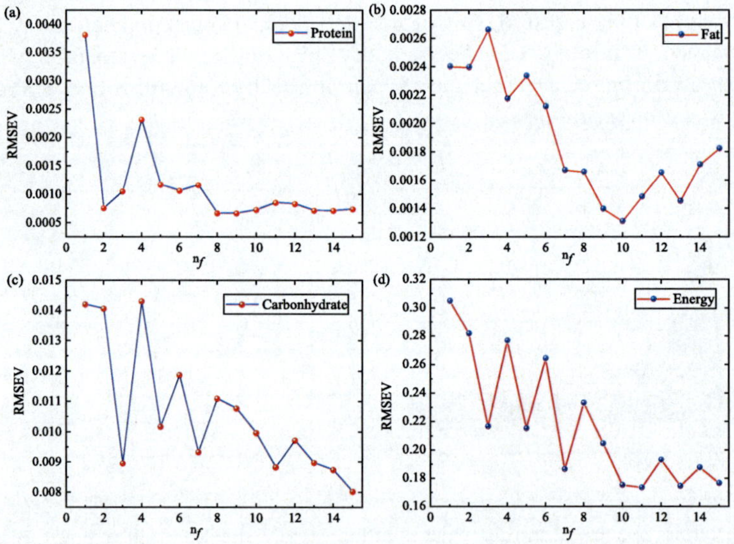

Fig. 2. Variations of RMSEVs with n_f

4.3 The Results of MLR Modeling

The four parameters were also predicted by MLR method, and its performances are also presented in Table 2. RMSEPs (0.0044–0.3115) and RPDs (3.2450–6.1120) were obtained. The values of R^2 are between 0.9002–0.9788, and p-values is taken as ≈0.

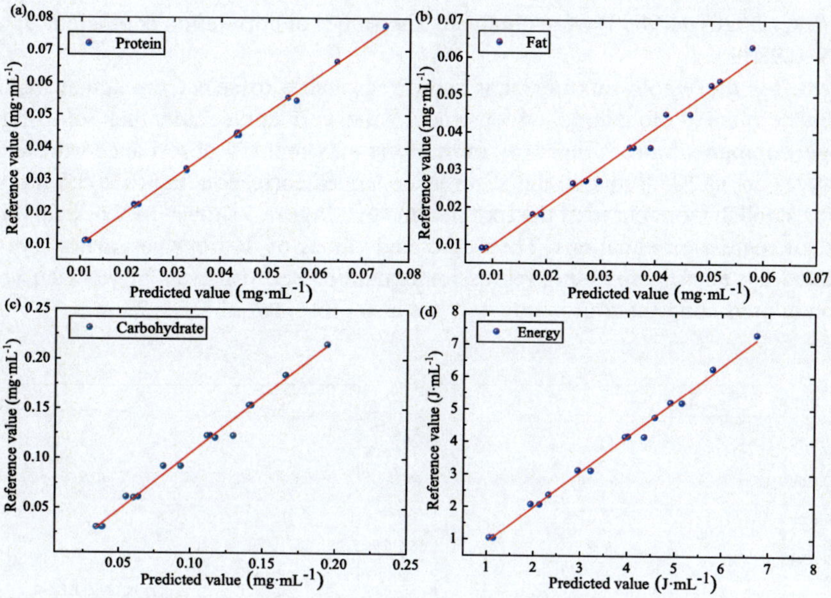

Fig. 3. The plots of the reference values versus the prediction values of PLS model

Through comparisons, clearly, the results of MLR are inferior to those of PLS, which illustrates the superiority of PLS modeling in dealing with multi-components complex system. Through comparisons, clearly, the results of MLR are inferior to those of PLS model. In PLS modeling, the principal components (latent variables) are simultaneously extracted from independent and dependent variables for regression, which shows the superiority of prediction accuracy and prediction stability in dealing with the multi-components complex system. Although RPDs of predicting the four parameters are all over 3.0 [26, 27], it can also meet the actual measure demand.

4.4 The Results of Work Curve Method

In the quantitative analysis by conventional spectrophotometry, work curve method is generally utilized, in which the maximal absorbance value with the corresponding concentration value will be selected to prepare a work curve. The wavelength corresponding to maximal absorbance value is called as maximal absorbance wavelength. Based on the description of 3.1 section, 280 nm, 208 nm and 280 nm are respectively selected as maximal absorbance wavelength of protein, fat and carbohydrate. While considering the energy value of milk powder is calculated based on values of protein, total carbohydrate and fat, 260 nm is proposed for energy [23]. The absorbance values corresponding to maximal absorbance wavelength were used to establish work curve. Figure 4 gives the work curves (I) with their regression equations. Then predicted concentrations of the four parameters are calculated by the corresponding regression equation, and performances of work curve (I) method in predicting the four parameters are presented in Table 2. It could be found that, in work curve (I) method, RMSEPs (0.0056–0.3816) and RPDs (2.5201–4.9893) were obtained. The values of correlation coefficient (R^2) are 0.9289–0.9859.

Next, for the whole mixture, it is more reasonable to select the actual maximal absorbance value of the sample for constructing the work curve rather than selecting that of single component. From Fig. 1(a), most of the maximal UV absorbance wavelengths of samples are at 205.5 nm, so the absorbance values corresponding to 205.5 nm were used to establish work curve of the four parameters. Figure 5 shows the work curves (II) with their regression equations. Then predicted values of the four parameters are also calculated by the corresponding regression equation. Performances of work curve (II) method in predicting the four parameters are also presented in Table 2.

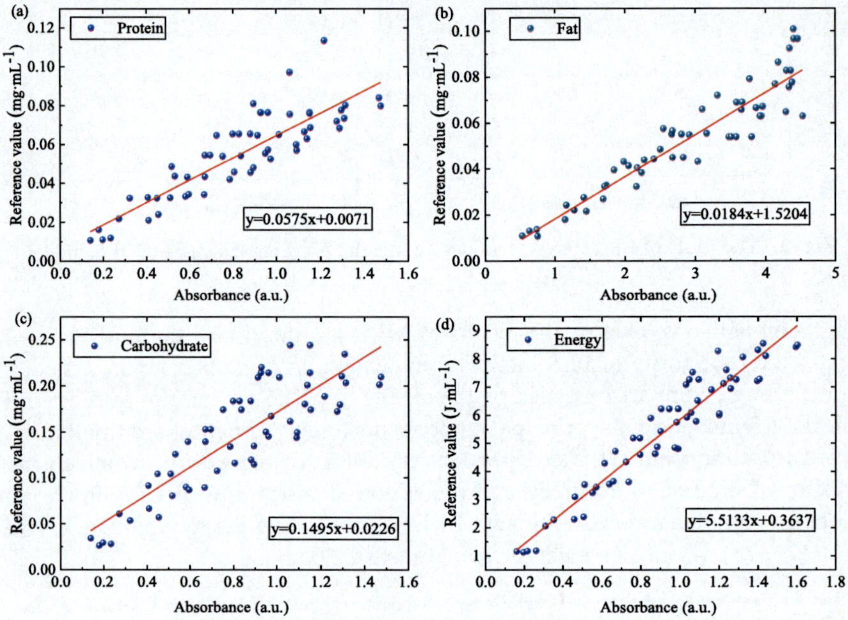

Fig. 4. Work curves (I) of four parameters

Clearly, most results of the work curve (II) method were better than those of the work curve (I) method, which is due to the absorption bands of multi-components in mixture overlap and interference, so the maximal absorbance wavelength of mixture is different with the single component. Therefore, the work curve should be established with the actual maximal absorbance value of the mixture.

Then, comparisons are made between PLS method and work curve methods. In Table 2, PLS performances larger correlation coefficient (R^2), smaller RMSEPs and high RPDs, which demonstrates high accuracy and high robustness of PLS model. In work curve methods, some RPD values are below 3.0, so the methods cannot meet the actual multi-components measure demand. A good prediction model is not only accurate but also robust, and PLS model meets the tradeoff of the two aspects.

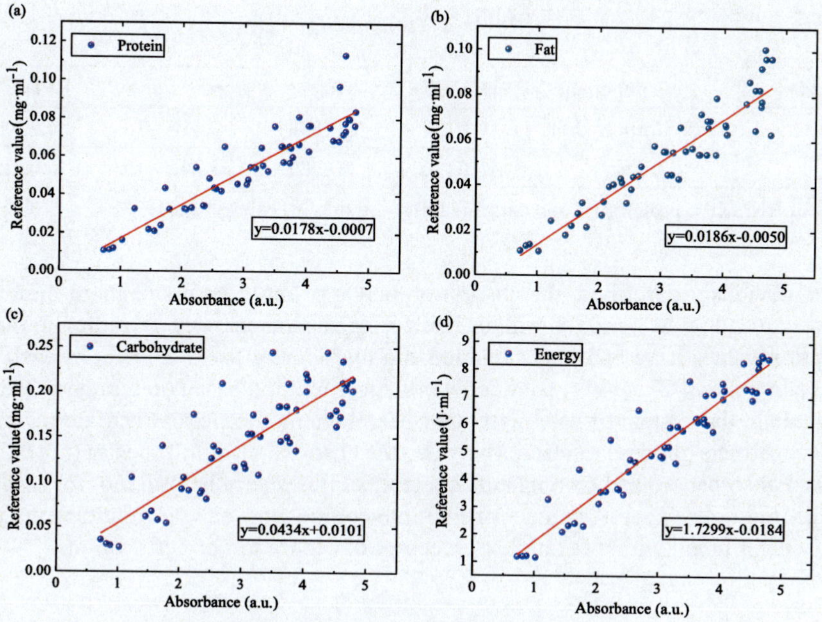

Fig. 5. Work curves (II) of four parameters

Table 2. The prediction results of PLS, MLR and work curve methods

Method	Component	RMSEP#	R^2	p	RPD
PLS	Protein	0.0011	0.9983	9.32E−21	18.2507
	Fat	0.0021	0.9863	1.94E−14	7.9554
	Carbohydrate	0.0099	0.9888	4.69E−15	5.6213
	Energy	0.2274	0.9939	6.70E−17	8.3734
MLR	Protein	0.0044	0.9788	4.12E−13	4.5846
	Fat	0.0035	0.9642	1.61E−11	4.8936
	Carbohydrate	0.0172	0.9002	2.17E−08	3.2450
	Energy	0.3115	0.9722	2.74E−12	6.1120
Work curve (I)	Protein	0.0066	0.9497	1.76E−10	3.0675
	Fat	0.0056	0.9450	3.26E−10	3.0426
	Carbohydrate	0.0222	0.9289	1.99E−09	2.5201
	Energy	0.3816	0.9859	2.31E−14	4.9893
Work curve (II)	Protein	0.0032	0.9884	6.07E−15	6.3052
	Fat	0.0059	0.9394	6.49E−10	2.8988

(continued)

Table 2. *(continued)*

Method	Component	RMSEP#	R^2	p	RPD
	Carbohydrate	0.0132	0.9889	4.41E−15	4.2399
	Energy	0.3156	0.9894	3.24E−15	6.0325

Unit of RMSEPs: protein, fat and carbohydrate: mg mL^{-1}, energy: J mL^{-1}.

As previously described, the absorption bands of multi-components in milk powder overlap and interference as well as interfering substances, which leading to the UV absorption values have bad linear relationship (association) with content of each individual component. Therefore, common work curve method based on Lambert-Beer law is suitable to the system of sole ingredient. Here, chemometric multivariate regression model constructs good relationships between the absorbance of full spectra (not only the maximal absorbance) and components content, so it is a very helpful tool for resolving complex mixtures. That is, content of multiple components can be simultaneously and quickly determined without complex processes of separation or purification.

5 Conclusion

Ultraviolet spectroscopy combined with PLS model was successfully used to determine the content of main ingredients including protein, fat, carbohydrate and energy in milk powder. Without any pretreatment process, PLS model can simultaneously predict the main parameters of milk powder, which presents a fast, simple, accurate and robust tool that could be studied further and extended to detect the composition of milk liquid and even other food. Furthermore, this study could promote the researches and applications of both chemometric quantitative models and UV spectroscopy.

Acknowledgments. This work is supported by the Fundamental Research Funds for the Central Universities (No. 2017MS135).

References

1. IDF: Dried milk, dried ice-mixes and processed cheese-determination of lactose content-Part 1: Enzymatic method utilizing the glucose moiety of the lactose. IDF 79-1. Brussels, Belgium: International Dairy Federation (2002)
2. IDF: Milk products and milk-based foods-determination of fat content by the Weibull-Berntrop gravimetric method (Reference method)-Part 1: Infant foods. IDF 124-1. Brussels, Belgium: International Dairy Federation (2005)
3. IDF: Milk and milk products-determination of lactose content by high-performance liquid chromatography (Reference method). IDF 198. Brussels, Belgium: International Dairy Federation (2007)
4. IDF: Milk-determination of fat content. IDF 226. Brussels, Belgium: International Dairy Federation (2008)

5. IDF: Milk and milk products-determination of nitrogen content-Part 4: Determination of protein and non-protein nitrogen content and true protein content calculation (Reference method). IDF 20-4. Brussels, Belgium: International Dairy Federation (2016)
6. Azad, T., Ahmed, S.: Common milk adulteration and their detection techniques. Int. J. Food Contam. **3**, 22–30 (2016)
7. Forcato, D.O., Carmine, M.P., Echeverria, G.E., Pecora, R.P., Kivatinitz, S.C.: Milk fat content measurement by a simple UV spectrophotometric method: an alternative screening method. J. Dairy Sci. **88**, 478–481 (2005)
8. Genisheva, Z., Quintelas, C., Mesquita, D.P., Ferreira, E.C., Oliveira, J.M., Amaral, A.L.: New PLS analysis approach to wine volatile compounds characterization by near infrared spectroscopy (NIR). Food. Chem. **246**, 172–178 (2018)
9. Li, P., et al.: Food science & nutrition, a simple and nondestructive approach for the analysis of soluble solid content in citrus by using portable visible to near-infrared spectroscopy. Food Sci. Nutr. **8**, 2543–2552 (2020)
10. Liu, C., Wang, Q.Y., Huang, W.Q., Chen, L.P., Yang, G.Y., Wang, X.B.: Measurement of light penetration depth through milk powder layer in Raman hyperspectral imaging system. Spectrosc. Spect. Anal. **37**, 3010–3017 (2017)
11. Pang, J.F., Tang, C., Li, Y.K., Xu, C.R., Bian, X.H.: Identification of melamine in milk powder by mid-infrared spectroscopy combined with pattern recognition method. Spectrosc. Spect. Anal. **40**, 3235–3240 (2020)
12. Bian, X.H., Lu, Z.K., Kollenburg, G.V.: Ultraviolet-visible diffuse reflectance spectroscopy combined with chemometrics for rapid discrimination of Angelicae Sinensis Radix from its four similar herbs. Anal. Methods **12**, 3499–3507 (2020)
13. Rukke, E.O., Olsen, E.F., Devold, T., Vegarud, G., Isaksson, T.: Comparing calibration methods for determination of protein in goat milk by ultraviolet spectroscopy. J. Dairy Sci. **93**, 2922–2925 (2010)
14. Ma, X.P., Pang, J.F., Dong, R.N., Tang, C., Shu, Y.X., Li, Y.K.: Rapid prediction of multiple wine quality parameters using Infrared spectroscopy coupling with chemometric methods. J. Food Compos. Anal. **91**, 103509 (2020)
15. Li, Y.K., Zeng, X.C.: Serum SELDI-TOF MS analysis model applied to benign and malignant ovarian tumor identification. Anal. Methods. **8**, 183–188 (2016)
16. Liu, Y., et al.: Discriminating geographic origin of sesame oils and determining lignans by near-infrared spectroscopy combined with chemometric methods. J. Food Compost Anal. **84**, 103327 (2019)
17. Han, L., Cui, X.Y., Cai, W.S., Shao, X.G.: Three–level simultaneous component analysis for analyzing the near–infrared spectra of aqueous solutions under multiple perturbations. Talanta **217**, 121036 (2020)
18. Yun, Y.H., Li, H.D., Deng, B.C., Cao, D.S.: An overview of variable selection methods in multivariate analysis of near-infrared spectra. Trac-Trend Anal. Chem. **113**, 102–115 (2019)
19. Li, Y.K., Jing, J.: Consensus PLS method based on diverse wavelength variables models for analysis of near-infrared spectra. Chemometr. Intell. Lab. Syst. **130**, 45–49 (2014)
20. Luke, B., Lisa, M., Angelo, S., MariaJose, O.C., Carol, W.: Analysis of seven salad rocket (Eruca sativa) accessions: the relationships between sensory attributes and volatile and non-volatile compounds. Food Chem. **218**, 181–191 (2017)
21. Wang, F., Zhao, C.J., Yang, G.J.: Development of a non-destructive method for detection of the juiciness of pear via VIS/NIR spectroscopy combined with chemometric methods. Foods **9**, 1778–1793 (2020)
22. Liang, Y.Z., Wu, H.L., Yu, R.Q.: Handbook of Analytical Chemistry: Chemometrics. Chemical Industry Press, Beijing (2016)

23. Brown, J., et al.: Metabolizable energy of high non-starch polysaccharide-maintenance and weight-reducing diets in men: experimental appraisal of assessment systems. J. Nutr. **128**, 986–995 (1998)
24. Wold, S., Albano, C., Dunll, W.J.I., Esbensen, K., Hellberg, S.: Pattern recognition: finding and using regularities in multivariate data food research, how to relate sets of measurements or observations to each other. Analysis Applied Science Publication, London (1983)
25. Lei, F., Zhu, S.S., Chen, S.S., Bao, Y., He, Y.: Combining fourier transform mid-Infrared spectroscopy with chemometric methods to detect adulterations in milk powder. Sensors **19**, 2934–2948 (2019)
26. Hamid, Z.A.: Evaluation of multivariate linear regression and artificial neural networks in prediction of water quality parameters. J. Environ. Health Sci. Eng. **12**, 1–8 (2014)
27. Williams, P.C., Sobering, D.: How do we do it: A brief summary of the methods we use in developing near infrared calibration. In: Near Infrared Spectroscopy: The Future Waves. NIR Publications, Chichester (1996)
28. Ferreira, D.S., Pallone, J.A.L., Poppi, R.J.: Fourier transform near-infrared spectroscopy (FT-NIRS) application to estimate Brazilian soybean [Glycine max (L.) Merril] composition. Food Res. Int. **51**, 53–58 (2013)

Nondestructive Analysis of Soluble Solids Content in Apple with a Portable NIR Spectrometer

Cheng Guo[1](✉), Cuiyan Han[1], Hui Yan[2], and Lei Li[3]

[1] School of Pharmacy, Jiangsu Vocational College of Medicine, Yancheng 224005, China
12263@jsmc.edu.cn
[2] School of Biotechnology, Jiangsu University of Science and Technology, Zhenjiang 212018, China
[3] Fiberhome Telecommunication Technologies Co., Ltd., Nanjing 210000, China

Abstract. Soluble solids content (SSC) in apples was determined by a portable near-infrared (NIR) spectrometer combined with chemometrics methods. To build a stable partial least squares (PLS) model, spectral pretreatment and variable selection methods were considered in this work. The result showed that the best spectral pretreatment method was the combination of Savitzky-Golay smoothing, first-order derivative, autoscale, standard normal variate, mean center. Variable selection method competitive adaptive reweighted sampling (CARS) achieved the best performance. Our work could be a useful tool for the fruit grading and the post-harvest management.

Keywords: Fruit · Nondestructive analysis · Portable spectrometer · Variable selection · Partial least squares

1 Introduction

Apple is a is a kind of fruit with different size, color, shape, and chemical composition. It is hard to determine the harvest time and sorting the fruit based on the quality using the conventional analytical techniques [1, 2]. Usually soluble solids content (SSC), as an important index to the ripeness of fruits, is the soluble sugar content including monosaccharide, oligosaccharide and polysaccharide. Therefore, SSC has been considered as the main index to evaluate the fruit and confirm the right harvest time. Ordinarily the SSC of the fruit was measured by using a destructive method. This method was accurate but it is time consuming and destructive.

Near-infrared (NIR) spectroscopy, as a rapid and nondestructive technology, has been demonstrated to be useful for quantitative analysis of complex samples in numerous fields, such as environmental, agricultural, and pharmaceutical analysis [3–5]. Partial least squares (PLS) is the most commonly used method in the NIR spectra modeling. However, uninformative variables may be contained in the spectral matrix and lead to

X. Chu et al. (Eds.): ICNIR 2021, *Sense the Real Change: Proceedings of the 20th International Conference on Near Infrared Spectroscopy*, pp. 157–161, 2022.
https://doi.org/10.1007/978-981-19-4884-8_16

inaccurate model performance. Therefore, variable selection methods are usually needed before building the model [6, 7].

In this work, we use a portable NIR spectrometer to measure the SSC value in apples rapidly and nondestructively. To build a robust model, spectral pretreatment and variable selection methods would be considered in the next section.

2 Materials and Methods

A total of 90 Fuji apples were purchased from the local supermarket in Zhenjiang. NIR spectra of apple samples were measured in the diffuse reflectance mode with DLP NIRScan Nano (Texas Instruments, USA). Each NIR spectrum was recorded in the wavelength range of 900–1650 nm. Each spectrum is composed of 228 data points. The SSC value was measured using a portable refractometer (WZ-103, TOP Instrument Co., Ltd, Zhejiang, China) based on National Standard Methods GB 12295-90. All samples were divided into calibration and validation set according to Kennard-Stone (KS) algorithm [8].

Spectral pretreatment can improve performance. In order to obtain reliable, accurate and stable calibration models, it is necessary to pretreat raw NIR spectra. In this work, Savitzky-Golay smoothing (SG), first-order derivative (1D), autoscale (A), standard normal variate (SNV), mean center (MC) and their combinations were studied.

Partial least squares (PLS) has been used widely in NIR spectra modeling. It takes the matrix **X** (spectra) and matrix **y** (properties such as Brix) into account, extracting maximum information from highly correlated raw data with a linear relationship and condensing them into potential variables. The valid information in spectra is very weak, some wavelength bands lack correlation with sample component. Even process does not work, therefore, optimized wavelength variables before modeling is necessary which could simplify models and improve prediction ability. In this work, random frog (RF) [9], uninformative variable elimination (UVE) [10] and competitive adaptive reweighted sampling (CARS) [11] were used to select the important variables for PLS modeling. Correlation coefficient (R for short) and root mean square error (RMSE) are indicators for evaluating performance of models. A good model should have higher R values and lower RMSE values.

3 Results and Discussion

In order to build a stable model, the data set should be partitioned as calibration and validation sets. Statistics of SSC in calibration and validation sets is shown in Table 1. Calibration and validation sets contain 60 and 30 samples respectively. The range of SSC in calibration covers the validation range. The mean values of SSC for calibration and validation set were 14.34 and 14.37 respectively. The standard deviation (S.D.) of calibration and validation set were 1.49 and 1.45 respectively. The coefficient of variance (CV, %) of calibration and validation set were 10.39 and 10.06 respectively.

As shown in Fig. 1, the peak around 950 nm was assigned to the second overtone of O-H functional group exists in H_2O and carbohydrates. The peak around 1180 nm was

Table 1. Statistical information of SSC (°Brix) in different sample sets.

Data	Number	Range (SSC, °Brix)	Mean (SSC, °Brix)	S.D	CV (%)
Calibration	60	9.10–17.80	14.34	1.49	10.39
Validation	30	10.50–17.70	14.37	1.45	10.06

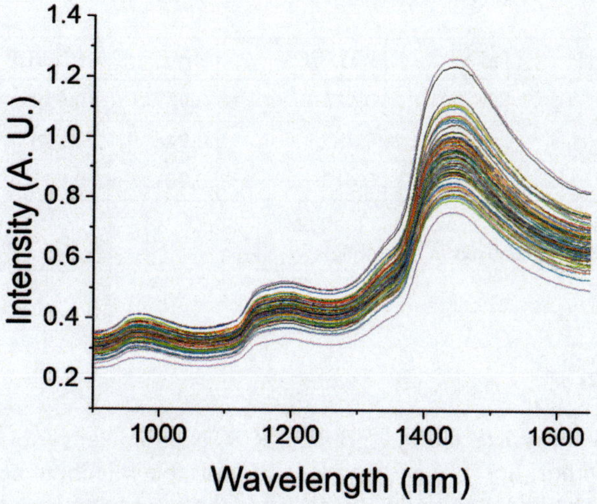

Fig. 1. Raw NIR spectra of apple samples

assigned to the second overtone of C-H functional group. The peak around 1450 nm was assigned to the first overtone of O-H functional group.

To improve the prediction results, five spectral pretreatment methods were compared in this work. PLS model was built with the pretreated spectra and the number of latent variables (*n*LV) was set according to the lowest RMSECV value. The results were shown in Table 2. Clearly, the spectral pretreatment method with the combination of SG, 1D, SNV, MC can improve the model performance.

Table 2. Results of PLS modeling with different spectral pretreatment methods.

Methods	nLV[a]	RMSECV	R_{CV}	RMSEP	R_P
SG, A	12	0.737	0.873	0.673	0.887
SG, 1D	8	0.713	0.879	0.651	0.894
SG, SNV	13	0.682	0.893	0.587	0.914
SG, MC	13	0.764	0.866	0.609	0.904
SG, 1D, SNV, MC	**11**	**0.641**	**0.885**	**0.569**	**0.917**

[a]Number of latent variables used in the models

To further improve the model performance, three variable selection methods were used in this work. PLS model was used in the calculation. The results of three variable selection methods were shown in Table 3. CARS shows the best performance with the lowest *n*LV and *n*Val, and achieves the lowest RMSECV, which means the model is more stable compared with the others.

Table 3. The results of RF, CARS and UVE.

Methods	*n*Val[a]	*n*LV[b]	RMSECV	R_{CV}	RMSEP	R_P
RF	50	9	0.516	0.937	0.539	0.929
CARS	**20**	**6**	**0.487**	**0.944**	**0.558**	**0.922**
UVE	99	6	0.642	0.901	0.631	0.902

[a]Number of spectral variables used in the models.
[b]Number of latent variables used in the models.

4 Conclusion

SSC value in apple was detected by a portable NIR spectrometer system combined with chemometric methods. Spectral pretreatment and variable selection methods should be considered before building the model. In this work, the best spectral pretreatment method was the combination of SG, 1D, SNV, MC and determined by lowest RMSECV. In the case of variable selection, CARS shows the best performance. Our work could be a useful tool for the fruit grading and the post-harvest management.

Acknowledgments. This work was supported by Scientific Research Start-up Fund of Jiangsu Vocational College of Medicine (No. 20216112).

References

1. Huang, H., Yu, H., Xu, H., Ying, Y.: Near infrared spectroscopy for on/in-line monitoring of quality in foods and beverages: a review. J. Food Eng. **87**, 303–313 (2008)
2. Alamar, M.C., Bobelyn, E., Lammertyn, J., Nicolaï, B.M., Moltó, E.: Calibration transfer between NIR diode array and FT-NIR spectrophotometers for measuring the soluble solids contents of apple. Postharvest Biol. Technol. **45**, 38–45 (2007)
3. Pasquini, C.: Near infrared spectroscopy: a mature analytical technique with new perspectives-a review. Anal. Chim. Acta **1026**, 8–36 (2018)
4. Shao, X.G., Bian, X.H., Liu, J.J., Zhang, M., Cai, W.S.: Multivariate calibration methods in near infrared spectroscopic analysis. Anal. Methods **2**, 1662–1666 (2010)
5. Kumar, N., Bansal, A., Sarma, G.S., Rawal, R.K.: Chemometrics tools used in analytical chemistry: an overview. Talanta **123**, 186–199 (2014)
6. Cai, W., Li, Y., Shao, X.: A variable selection method based on uninformative variable elimination for multivariate calibration of near-infrared spectra. Chemom. Intell. Lab. Syst. **90**, 188–194 (2008)

7. Yun, Y.H., Li, H.D., Deng, B.C., Cao, D.S.: An overview of variable selection methods in multivariate analysis of near-infrared spectra. TrAC Trends Anal. Chem. **113**, 102–115 (2019)
8. Kennard, R.W., Stone, L.A.: Computer aided design of experiments. Technometrics **11**, 137–148 (1969)
9. Yun, Y.H., et al.: An efficient method of wavelength interval selection based on random frog for multivariate spectral calibration. Spectrochim. Acta A Mol. Biomol. Spectrosc. **111**, 31–36 (2013)
10. Centner, V., Massart, D.L., de Noord, O.E., de Jong, S., Vandeginste, B.M., Sterna, C.: Elimination of uninformative variables for multivariate calibration. Anal. Chem. **68**, 3851–3858 (1996)
11. Li, H., Liang, Y., Xu, Q., Cao, D.: Key wavelengths screening using competitive adaptive reweighted sampling method for multivariate calibration. Anal. Chim. Acta **648**, 77–84 (2009)

Aquaphotomics

The Aquaphotomics and E-nose Approaches to Evaluate the Shelf Life of Ready-To-Eat Rocket Salad

L. Marinoni[1(✉)], G. Bianchi[1], and T. M. P. Cattaneo[1,2]

[1] Research Centre for Engineering and Agro-Food Processing, Council for Agricultural Research and Economics, 20133 Milan, Italy
laura.marinoni@crea.gov.it
[2] DAFNE, Tuscia University, 01100 Viterbo, Italy

Abstract. The shelf life of ready-to-eat rocket salad packed under modified atmospheres was evaluated. Freshly cut rocket salad was packed in plastic bags under three modified atmospheres (A = atmospheric air; B = 30% O_2, 70% N_2; C = 10% CO_2, 5% O_2, 85% N_2). At t = 0 and after 0, 1, 4, 7, 11 and 13 days NIR spectra were collected with a microNIR OnSite-W spectrometer (VIAVI Srl, Italy) in reflectance mode, over the range 900–1600 nm (50 scans; 125 reading points). Aquaphotomic approach was used to evaluate the maintenance of product freshness studying the changes in the water absorption profile. Samples were also analyzed by a Portable Electronic Nose PEN3 (AIRSENSE Analytics GmbH, Germany) with a sensor array composed of 10 metal oxide semiconductor (MOS) type chemical sensors. The PCA, applied 1300–1600 nm region, allowed the samples grouping according to the storage time. The obtained Aquagrams showed shifts of the selected water absorptions bands during the shelf life, estimating a first loss of freshness after 4–7 days from packaging for A and C theses, and after 7–11 days for B thesis. Similarly, E-nose detected important variations in MOS sensors data after 7 days for A and C theses, and after 11 days for the B one. NIR and E-nose results agreed in identifying the B modified atmosphere as the best for maintaining the product freshness. The B composition, characterized by a high O_2 concentration, seemed to be able to lengthen by about 3 days the shelf life of the ready-to-eat rocket salad.

Keywords: Shelf life · Rocket salad · Freshness · Water · E-nose · Portable instrument

1 Introduction

Consumption of ready-to-eat (RTE) leafy vegetables has increased rapidly due to changes in consumer behavior. RTE products are perceived as natural, fresh, aromatic, convenient, high-quality and with health benefits [1, 2]. However, the processing steps, such as cutting and peeling, involved in preparing the fresh-cut products can cause severe tissue damage, leading to rapid quality deterioration [3]. Physical damage caused by

X. Chu et al. (Eds.): ICNIR 2021, *Sense the Real Change: Proceedings of the 20th International Conference on Near Infrared Spectroscopy*, pp. 165–173, 2022.
https://doi.org/10.1007/978-981-19-4884-8_17

postharvest processing brings increase in the microbial spoilage, biochemical changes, and respiration rates. These changes imply a degradation of color, texture, and flavor of the RTE product [2].

Many post-harvest technologies, such as refrigeration and modified atmosphere packaging (MAP), are intended to delay senescence [4]. They are successful in preserving visual quality and microbial safety of minimally processed fruit and vegetables during the supply chain [5]. Maintaining low temperatures −5 °C or below - is recommended for packaged fresh-cut leafy green vegetables (such as rocket) during transport, storage and retail display [6]. MAP is effective in prolonging the shelf-life of fresh-cut produces by modifying the ratios between gases within the packaging. Several factors contribute to the modification of the atmosphere: the composition of the gas mixture injected into the package; the produce respiration rate (which itself is affected by temperature, produce type, variety, size, maturity and severity of preparation); the packaging film permeability [2, 7]. Successful applications of MAP with low O_2 and high CO_2 for minimally processed fruits and vegetables have been extensively reported in the literature [2] as low O_2 concentrations decrease respiration, inhibit the growth of postharvest pathogens and therefore decrease the rate of deterioration. However, very low levels of O_2 may induce anaerobic fermentation with the corresponding accumulation of unpleasant odors undesirable tastes and damage on tissues [2, 4, 8]. Moreover, the presence of a very high CO_2 concentration (25%) in the storage atmosphere has been found to be deleterious for fresh-cut artichokes, while only slight beneficial effects were observed at lower concentrations (5 and 15%) [9].

The monitoring of the quality of RTE vegetables is crucial, especially in relation to safety issues [10]. Not only the quality of the product at the end of its shelf life, but also during storage is appropriate to evaluate. The conventional techniques used to monitor and extend the shelf life of fresh-cut fruit and vegetables involve destructive measurements, such as texture evaluation; respiration rate and pH determinations; microbiological and sensory analyses; nutritional and antioxidants status determinations [11]. These methods are generally expensive, slow, and are not suited to automation. Near infrared (NIR) spectroscopy in combination with chemometrics and Aquaphotomics represents a powerful, rapid, and non-destructive analytical tool to monitor the quality of packaged foods by evaluating the changes occurring during the storage [12].

Aquaphotomics is a recent scientific discipline based on NIR measurements and multivariate spectral analysis that investigates the water–light interactions in biological systems, exploiting the fact that changes in the water matrix reflect, like a mirror, the rest of the molecules the water surrounds [13, 14]. The Aquaphotomics approach is based on the high sensitivity of water's hydrogen bonds that reflect any change of the aqueous system highlighting perturbations that can be observed, measured, analysed and interpreted [14]. According to this approach, the NIR spectra acquired in living systems under various perturbations (temperature, ion concentrations, oxidative stress, illumination, disease, and damage) are characterized by 12 water absorption ranges (6–20 nm width each) in the spectral region of the first overtone of water (1300–1600 nm). Such spectral ranges have been called Water Matrix Coordinates (WAMACs) and labelled Ci, i = 1–12. Within the WAMACs, specific water absorbance bands are related to specific water molecular conformations (water species and water molecular structures) [13].

When a certain perturbation of interest is shown to produce the changes at specific water absorbance bands, and when this is determined consistently and repeatedly throughout the Aquaphotomics analysis, these water absorbance bands (WABS) are considered 'activated' by the respective perturbation. The selected WABS are plotted in spider charts, named 'Aquagrams', to observe the Water Absorbance Spectral Patterns (WASPs) [15].

The aim of this work was to evaluate the maintenance of product freshness of ready-to-eat rocket salad packed under modified atmospheres during the shelf life. Rapid and non-destructive techniques were employed, such as NIR spectroscopy to study the changes in the water absorption profile in the range 1300–1600 nm (Aquaphotomics) and a Portable Electronic Nose. The influence of modified atmospheres on the product shelf life was also investigated.

2 Materials and Methods

Freshly harvested rocket leaves (*Diplotaxis tenuifolia* L.) were purchased from a local grower and immediately transported to the CREA IT lab. Rocket leaves were then inspected for impurity and visual defects, discarding damaged and dehydrated leaves. The product was then dipped in a 1% sodium hypochlorite solution for 1 min, rinsed with cold water, and centrifuged to remove the remaining water. Rocket leaves (ca. 60 g) were packed in polypropylene film bags, and each bag was sealed under atmospheric and modified headspace conditions. The following gas mixtures were used: thesis A, atmospheric air (21% O_2, 78% N_2); thesis B, 30% O_2, 70% N_2; thesis C, 10% CO_2, 5% O_2, 85% N_2. Three bags were prepared for each evaluation day for each thesis. All samples were stored at 4 °C for 13 days and subjected to NIR and E-nose analysis at scheduled sampling points: 0, 1, 4, 7, 11 and 13 days for NIR analyses; 1, 4, 7, 11 and 13 days for E-nose analyses.

The E-nose measurements were performed in triplicate on 3 bags for each thesis trough a pierceable Silicon/Teflon disk, using a commercial portable electronic nose (PEN3, Win Muster Airsense Analytic Inc., Schwerim, Germany). The instrument is made up of a sampling apparatus, a detector unit containing the sensor array and pattern-recognition software (Win Muster v.16) for data recording. The sensor array is composed of 10 metal oxide semiconductor (MOS) type chemical sensors: W1C (aromatic compounds), W5S (broad range), W3C (aromatic compounds), W6S (hydrogen), W5C (alkanes, aromatic compounds, less polar compounds), W1S (broad range, methane), W1W (terpenes, sulfur organic compounds, limonene, pyrazine), W2S (alcohols, broad range), W2W (aromatic compounds, sulfur organic compounds), and W3S (methane aliphatics, reacts on high concentrations >100 mg/kg). The sensor response is given by the ratio between the conductivity response of the sensors to both the sample gas (G) and the carrier gas (G0) over time (G/G0). The E-nose analyses were performed following the conditions reported by Vanoli et al. [16]. For each E-nose run, the conductivity G/G0 of the 10 sensors at the time corresponding to the normalized maximum of all signals was taken as the vector of sensors signal. The average of the runs of each replicate was used for statistical analysis. Data were subjected to one-way analysis of variance (ANOVA) for means comparison using Statgraphics ver. 5.1 (Manugistic Inc, Rockville, MD, USA) software package. The means were separated using a Tukey's HSD test, and their statistical significance was determined at 5% ($p < 0.05$) level. The E-nose data were also analyzed by principal component analysis (PCA) with the Unscrambler software package (v 9.7, Camo, Inondhcim, Norway).

Near infrared spectra were collected in reflectance mode using the MicroNIR OnSite-W (VIAVI Solutions Italia S.r.l., Monza, Italy) portable spectrometer. Spectra acquisition took place in the spectral region between 908 and 1676 nm (125 data points per spectrum with pixel-to-pixel interval of 6.2 nm; each spectrum given by the software is the average of 50 scans) on 3 bags for each thesis, on both the two sides of the bags. Ten replicates for each side were collected, for a total of 60 spectra for each thesis. Spectra of PET packaging (#10) were also collected, and the average spectrum was subtracted from each sample spectrum. In order to avoid temperature-related spectral differences, prior to analyses samples were kept at room temperature for 15 min. Before the analysis, the instrument underwent calibration for the black, on the air, and for the white, on the supplied standard tile.

Chemometric analysis on NIR data was performed using The Unscrambler software package (v 9.7, Camo, Inondhcim, Norway). Spectra were pre-processed using the Moving Average smoothing (gap size 15 points) and the first derivative Norris Gap transformation (gap size 21 points). An explorative PCA was applied on the averaged pretreated spectra in the range of the first overtone of water (1300–1600 nm). Spectra were also pretreated according to Tsenkova et al. [15]. The second derivative Savitzky–Golay filter (second order polynomial fit and 21 points) and multiplicative scatter correction (MSC) were applied to absorbance spectra to remove potential scatter effects. The transformed spectra were then normalized by applying the following formula:

$$(A\lambda - \mu\lambda)/\sigma\lambda$$

where $A\lambda$ is the transformed absorbance, $\mu\lambda$ is the mean value of all spectra and $\sigma\lambda$ is the standard deviation of all spectra at wavelength λ, and properly averaged using MS Excel®.

The Aquagrams were then built up using 12 wavelengths selected from the normalized spectra: 1342 nm, 1366 nm, 1373 nm, 1385 nm, 1397 nm, 1410 nm, 1435 nm, 1447 nm, 1466 nm, 1472 nm, 1490 nm, and 1515 nm.

3 Results and Discussion

The E-nose was applied in order to evaluate the evolution of the volatile compounds profile of rocket salad during storage. In order to evaluate the ability of the electronic nose to discriminate among the different samples during the shelf life, the sensor responses were elaborated by PCA. The first two principal components accounted for 99% of the total variance. The biplot for scores and factor loadings is reported in Fig. 1, which illustrates the mutual relationships between samples and sensors.

From the figure it was evident the important influence of the three wide range sensors, W5S, W1S and W2S, on the characterization of the samples. In particular, the W5S sensor mainly characterized the samples of thesis C in the last checkpoints; W1S and W2S mostly characterized the samples of theses A and B at the end of the shelf life.

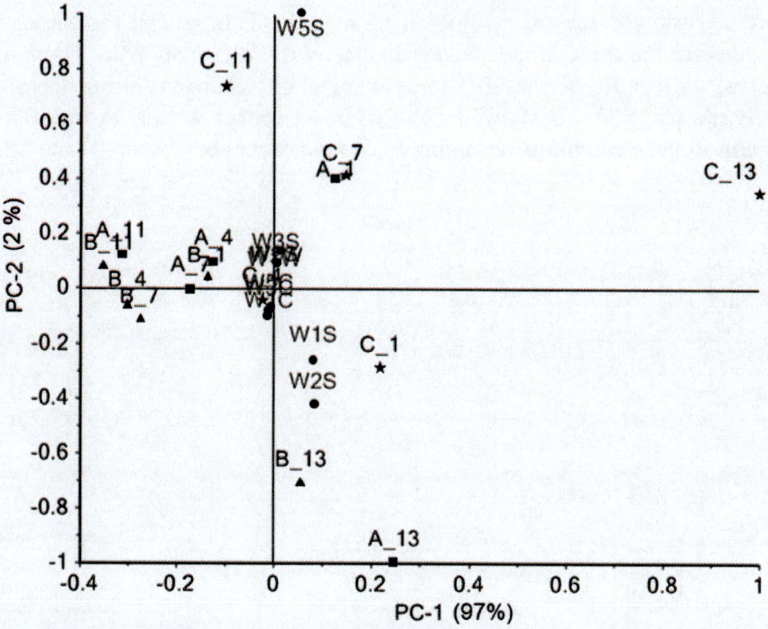

Fig. 1. Biplot of E-nose sensor responses (loadings; dots) and samples scores (average values; A treatment = squares; B treatment = triangles; C treatment = stars) during the storage of rocket salad.

In general, thesis C samples showed a peculiar pattern and seemed to be more characterized by these three sensors at all sampling points. These sensors are sensitive towards a wide range of compounds, methane, and alcohols, indicating the presence of such compounds in the atmosphere of thesis C. Among alcohols, ethanol, for example, is considered a major fermentative metabolite It is reported to accumulate in response to low O_2 and/or high CO_2 treatment during controlled atmosphere storage of broccoli florets [17] and modified atmosphere packaged fresh-cut artichoke [8]. Methanol has also been reported to be released under restricted O_2 conditions. Its formation could derive from enzymatic degradation of pectin by the enzyme pectin methyl esterase into methanol and other constituents [18]. The PCA interpretation, where the thesis C was mainly described by W5S and characterized by low oxygen concentrations, was in agreement with literature data reported above. It can be assumed that under these conditions, the lack of O_2 for aerobic respiration led to a switch to anaerobic respiration, with release of alcohols.

All other sensors were located close to the axes intersection and described samples during the first days of storage. Such sensor is sensitive to aromatic compounds, alkanes, terpenes, and sulphur organic compounds. These findings were in accordance with Mastrandrea et al. who studied the changes in volatile compounds responsible for flavour in wild rocket stored in MAP conditions. They found volatile compounds included sulphur, C6 and C5 compounds, acetaldehyde, isothiocyanate and thiocyanate derivatives. These compounds are responsible for the typical odour and flavour notes of fresh rocket [19].

Figure 2a) showed the profiles obtained with the E-nose. All the theses showed higher values for the three broad range sensors, W5S, W1S and W2S. Comparing the three theses, the profile of thesis C underwent major changes during the shelf life. Conversely, theses A and B showed more constant sensor values. Profile B reflected minimal atmosphere variations occurring inside the rocket bag.

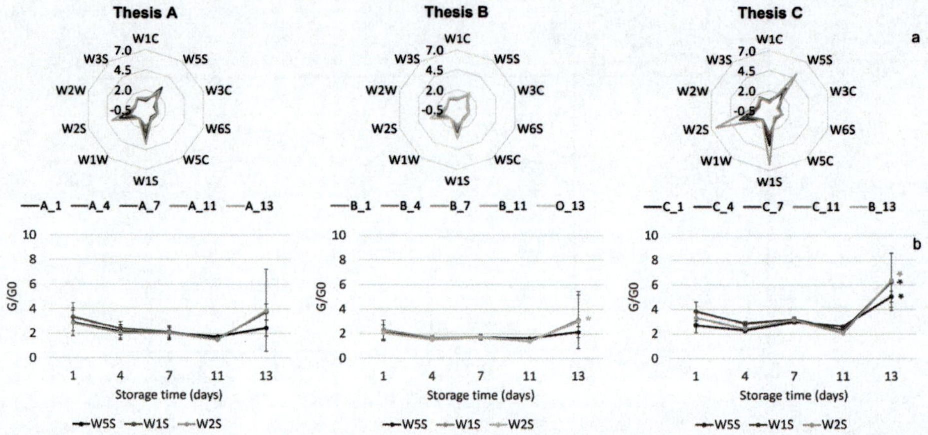

Fig. 2. E-nose profiles (a) and trend curves of W5S, W1S and W2S sensors (b) of the three theses during shelf life. Data are expressed as mean value of three replicates. Coloured asterisks indicate significant differences in the corresponding sensor signals according to ANOVA and Tukey's post hoc test ($p < 0.05$).

Observing in detail the trend of only the three most involved sensors (Fig. 2b), thesis A and B showed a similar trend, with constant values until 11 days followed by an increase at 13 days. This increase was significant only for the W2S sensor in thesis B. Thesis C showed more fluctuating trends for all the three sensors, with a marked and significant increases at 13 days and higher signal values during the whole storage. Thesis B was also characterized by the lowest signal values and therefore seemed to be the best solution to minimize the metabolic variations occurring during the shelf life.

Figure 3 depicted the NIR spectra of the packed rocket salad samples. The raw spectra (Fig. 3a) showed broad and overlapped absorption bands detecting predominant water absorptions at 960–990 (second overtone –OH stretching), 1150–1170 (combination of first overtone –OH stretching and OH bending) and 1430–1440 nm (first overtone –OH stretching) [15]. The application of pre-treatments made the peaks narrower and more defined, allowing their easier identification. In particular, Fig. 3b highlighted the presence of wavelength shifts with a high variability in the 1100–1500 nm NIR range. These findings supported the hypothesis that Aquaphotomics could be helpful to discriminate between the different theses.

The explorative PCA explained 98% of the total variance showing a sample grouping according to the storage time (Fig. 4): i) t0 and t1 samples; ii) t4 samples; iii) t7 to t13 samples. Noteworthy is that the sample B at 7 days located together with t4 samples.

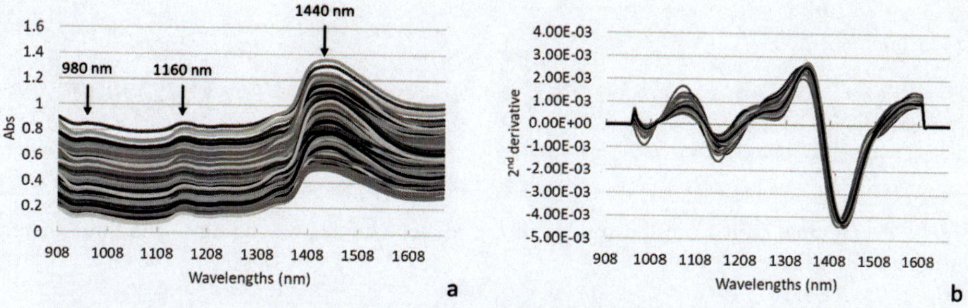

Fig. 3. Examples of NIR raw (a) and pre-treated (b) spectra of packed rocket salad.

This suggested that the characteristics of the biosystem of the 4 days samples were also maintained for 7 days rocket samples stored in atmosphere B.

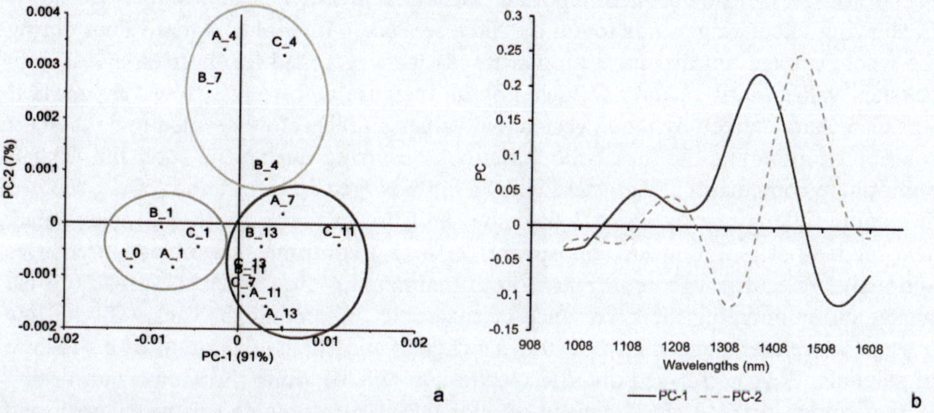

Fig. 4. PCA score (a) and loading (b) plots of pre-treated NIR spectra of rocket salads (each point corresponded to the mean spectra for each checkpoint).

According to the loadings plot (Fig. 4b), the separation along PC1 between fresh and longer-stored samples was mainly based on wavelength at 1376 nm while the separation along PC2 was due to 1460 nm. According to Tsenkova et al. [15], these wavelengths correspond to the absorbance of the free OH stretch and of strongly hydrogen bonded water, respectively.

The Aquagrams highlighted differences among the samples, giving rise to different WASPs for each sampling time and for each thesis (Fig. 5).

For theses A and C, a shift from left to right occurred up to the 4th day, then the graph went back to the right. For thesis B, the inversion occurred on the seventh day. From these premises, it can be assumed that Aquaphotomics estimated a first loss of freshness after 4 days from packaging for A and C theses, and after 7 days for B thesis. This suggested that thesis B was the best at preserving the freshness of rocket salad, confirming the E-nose findings.

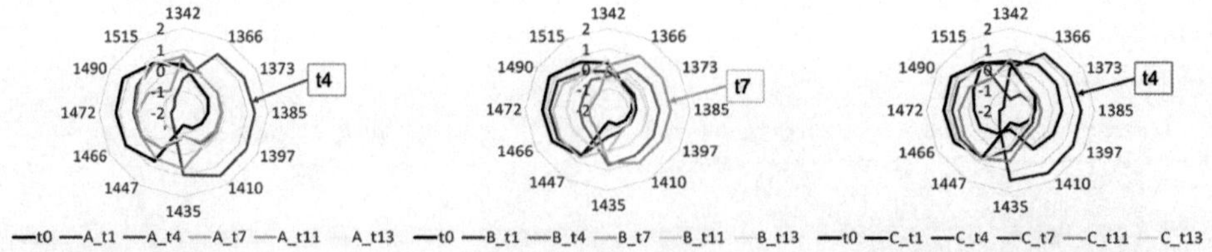

Fig. 5. Aquagrams built from the spectra of the rocket samples belonging to the three theses during the shelf life

4 Conclusion

NIR and E-nose results agreed in identifying the B atmosphere as the best for maintaining the product freshness. Three E-nose sensors - more sensitive towards a wide range of compounds, methane, and alcohols – characterized and described the samples at the end of the storage. E-nose detected important variations in MOS sensors signals for thesis C, showing fluctuating trends for all the three sensors, with higher signal values during the whole storage. Furthermore, similar trends were recorded for thesis A and B with constant values until 11 days followed by an increase at 13 days. However, thesis B was also characterized by the lowest signal values and therefore seemed to be the best solution to minimize the metabolic variations occurring during the shelf life. The B atmosphere composition, characterized by a high oxygen concentration (30%), seemed to be able to lengthen by about 3 days the shelf life of the ready-to-eat rocket salad. The second-best solution (A atmosphere, 21% O_2) confirmed the plausible positive and active role of oxygen concentration in maintaining the freshness of ready to eat rocket and minimizing the occurrence of anaerobic fermentations. Thesis C with low oxygen concentration seemed to lead to a switch to anaerobic respiration with a release of alcohols. PCA performed on NIR spectra was able to group the samples according to the storage time. A good separation along PC1 was obtained between fresh and longer-stored samples, mainly based on wavelength at 1376 nm, corresponding to the absorbance of the free OH stretch. The Aquagrams highlighted differences among the samples, giving rise to different WASPs for each sampling time and for each thesis. From these preliminary findings, it can be assumed that Aquaphotomics estimated a first loss of freshness after 4 days from packaging for A and C theses, and after 7 days for B thesis. This suggested that thesis B was the best at preserving the freshness of rocket salad, according with the E-nose results. The results suggest that Aquaphotomics is more sensitive than other methods for detecting minimal qualitative changes in the product in advance (4–7 days vs 11 days of E-nose), allowing for early detection of any anomalies. Some other destructive parameters, such as consistency, loss of electrolytes, fluorescence, ripening index, are under evaluation to confirm these results obtained using rapid and non-destructive techniques only.

Acknowledgments. Authors thank the Italian Ministry of Agriculture for the funding of the Agridigit project, sub-project Agrofiliere (D.M. 36503/7305/2018, 20/12/2018).

References

1. Pereira, M.J., Amaro, A.L., Oliveira, A., Pintado, M.: Bioactive compounds in ready-to-eat rocket leaves as affected by oxygen partial pressure and storage time: a kinetic modelling. Postharvest Biol. Technol. **158**, 110985 (2019)
2. Ghidelli, C., Pérez-Gago, M.B.: Recent advances in modified atmosphere packaging and edible coatings to maintain quality of fresh-cut fruits and vegetables. Crit. Rev. Food Sci. Nutr. **58**, 662–679 (2018)
3. Kim, J.G., Luo, Y., Tao, Y., Saftner, R.A., Gross, K.C.: Effect of initial oxygen concentration and film oxygen transmission rate on the quality of fresh-cut romaine lettuce. J. Sci. Food Agric. **85**, 1622–1630 (2005)
4. Torales, A.C., Gutiérrez, D.R., Rodríguez, S.C.: Influence of passive and active modified atmosphere packaging on yellowing and chlorophyll degrading enzymes activity in fresh-cut rocket leaves. Food Packag. Shelf Life **26**, 100569 (2020)
5. Waghmare, R.B., Mahajan, P.V., Annapure, U.S.: Modelling the effect of time and temperature on respiration rate of selected fresh-cut produce. Postharvest Biol. Technol. **80**, 25–30 (2013)
6. Yahya, H.N., Lignou, S., Wagstaff, C., Bell, L.: Changes in bacterial loads, gas composition, volatile organic compounds, and glucosinolates of fresh bagged Ready-To-Eat rocket under different shelf life treatment scenarios. Postharvest Biol. Technol. **148**, 107–119 (2019)
7. Day, B.P.F.: Modified atmosphere packaging of fresh fruit and vegetables – an overview. Acta Hortic. **553**, 585–590 (2001)
8. la Zazzera, M., Amodio, M.L., Colelli, G.: Designing a modified atmosphere packaging (MAP) for fresh-cut artichokes. Adv. Hortic. Sci. **29**, 24–29 (2015)
9. la Zazzera, M., Rinaldi, R., Amodio, M.L., Colelli, G.: Influence of high CO_2 atmosphere composition on fresh-cut artichoke quality attributes. Acta Hortic. **934**, 633–640 (2012)
10. Castro-Ibáñez, I., Gil, M.I., Allende, A.: Ready-to-eat vegetables: current problems and potential solutions to reduce microbial risk in the production chain. LWT Food Sci. Technol. **85**, 284–292 (2017)
11. Rico, D., Martín-Diana, A.B., Barat, J.M., Barry-Ryan, C.: Extending and measuring the quality of fresh-cut fruit and vegetables: a review. Trends Food Sci. Technol. **18**, 373–386 (2007)
12. Sánchez, M.-T., Pérez-Marín, D., Flores-Rojas, K., Guerrero, J.-E., Garrido-Varo, A.: Use of near-infrared reflectance spectroscopy for shelf-life discrimination of green asparagus stored in a cool room under controlled atmosphere. Talanta **78**, 530–536 (2009)
13. Tsenkova, R.: Aquaphotomics: dynamic spectroscopy of aqueous and biological systems describes peculiarities of water. J. Near Infrared Spectrosc. **17**, 303–313 (2009)
14. Tsenkova, R.: Aquaphotomics: water in the biological and aqueous world scrutinised with invisible light. Spectrosc. Eur. **22**, 6–10 (2010)
15. Tsenkova, R., Munćan, J., Pollner, B., Kovacs, Z.: Essentials of Aquaphotomics and its chemometrics approaches. Front. Chem. **6**, 363 (2018)
16. Vanoli, M., Grassi, M., Buccheri, M., Rizzolo, A.: Influence of edible coatings on postharvest physiology and quality of Honeydew melon fruit. Adv. Hortic. Sci. **29**, 65–74 (2015)
17. Hansen, M.E., Sørensen, H., Cantwell, M.: Changes in acetaldehyde, ethanol and amino acid concentrations in broccoli florets during air and controlled atmosphere storage. Postharvest Biol. Technol. **22**, 227–237 (2001)
18. Luca, A., Mahajan, P.V., Edelenbos, M.: Changes in volatile organic compounds from wild rocket (*Diplotaxis tenuifolia* L.) during modified atmosphere storage. Postharvest Biol. Technol. **114**, 1–9 (2016)
19. Mastrandrea, L., Amodio, M.L., Pati, S., Colelli, G.: Effect of modified atmosphere packaging and temperature abuse on flavor related volatile compounds of rocket leaves (*Diplotaxis tenuifolia* L.). J. Food Sci. Technol. **54**, 2433–2442 (2017)

Check for updates

Near Infrared Aquaphotomics Evaluation of Nasal Secretions as a Potential Diagnostic Tool for Bovine Respiratory Syncytial Virus (BRSV) Infection

M. Santos-Rivera[1](✉), A. R. Woolums[2], M. Thoresen[2], F. Meyer[1], and C. K. Vance[1]

[1] Department of Biochemistry, Molecular Biology, Entomology and Plant Pathology, Mississippi State University, Mississippi State, MS 39762, USA
jms2033@msstate.edu

[2] Department of Pathobiology and Population Medicine, College of Veterinary Medicine, Mississippi State University, Mississippi State, MS 39762, USA

Abstract. This study evaluated near infrared (NIR) spectra (n = 970) of nasal secretions (NS) from dairy calves (n = 5) challenged with bovine respiratory syncytial virus (BRSV). This pathogen is a common cause of respiratory disease in young calves worldwide and is typically diagnosed by evaluation of the clinical signs, followed by time-consuming serological and molecular methods. More rapid diagnostic methods could improve outcomes for infected calves. The near infrared aquaphotomics evaluation of this biofluid unveiled changes between the spectra (1300–1600 nm) of samples collected during the uninfected (n = 200) and infected (n = 200) stages, specifically identified in the WAMACS (water matrix coordinates) C1, C9, C10, and C11, where water molecules are highly associated with chaotropic solutes in water asymmetrical stretching vibrations (ν_3) and with kosmotropic solutes in water clusters with 2, 3, and 4 hydrogen bonds (S_2, S_3, S_4). These chemical differences were discriminated by PCA-LDA using a leave-one-animal-out approach with averaged percentages of accuracy, sensitivity, and specificity of 90.1 ± 4.3, 88.1 ± 3.8, 92.0 ± 5.5 in the calibration process, respectively. By collecting spectra from nasal secretions, we revealed the potential of NIR spectroscopy in combination with aquaphotomics and chemometrics for the detection of this viral infection in-vivo; as a first step toward developing a rapid in-field diagnostic tool for BRSV infection.

Keywords: Absorbance · Cattle · Early disease diagnostics · Near infrared spectroscopy · Viral infection

1 Introduction

Bovine respiratory syncytial virus (BRSV) is a *Pneumovirus* causing respiratory disease in cattle, contributing to economic losses in this sector worldwide [1–3]. The BRSV infection can be asymptomatic or can present with clinical signs such as fever and cough

X. Chu et al. (Eds.): ICNIR 2021, *Sense the Real Change: Proceedings of the 20th International Conference on Near Infrared Spectroscopy*, pp. 174–183, 2022.
https://doi.org/10.1007/978-981-19-4884-8_18

with a seromucoid nasal and ocular discharge [2]. The clinical diagnosis of respiratory infection is traditionally carried out by use of disease scoring systems that provide on average 61% screening sensitivity, 88% diagnostic sensitivity, and 69% specificity in dairy cattle [4–7]. Although clinical signs may suggest BRSV infection, only laboratory diagnostic tools can confirm the diagnosis [3]. Diagnostic techniques such as virus isolation, immunofluorescent antibody (IFA), enzyme-linked immunosorbent assay (ELISA), and reverse transcription-polymerase chain reaction (RT-PCR) have been used to detect BRSV in biofluids collected from the lungs, trachea, or nostrils with sensitivities and specificities similar to, or higher than the clinical diagnosis [2, 8, 9]. However, the results from these methods take multiple days to acquire. In previous attempts to create new alternatives for the diagnosis of bovine respiratory infections, a comparison of nasal secretions collected from healthy and sick animals was made using gas chromatography-based mass spectrometry (GC-MS), ELISA, and PCR [10–14]. However, inefficiencies arise with the labor of sample preparation and cost of analysis, as well as the time delay in acquiring diagnostic results, creating a risk of disease spread before mitigating action can be implemented. Thus, there is a need for rapid testing in the field that would support more specific treatments, improving cattle health and overall enhanced biosecurity of the food chain.

Near Infrared Spectroscopy (NIRS) allows fast, non-invasive assessment of biofluids [15, 16], whereby the biochemical profile is described by the intensity and structure of sample transmittance spectra in the NIRS region (750–2500 nm) [17, 18]. This method provides a potential alternative to the current diagnostic tools and also reveals new information regarding the biochemistry of bovine nasal secretions, which has not been thoroughly described, probably due to the perceived difficulties in collecting a significant amount for examination [14, 19]. Here, aquaphotomics and chemometrics were applied to characterize the NIR spectral profile of nasal secretions (NS) from dairy calves intentionally infected with BRSV in a controlled challenge study in order to determine if the molecular organization of the aqueous phase of this biofluid is suitable as a potential source of detecting this viral infection in real-time in the early stages of the disease.

2 Materials and Methods

2.1 Animals and Controlled Challenge

The BRSV strain GA-1, passage 5 was prepared (5 ml at 1×10^5 $TCID_{50}$ units/ml) and administered via a nebulizer (DeVilbiss Pulmo-Neb) through a custom-made face mask to five healthy non-vaccinated Holstein steers (approx. 3 months old, 130 kg) housed in outdoor research pens at Mississippi State University (MSU). Animals were devoid of maternally derived anti-BRSV antibodies before being challenged with the virus. A standard clinical evaluation was used to detect signs of viral infection and included rectal temperature, heart rate, respiratory rate, and assessment of overall airway health pre- and post-challenge, including spontaneous or induced cough, nasal and ocular discharge, submandibular lymphadenopathy, difficult breathing, and abnormal lung sounds (crackles, wheezes, large airway sounds, or absence of sounds) [20, 21]. The NS were collected daily for four days prior to the BRSV challenge, daily for 11 days after challenge, then every other day until Day 23 post-challenge. To collect the NS (Fig. 1), a string-secured sponge (3 cm × 2 cm) was inserted into one nostril approximately 2 cm deep for 2 min (Fig. 1A), and nasal fluids (approx. 1 ml) were pressed out of the sponge using a 30 ml syringe and stored at –80 °C (Fig. 1B). The experiment was conducted in accordance with MSU-Institutional Animal Care and Use Committee guidelines and regulations (IACUC-19-037).

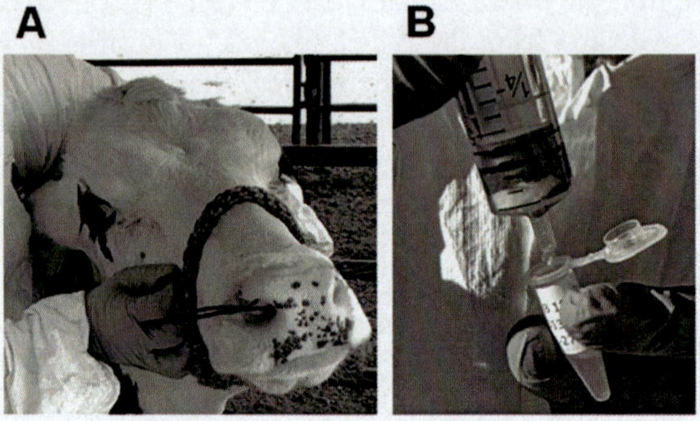

Fig. 1. Bovine nasal secretion (NS) collection. A) string-secured sponge inside the calf's nostril. B) mechanism of extraction of nasal fluid for storage.

2.2 NIR Spectra Acquisition and Analysis

Transmittance NIR spectra (n = 970) were collected from 300 μl of thawed nasal secretions (NS) as previously described [22] using a portable spectrophotometer ASD FieldSpec®3 + Indico®Pro software (Malvern Panalytical, Cambridge, United Kingdom). The NIR absorbance was transformed by applying the mathematical pre-treatments: SNV (Standard Normal Variate) with de-trending (polynomial order: 2), Baseline Offset & Linear Baseline Correction, and a 2nd derivative (polynomial order: 2, Gap Size: 25, Segment Size: 19, Savitzky-Golay smoothing points: 25). The NS samples (n = 97) were categorized as having been obtained during the Pre-challenge (healthy before the challenge), Asymptomatic (post-challenge, but not yet showing clinical signs of respiratory disease), Symptomatic (showing signs of respiratory disease after challenge) or Recovered (returned to apparently healthy after being symptomatic) stages based on the clinical signs evaluated before and after the BRSV challenge. To prevent compromising the interpretation of the aquaphotomics profiles due to the biochemical changes happening during the Asymptomatic phase or the Recovery process, only samples labeled as Pre-challenge or Symptomatic were selected for the aquaphotomics evaluation and the chemometrics, and models were built using an equal number of each. Principal Component Analysis (PCA) of the general database (n = 400) and Linear Discriminant Analysis (PCA-LDA) using a leave-one-animal-out approach were applied on the first overtone region of the near infrared spectrum in the vibrational combination band between 1300–1600 nm using Unscrambler® X v.11 (Aspen Technology Inc, MA, USA) [20, 22]. The PCA-LDA was reported from the confusion matrix using quality parameters as a percentage (%) of accuracy (correctly classified), sensitivity (true positives), and specificity (true negatives) to assess the performance of the classification method. The aquaphotomics evaluation, including absorbance normalization, barcodes, and aquagrams based on Water Absorbance Bands (WABS) and 12 Water Matrix Coordinates (WAMACS), were created following published procedures [20, 22–24].

3 Results and Discussion

The raw NIR spectra collected for NS from dairy calves challenged with BRSV indicated a predictable spectral water pattern in the wavelength range 1300–1600 nm (Fig. 2A,

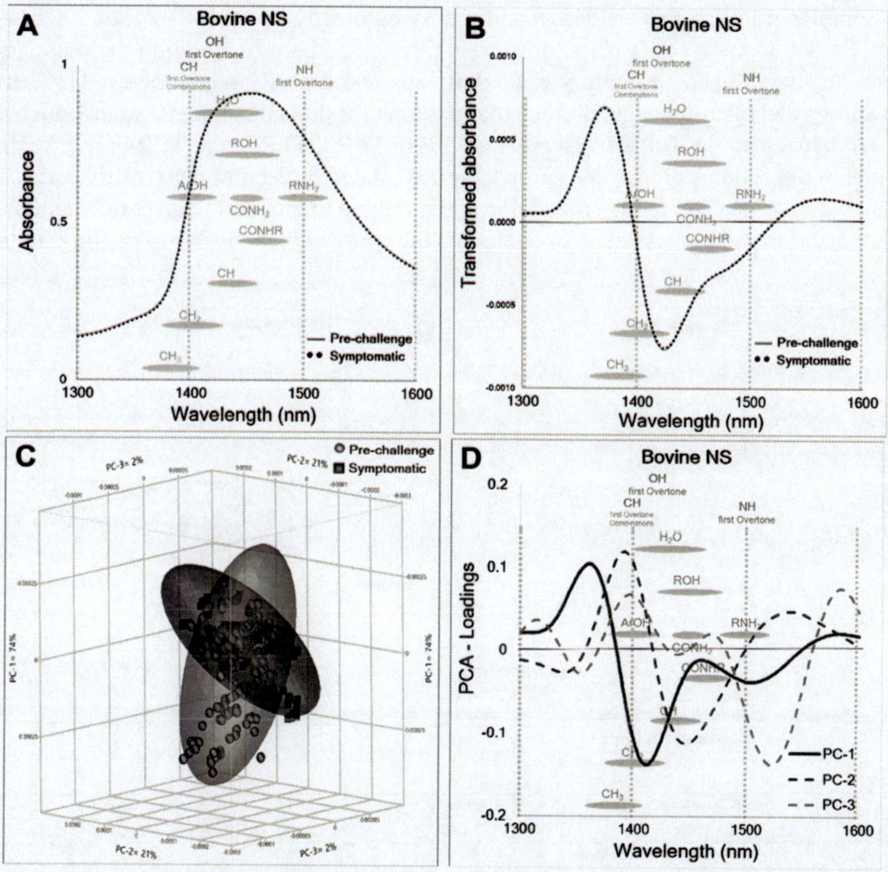

Fig. 2. NIR absorbance from Bovine NS collected before and after the BRSV challenge. A) Raw NIR absorbance from the Pre-challenge (n = 200) and the Symptomatic (n = 200) stages, revealing the distinctive water spectral pattern. B) Transformed absorbance from the Pre-challenge (n = 200) and the Symptomatic (n = 200) stages showing two prominent features at 1375 and 1427 nm and the functional groups interacting with NIR light in this wavelength range. C) PCA scores plot containing samples from all the calves (n = 5) challenged in this study. D) PCA loadings plot showing the dominant peaks influencing the trends in the scores plot: PC-1 = 74%, PC-2 = 21%, PC-3 = 2%.

B). In the PCA scores plot (Fig. 2C), both assessed stages overlap but are defined by the 95% coverage of the normal contour ellipsoids, with the first three PCs explaining 97% of the variance of the overall spectral database encompassing samples from all the dairy calves. The positive and negative dominant peaks or eigenvectors associated with functional groups interacting with NIR light and found in the PCA loadings plot (Fig. 2D) explained these pattens (PC-1 = 1365, 1414, 1463, 1511, 1587 nm; PC-2 = 1340, 1397, 1449, 1542 nm; PC-3 = 1318, 1356, 1400, 1473, 1526, 1580 nm). In addition, no outliers were observed in the Hotelling's T^2 influence plot (not shown).

The discoveries in the spectral signatures and the PCA emphasize the need to employ both aquaphotomics and chemometrics to identify, and discriminate, distinguishing biochemical profiles of BRSV infection in this biofluid. Biochemical differences between NS samples from the Pre-challenge and the Symptomatic stages are expected because once BRSV enters a calf's respiratory tract through aerosols containing the pathogen, it reaches the ciliated respiratory epithelial cells and type II pneumocytes [1]. There, the known pathogen-associated molecular patterns for this virus (single-stranded RNA) are recognized by the Toll-like receptors (TLRs), TLR2, TLR3, and TLR4 [1, 25]. This results in the cellular production of damage-associated molecular patterns that activate signaling cascades and induce metabolic shifts related to glucose, triglyceride, and protein metabolic pathways needed to overcome the energy demands due to the production

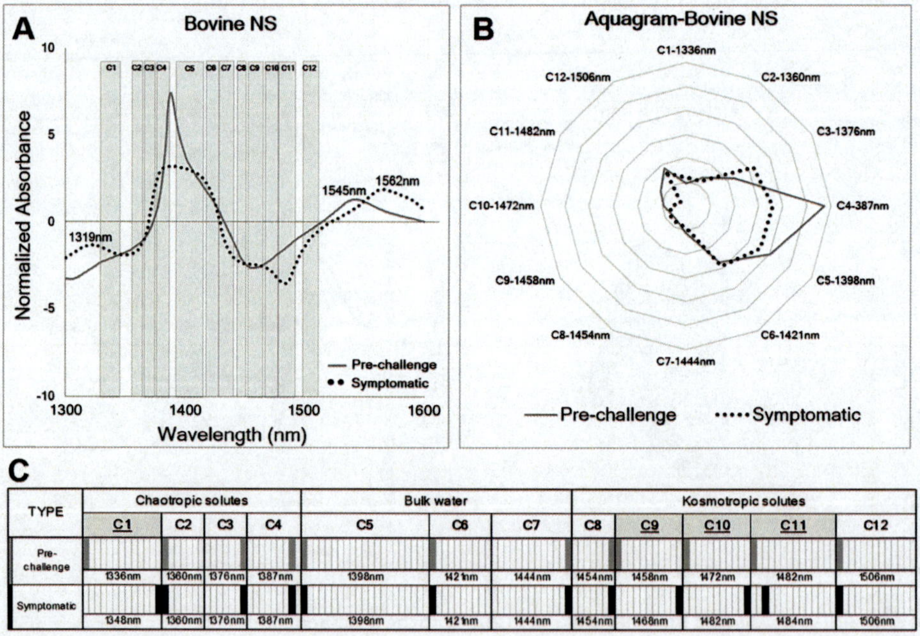

Fig. 3. Aquaphotomics of NS from dairy calves challenged with BRSV. A) Normalized absorbance for samples from the Pre-challenge (n = 200) and the Symptomatic (n = 200) stages from all the calves (n = 5). B) Aquagram generated with the key WABS from the Pre-challenge stage. C) NS WAMACS barcode showing chemical shifts and their association with the types of solutes.

of inflammatory cytokines and chemokines that are required for the activation of the innate immune response [21, 26–28].

When comparing the water spectral patterns of NS from both stages (Fig. 3), the normalized absorbance highlighted distinct trends for each category (Fig. 3A), demonstrating variations in the chemical makeup of NS from healthy and sick dairy calves. Furthermore, significant absorbance values at 1319, 1545, and 1562 nm which are associated with free OH, free water, bulk water [29] and hydrogen bonded water around large macromolecules [18] were also shown to be critical in the differentiation of both clinical stages but are not included in the currently stated WAMACS by Dr. Tsenkova, opening the opportunity to this discipline to develop more WAMACS in this wavelength range. The aquagram displaying the WABS selected from the normalized absorbance (Fig. 3B) revealed different Water Spectral Patterns (WASPS) for both groups, with the highest absorbance points at C4 (water shells) for the Pre-challenge stage, and at C3, C4, C5, and C6, for the Symptomatic stage. This suggests that in the NS from healthy calves, the water molecules associated with chaotropic solutes, such as the dilute ions in the inorganic salts (KCl, $CaCl_2$, NaCl, and $PO_4{}^{3-}$) present in this biofluid strongly absorb NIR light [23, 24, 30]. In contrast, the water molecules for NS from Symptomatic calves were not only associated with chaotropic solutes but also with bulk water, likely because of the inflammation of the sinuses caused by the virus preventing them from draining normally, leading to a build-up of mucus (~95% water) and the increase of S_0: free water [23, 24, 31]. The WAMACS barcode (Fig. 3C) exhibited shifts between NS samples collected in both stages in four of the 12 established WAMACS in the coordinates C1, C9, C10, and C11, whereas the samples from the Symptomatic stage were right-shifted at 1348, 1468, 1482, and 1484 nm in comparison to Pre-challenge peaks at 1336, 1458, 1472, and 1482 nm. This suggests a shift towards water molecules highly associated with kosmotropic solutes, such as macromolecules which increase H-bonding numbers and strength by forming water clusters with 2, 3, and 4 hydrogen bonds (S_2, S_3, S_4) during thc infcction [23, 24].

This is the first study to employ NIRS to identify and distinguish BRSV infection in nasal secretions from dairy calves. The average percentage of classification accuracy, sensitivity, and specificity during the calibration and internal validation of the PCA-LDA using 9 ± 2 PCs explaining $99.9 \pm 0.09\%$ of the variance of the spectral database (1300–1600 nm) was 90.1 ± 4.3, 88.1 ± 3.8, 92.0 ± 5.5, and 90.1 ± 1.3, 91.9 ± 2.8, 90.0 ± 2.6, respectively (Table 1). The external validation yielded percentages of 79.0 ± 3.2, 72.0 ± 16.6, and 86.0 ± 18.4 of accuracy, sensitivity, and specificity for classifying Pre-challenge and Symptomatic bovine nasal secretions, correspondingly. The PCA-LDA plot (Fig. 4) from model 4 (M4) is shown as a representation of the defined trends obtained for all the evaluated models. Previous studies used visible and near-infrared spectroscopy (VIS-NIRS) in the region 600–1100 nm, combined with soft modeling of class analogy (SIMCA) to discriminate influenza infection in humans in 64 samples of nasal secretions (34 infected, 33 healthy), finding a sensitivity of 96.7% and a specificity of 100% [32]. More recently, the application of infrared spectroscopy techniques in human nasal secretions has been proposed for the detection of SARS-CoV-2 (COVID-19) [33, 34].

Table 1. Classification of bovine nasal secretions (NS) collected before and after the BRSV challenge by PCA-LDA (1300–1600 nm). No significant differences ($p < 0.05$) were found between models after the application of ANOVA and Tukey-Kramer HSD (honestly significant difference).

Model	#Selected PCs	%Explained variance	Category and quality	% PCA-LDA Mahalanobis		
				Cal 80%	Val 20%	External validation
1 (Calf 1 out)	8	99.85	Pre-challenge	111/128	28/32	40/40
			Symptomatic	114/128	31/32	27/40
2 (Calf 2 out)	9	99.93	Pre-challenge	119/128	29/32	35/40
			Symptomatic	110/128	29/32	25/40
3 (Calf 3 out)	7	99.77	Pre-challenge	110/128	28/32	35/40
			Symptomatic	106/128	29/32	29/40
4 (Calf 4 out)	12	99.99	Pre-challenge	125/128	30/32	40/40
			Symptomatic	118/128	29/32	23/40
5 (Calf 5 out)	10	99.96	Pre-challenge	124/128	29/32	22/40
			Symptomatic	116/128	29/32	40/40
Mean ± SD	9 ± 2	99.9 ± 0.09	% Accuracy	90.1 ± 4.3	90.1 ± 1.3	79.0 ± 3.20
			% Sensitivity	88.1 ± 3.8	91.9 ± 2.8	72.0 ± 16.6
			% Specificity	92.0 ± 5.5	90.0 ± 2.6	86.0 ± 18.4

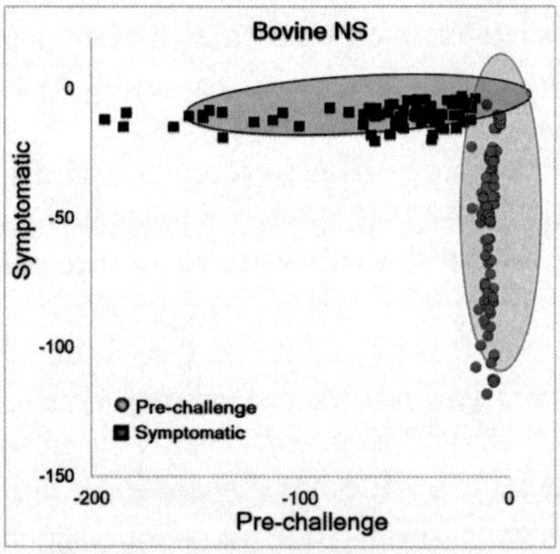

Fig. 4. PCA-LDA plot for the calibration of model M4 developed with the transformed absorbance (1300–1600 nm) from NS collected from the Pre-challenge (n = 128) and Symptomatic (n = 128) stages.

The described WASPS and the PCA-LDA discrimination values for NS collected during the Pre-challenge and the Symptomatic stages can be associated with chemical changes previously reported in nasal secretions from healthy and sick cattle with respiratory disease. The biochemical and immunological composition of NS from 38 healthy

Holstein-Friesian cows (2–5 years old) was reported to contain lower concentrations of albumin, calcium, chloride, phosphate, sodium, and total protein than the bovine serum reference range, whereas aspartate transaminase, bilirubin, creatinine, immunoglobulin A (IgA), IgG, and urea levels were comparable between both biofluids. Alkaline phosphatase and gamma-glutamyltransferase activity, on the other hand, were greater in NS than in the serum reference [14, 19]. The concentration of the volatile compounds phenol (C_6H_6O), benzothiazole (C_7H_5NS), *p*-Cresol (C_7H_8O), and 5-Octadecenal ($C_{18}H_{34}O$) were found to decrease in the nasal secretions from sick cattle with respiratory disease (n = 50) in comparison to healthy animals (n = 50) when analyzed by GC-MS [11]. The volatile compounds cyclohexanone ($C_6H_{10}O$), 2-butanone ($C_2H_5COCH_3$), and 4-methyl-2-pentanone ($C_6H_{12}O$) were correlated with the clinical state of five calves by PCA using a gas sensor array in a portable electronic nose [13]. By using the same approach in combination with PCA and PCA-LDA, a group of 20 sick calves with respiratory disease was discriminated from a group of 20 healthy calves with an accuracy of 100%, finding a change in the concentration of alcohols, aldehydes, amines, ketones, and organic carboxylic acids, when comparing the nasal secretions from both groups [12].

4 Conclusions

Our findings suggest that the biochemical differences discovered in the aquaphotomics evaluation due to changes in NS composition during BRSV infection can be detected using chemometrics-based MVA methods with percentages of accuracy (correctly classified), sensitivity (true positives), and specificity (true negatives) that are higher than traditional diagnostic methods. By collecting spectra from nasal secretions, we demonstrated the potential of NIRS in combination with aquaphotomics for the detection of this viral infection in-vivo, as a first step toward developing a rapid in-field diagnostic tool for BRSV infection.

Acknowledgments. The authors thank Ellianna Blair, Amanda Free, Hannah Bostick, Matt Harjes, Victoria Jefferson, and Matt Scott (DVM) for their help with the sample collection. This project was supported by the Mississippi Agricultural and Forestry Experiment Station, the National Institute of Food and Agriculture, the U.S. Department of Agriculture, Hatch project under accession number W3173, and the U.S. Department of Agriculture, Agricultural Research Service, Biophotonics project # 6066-31000-015-00D.

References

1. Valarcher, J.F., Taylor, G.: Bovine respiratory syncytial virus infection. Vet. Res. **38**, 153–180 (2007)
2. Larsen, L.E.: Bovine respiratory syncytial virus (BRSV): a review. Acta Vet. Scand. **41**(1), 1–24 (2000)
3. Makoschey, B., Berge, A.C.: Review on bovine respiratory syncytial virus and bovine parainfluenza – usual suspects in bovine respiratory disease – a narrative review. BMC Vet. Res. **17**(261), 1–18 (2021)

4. Love, W.J., et al.: Sensitivity and specificity of on-farm scoring systems and nasal culture to detect bovine respiratory disease complex in preweaned dairy calves. J. Vet. Diagnostic Investig. **28**(2), 119–128 (2016)
5. Maier, G.U., et al.: Development of a clinical scoring system for bovine respiratory disease in weaned dairy calves. J. Dairy Sci. **102**, 7329–7344 (2019)
6. Buczinski, S., Ollivett, T.L., Dendukuri, N.: Bayesian estimation of the accuracy of the calf respiratory scoring chart and ultrasonography for the diagnosis of bovine respiratory disease in pre-weaned dairy calves. Prev. Vet. Med. **119**, 227–231 (2015)
7. Woolums, A.R., et al.: Effects of a single intranasal dose of modified-live bovine respiratory syncytial virus vaccine on cytokine messenger RNA expression following viral challenge in calves. Am. J. Vet. Res. **65**(3), 363–372 (2004)
8. Brodersen, B.W.: Bovine respiratory syncytial virus. Vet. Clin. North Am. - Food Anim. Pract. **26**, 323–333 (2010)
9. Nefedchenko, A.V., Glotov, A.G., Koteneva, S.V., Glotova, T.I.: Developing and testing a real-time polymerase chain reaction to identify and quantify bovine respiratory syncytial viruses. Mol. Genet. Microbiol. Virol. **35**(3), 168–173 (2020). https://doi.org/10.3103/S0891416820030052
10. Ellis, J.A., Chamorro, M.F., Lacoste, S., Gow, S.P., Haines, D.M.: Bovine respiratory syncytial virus-specific lgG-1 in nasal secretions of colostrum-fed neonatal calves. Can. Vet. J. **59**, 505–508 (2018)
11. Maurer, D.L., Koziel, J.A., Engelken, T.J., Cooper, V.L., Funk, J.L.: Detection of volatile compounds emitted from nasal secretions and serum: towards non-invasive identification of diseased cattle biomarkers. Separations **5**, 1–18 (2018)
12. Kuchmenko, T., Shuba, A., Umarkhanov, R., Chernitskiy, A.: Portable electronic nose for analyzing the smell of nasal secretions in calves: toward noninvasive diagnosis of infectious bronchopneumonia. Vet. Sci. **8**(74), 1–16 (2021)
13. Kuchmenko, T., Shuba, A., Umarkhanov, R., Lvova, L.: The new approach to a pattern recognition of volatile compounds: the inflammation markers in nasal mucus swabs from calves using the gas sensor array. Chemosensors **9**(116), 1–16 (2021)
14. Ghazali, M.F., Koh-Tan, H.C., McLaughlin, M., Montague, P., Jonsson, N.N., Eckersall, P.D.: Alkaline phosphatase in nasal secretion of cattle: biochemical and molecular characterization. BMC Vet. Res. **10**(204), 1–8 (2014)
15. Shaw, R.A., Mantsch, H.H.: Infrared Spectroscopy of Biological Fluids in Clinical and Diagnostic Analysis. Encyclopedia of Analytical Chemistry. Wiley (2008)
16. Krzysztof, B., Grabska, J., Huck, C.W.: Near-infrared spectroscopy in bio-applications. Molecules **25**, 1–37 (2020)
17. Pasquini, C.: Near infrared spectroscopy: fundamentals, practical aspects and analytical applications. J. Braz. Chem. Soc. **14**, 198–219 (2003)
18. Williams, P., Antoniszyn, J., Manley, M.: Near Infrared Technology. Getting the Best Out of Light, 1st edn. AFRICAN SUN Media, Stellenbosch (2019)
19. Ghazali, M.F.: Biochemical and proteomic investigation of bovine nasal secretion, University of Glasgow (2015)
20. Santos-Rivera, M., Woolums, A.R., Thoresen, M., Meyer, F., Vance, C.K.: Bovine respiratory syncytial virus (BRSV) infection detected in exhaled breath condensate of dairy calves by near- infrared aquaphotomics. Molecules **27**, 1–13 (2022)
21. Santos-Rivera, M., et al.: NMR-based metabolomics of plasma from dairy calves infected with two primary causal agents of Bovine Respiratory Disease (BRD). Sci. Reports (2022, in Revision)
22. Santos-Rivera, M., et al.: Profiling Mannheimia haemolytica infection in dairy calves using near infrared spectroscopy (NIRS) and multivariate analysis (MVA). Sci. Rep. **11**, 1–13 (2021)

23. Muncan, J., Tsenkova, R.: Aquaphotomics-from innovative knowledge to integrative platform in science and technology. Molecules **24**, 1–26 (2019)
24. Tsenkova, R., Munćan, J., Pollner, B., Kovacs, Z.: Essentials of aquaphotomics and its chemometrics approaches. Front. Chem. **6**, 1–25 (2018)
25. McGill, J.L., Sacoo, R.E.: The immunology of bovine respiratory disease. Vet. Clin. Food Anim. **36**, 333–348 (2020)
26. Werling, D., Coffey, T.J.: Pattern recognition receptors in companion and farm animals - the key to unlocking the door to animal disease? Vet. J. **174**, 240–251 (2007)
27. Kominsky, D.J., Campbell, E.L., Colgan, S.P.: Metabolic shifts in immunity and inflammation. J. Immunol. **184**(8), 4062–4068 (2010)
28. Gleeson, L.E., Sheedy, F.J.: Metabolic reprogramming & inflammation: fuelling the host response to pathogens. Semin. Immunol. **28**, 450–468 (2016)
29. Xantheas, S.S.: Theoretical study of hydroxide ion—water clusters. J. Am. Chem. Soc. **117**, 10373–10380 (1995)
30. Burke, W.: The ionic composition of nasal fluid and its function. Health (Irvine. Calif) **6**, 720–728 (2014)
31. Divers, T.: Respiratory diseases, chapter 4. In: Rebhun's Diseases of Dairy Cattle, 2nd edn., vol. 19, pp. 79–129 (2008)
32. Sakudo, A., Baba, K., Ikuta, K.: Discrimination of influenza virus-infected nasal fluids by Vis-NIR spectroscopy. Clin. Chim. Acta **414**, 130–134 (2012)
33. Khan, R.S., Rehman, I.U.: Spectroscopy as a tool for detection and monitoring of Coronavirus (COVID-19). Expert Rev. Mol. Diagn. **20**, 1–3 (2020)
34. de Luque Ripoll, D.M.: COVINIRs Boscalia Technologies, Boscalia Technologies (2021). http://boscalia.org/en/covinirs-2

Biomedicine, Environment, and fNIR

Vis-NIR Spectroscopic Discriminant Analysis Applied to Serum Breast Cancer Screening

Lu Yuan[1], Jing Zhang[1,2], Jianhua Xu[3], Lijun Yao[1], Dawei Wang[3], and Tao Pan[1](✉)

[1] Department of Optoelectronic Engineering, Jinan University, Guangzhou 510632, China
tpan@jnu.edu.cn

[2] School of Dongguan, University of Technology, Dongguan 523000, China

[3] Shunde Hospital Guangzhou University of Chinese Medicine, Foshan 528000, China

Abstract. The research and development of fast, simple and accurate technology for breast cancer screening has important application value. In this paper, the discriminant analysis models for breast cancer and normal control samples were established using serum Vis-NIR spectroscopy combined with the equidistant combination-partial least squares-discriminant analysis (EC-PLS-DA) method. Standard normal variable (SNV) method was adopted for the spectral pretreatment of serum samples to improve spectral prediction.

The parameters of the selected optimal EC-PLS-DA model were initial wavelength (I) = 1976 nm, ending wavelength (E) = 2396 nm, number of wavelengths (N) = 31, number of wavelength gaps (G) = 7 and the number of PLS latent variables (LV) = 10, respectively. In modelling, the calibration, prediction and total recognition accuracy rates were 96.0%, 97.5%, and 96.7%, respectively. Using independent validation samples not involved in modelling, the positive, negative, and total recognition accuracy rates were 85.0%, 90.0%, and 87.5%, respectively. The results showed the feasibility of serum Vis-NIR spectroscopic applied to discriminant analysis of breast cancer and normal control samples. The EC-PLS-DA method can extract information wavelengths, improve the recognition accuracy of discriminant analysis and reduce wavelength model complexity. The relevant wavelength model can provide valuable references for specialized spectrometer design and clinical application. The analytical technique is simple and novel, and has potential application in breast cancer screening.

Keywords: Serum · Breast cancer screening · Equidistant combination-partial least squares-discriminant analysis · Discriminant analysis model · Recognition accuracy rate

1 Introduction

Breast cancer is a serious disease with high incidence. And breast cancer screening is very frequent in clinic. Existing diagnostic methods are traumatic to the body and not suitable for frequent screening of large populations. There are specific protein tumor markers in the serum of breast cancer patients, which cause changes in serum components and further reflected in the overall changes in the serum spectrum. The qualitative

X. Chu et al. (Eds.): ICNIR 2021, *Sense the Real Change: Proceedings of the 20th International Conference on Near Infrared Spectroscopy*, pp. 187–192, 2022.
https://doi.org/10.1007/978-981-19-4884-8_19

discriminant analysis of spectroscopy is based on the spectral similarity of samples in the same type and the spectral differences of different types of samples to distinguish and classify samples. This study explored the feasibility of serum Vis-NIR spectroscopic applied to discriminant analysis of breast cancer and normal control samples. The equidistant combination-partial least squares-discriminant analysis (EC-PLS-DA) method [1] was performed to improve the recognition accuracy.

2 Materials and Methods

2.1 Experimental Materials, Instruments and Measurement

A total of 130 serum samples of breast cancer and normal control subjects were collected from the hospital, and were used for modelling and validation of the Vis-NIR spectral discrimination method. The serum types of the subjects were accurately determined by the hospital using the biopsy examination of clinical gold standard method. Among them, there are 65 samples of breast cancer and normal control subjects each. Informed consent was obtained from all individual participants as human serum samples were collected and used in this work. The experiment was performed in accordance with relevant laws and institutional guidelines and approved by local medical institutions.

The XDS Rapid Content™ Liquid Grating Spectrometer (FOSS, Denmark) and a transmission accessory with 0.8 mm cuvette were used for spectral measurement. Spectral scope ranged as 400–2498 nm with a 2-nm wavelength interval. Wavebands of 400–1100 nm and 1100–2498 nm were used for Si and PbS detection, respectively. Each sample was measured three times and the average spectrum was used. The experimental temperature and humidity were 25 ± 1 °C and $45 \pm 1\%$, respectively.

2.2 Experimental Design of Calibration-Prediction-Validation and Evaluation Indicators

In chronological order, the randomly collected serum samples of breast cancer (positive, 65) and normal control (negative, 65) were divided into calibration (positive 25, negative 25, total 50), prediction (positive 20, negative 20, total 40), and validation (positive 20, negative 20, total 40) sets. Among them, the calibration and prediction sets were used for modelling and parameter optimization, and the validation set that does not participate in the modelling was used to test the established model for obtaining objective evaluation results.

According to the true type and prediction type of the samples, the eight recognition accuracy rates (RAR_C^+, RAR_C^-, RAR_C, RAR_P^+, RAR_P^-, RAR_P, RAR^+, RAR^-) of the positive (breast cancer) and negative (normal control) samples in calibration and prediction sets, the total recognition accuracy rate (RAR_{Total}) and their standard deviations (RAR_{SD}) were calculated respectively for modelling effect evaluation. The three recognition accuracy rates (RAR_V^+, RAR_V^-, RAR_V) of the positive and negative validation samples were also calculated respectively for validation effect evaluation.

2.3 EC-PLS-DA

The EC-PLS-DA method used the cycle parameters of initial wavelength (I), number of wavelengths (N), and number of wavelength gaps (G) to perform wide range of wavelength selection for improving recognition accuracy [1–4]. The optimal EC-PLS-DA model was determined according to the optimal total modelling accuracy. Corresponding to the wavelength-screening range of 400–2498 nm, the parameters I, N, G, and the number of PLS latent variables (LV) were set as $I \in \{400, 402, \cdots, 2498\}$, $N \in \{1, 2, \cdots, 1050\}$, $G \in \{1, 2, \cdots, 10\}$, and $\text{LV} \in \{1, 2, \cdots, 20\}$, respectively. Notably, when $G = 1$, the equidistant wavelength combination was corresponded to a continuous waveband. Therefore, the wavelength selection method of equidistant combination covers the well-known moving-window waveband selection method.

3 Results and Discussion

3.1 PLS-DA Models Without and with SNV

The commonly used standard normal variable (SNV) method was adopted for the spectral pretreatment of serum samples to improve spectral prediction. Based on the spectra in whole scanning region without and with SNV pretreatment, the PLS-DA models were initially established. The corresponding nine RARs and RAR_{SD} are summarized in Table 1, in which the recognition rate and indicator balance of SNV model both were improved. And the further wavelength model optimization was based on the SNV pretreatment.

Table 1. Recognition-accuracy rates of full-spectrum PLS-DA models without and with SNV pretreatment in modelling (%)

Method	LV	RAR_C^-	RAR_C^+	RAR_P^-	RAR_P^+	RAR_C	RAR_P	RAR^-	RAR^+	$\text{RAR}_{\text{Total}}$	RAR_{SD}
without SNV	16	100	100	70.0	70.0	100	70.0	86.7	86.7	86.7	13.0
with SNV	15	100	100	70.0	75.0	100	72.5	86.7	88.9	87.8	12.0

3.2 EC-PLS-DA Model

The EC-PLS-DA method was used for further wavelength selection and modelling optimization. The parameters of the optimal model were $I = 1976$ nm, $N = 31$, $G = 7$ and $\text{LV} = 10$. The corresponding ending wavelength (E) was 2396 nm. The corresponding modelling effect is summarized in Table 2. The results show that compared with the full-spectrum model ($N = 1050$), the total accuracy ($\text{RAR}_{\text{Total}}$) and indicator balance (RAR_{SD}) both were significantly improved. Moreover, the wavelength model complexity was also significantly reduced ($N = 31$). The 31 wavelengths of the optimal EC-PLS-DA model were located in the NIR combined frequency region, their positions (hollow circles) of in the average spectrum are shown in Fig. 1.

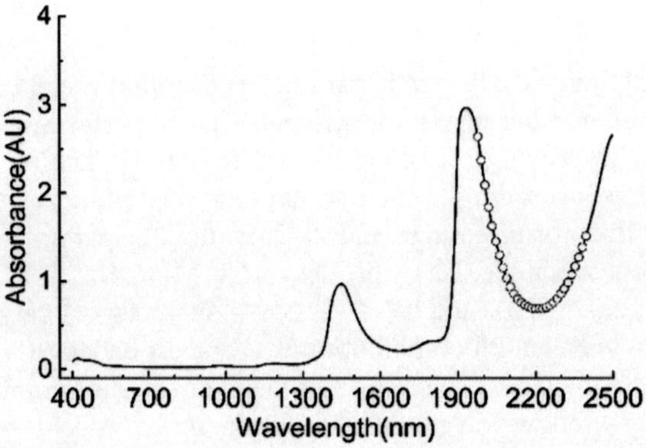

Fig. 1. Position of the wavelength combination of the optimal EC-PLS-DA model in the average spectrum

Table 2. Recognition-accuracy rates of the optimal EC-PLS-DA model in modelling (%)

I	*N*	*G*	LV	RAR_C^-	RAR_C^+	RAR_P^-	RAR_P^+	RAR_C	RAR_P	RAR^-	RAR^+	RAR_{Total}	RAR_{SD}
1976	31	7	10	96.0	96.0	100	95.0	96.0	97.5	97.8	95.6	96.7	1.5

3.3 Independent Validation

A total of 45 validation samples not involved in modelling were used to validate the optimal EC-PLS-DA model ($I = 1976$nm, $N = 31$, $G = 7$, LV = 10). The recognition accuracy rates of independent validation is summarized in Table 3. The results indicated that the optimal EC-PLS-DA model has achieved relatively good discrimination accuracy. The recognition effect of the optimal EC-PLS-DA model for the validation samples is shown in Fig. 2. Among them, hollow dots indicated correct recognition (true negative, true positive), and solid dots indicated wrong recognition (false negative, false positive). Confusion matrix of recognition effect of the optimal EC-PLS-DA model in validation is also shown Table 4.

Table 3. Recognition-accuracy rates of the optimal EC-PLS-DA model in validation (%)

I	*E*	*N*	*G*	LV	RAR_V^-	RAR_V^+	RAR_V
1976	2396	31	7	10	90.0	85.0	87.5

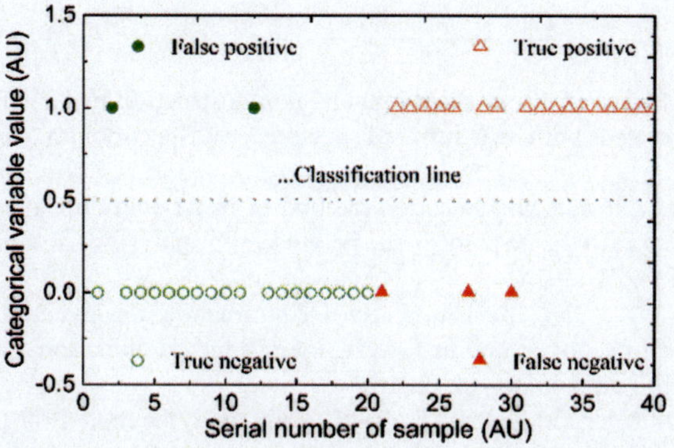

Fig. 2. Recognition effect of the optimal EC-PLS-DA model for the validation samples

Table 4. Confusion matrix of recognition effect of the optimal EC-PLS-DA model in validation

Actual\Predictive	Breast cancer	Normal control	Recognition-accuracy rates
Breast cancer	17	3	85.0%
Normal control	2	18	90.0%

4 Conclusion

The discriminant analysis models for breast cancer and normal control samples were established using serum Vis-NIR spectroscopy combined with the EC-PLS-DA method. The selected optimal EC-PLS-DA model used only 31 wavelengths and achieved significantly better discriminant analysis modelling effect than the full-spectrum model. Using independent validation samples not involved in modelling, the positive, negative, and total recognition accuracy rates were 85.0%, 90.0%, and 87.5%, respectively. The results showed the feasibility of serum Vis-NIR spectroscopic applied to discriminant analysis of breast cancer and normal control samples. The EC-PLS-DA method can extract information wavelengths, improve the recognition accuracy of discriminant analysis and reduce wavelength model complexity. The relevant wavelength model can provide valuable references for specialized spectrometer design and clinical application. The analytical technique is simple and novel, and has potential application in breast cancer screening. The proposed method framework is also expected to be applied to other application fields.

Acknowledgments. This work was supported by National Natural Science Foundation of China (No. 61078040) and Guangdong Province Project of China (No. 2014A020213016, No. 2014A020212445).

References

1. Chen, J.M., Li, M.M., Pan, T., et al.: Rapid and non-destructive analysis for the identification of multi-grain rice seeds with near-infrared spectroscopy. Spectrochim. Acta. A. **219**, 179–185 (2019)
2. Pan, T., Li, M.M., Chen, J.M.: Selection method of quasi-continuous wavelength combination with applications to the near-infrared spectroscopic analysis of soil organic matter. Appl. Spectrosc. **68**, 263–271 (2014)
3. Han, Y., Chen, J.M., Pan, T., Liu, G.S.: Determination of glycated hemoglobin using near-infrared spectroscopy combined with equidistant combination partial least squares. Chemometr. Intell. Lab. **145**, 84–92 (2015)
4. Yao, L.J., Lyu, N., Chen, J.M., Pan, T., Yu, J.: Joint analyses model for total cholesterol and triglyceride in human serum with near-infrared spectroscopy. Spectrochim. Acta. A. **159**, 53–59 (2016)

Grouping Modeling Strategy for Hematocrit Analysis with Blood Vis-NIR Spectroscopy

Zeqi Chen[1], Yan Tang[1], Haoran Lin[1], Zhiyuan Yin[2], Junyu Fang[3], and Tao Pan[1](✉)

[1] Department of Optoelectronic Engineering, Jinan University, Guangzhou 510632, China
tpan@jnu.edu.cn, 466945939@qq.com

[2] Department of Software Engineering, University of New South Wales, Sydney, NSW 1466, Australia

[3] Dongguan Public Hospital Operation Service Center, Dongguan 523000, China

Abstract. Vis-NIR spectroscopy combined with equidistant combination-PLS (EC-PLS) method was applied for the rapid and reagent-free analysis of blood Hematocrit (HCT). The multi-parameter optimization platform based on Norris derivative filter (NDF) was constructed to select appropriate spectral preprocessing. Multi-partition modeling and independent validation in calibration-prediction-validation design were adopted to ensure the stability of parameter selection and the objectivity of modeling effect. For male and female groups, the optimal EC-PLS models of the grouping modeling were selected and achieved significantly better validation effects than hybrid modeling. In independent validation, the root mean square error of prediction (SEP) of male, female and mixed sample groups were decreased by 12.7%, 32.4% and 20.4%, respectively. The results showed that the predicted and clinical actual values of the all validation samples have high correlation coefficient of prediction ($R_P = 0.93$) and low prediction error (SEP = 1.21%), and thus have potential for clinical application.

Keywords: Blood hematocrit · Equidistant combination-PLS · Norris derivative filter · Multi-partition modeling · Grouping modeling

1 Introduction

Hematocrit (HCT) refers to the volume ratio of red blood cells in a certain volume of whole blood, and is often used for the diagnosis and classification of anemia. HCT is related to blood viscosity and has reference value for cardiovascular and cerebrovascular diseases. It is one of the important indicators included in blood routine and hemrheology testing. Existing detection methods either require reagents or are complicated to operate, and which are not suitable for large population screening and real-time monitoring during surgery. In this paper, the Vis-NIR spectroscopy combined with equidistant combination-PLS (EC-PLS) [1, 2] was used to establish the quantitative models for rapid detection of HCT in blood without reagents. Considering the homogeneity of the samples helped to improve the analysis accuracy, the spectral analysis models of male, female and mixed groups were established respectively, and the parameters were optimized, and a comparative analysis of modeling effects was given.

X. Chu et al. (Eds.): ICNIR 2021, *Sense the Real Change: Proceedings of the 20th International Conference on Near Infrared Spectroscopy*, pp. 193–198, 2022.
https://doi.org/10.1007/978-981-19-4884-8_20

2 Materials and Methods

2.1 Experimental Materials, Instruments and Measurement

A total of 450 human peripheral blood samples (male 240, female 210) were collected from the hospital, and were used for modelling and validation of the Vis-NIR spectral method. The HCT index (%) of the blood samples was detected by hemorheology method. The measuring instrument was an automatic blood rheometer with enhanced dual system (LBY-N7500) (Beijing Precil Instrument Co., Ltd., China). Informed consent was obtained from all individual participants as the human peripheral blood samples were collected and used in this work. The experiment was performed in accordance with relevant laws and institutional guidelines and approved by local medical institutions.

The XDS Rapid Content™ Liquid Grating Spectrometer (FOSS, Denmark) and a transmission accessory with 0.8 mm cuvette were used for spectral measurement. Spectral scope ranged as 400–2498 nm with a 2-nm wavelength interval. Wavebands of 400–1100 nm and 1100–2498 nm were used for Si and PbS detection, respectively. Each sample was measured six times and the average spectrum was used. The experimental temperature and humidity were 25 ± 1 °C and $45 \pm 1\%$, respectively.

2.2 Multi-partition Modeling in Calibration-Prediction-Validation and Evaluation Indicators

The spectral analysis models of the men, women and mixed groups were established respectively. In chronological order, the randomly collected 240 male samples were divided into modeling (160) and validation (80) sets, and the 160 modeling samples were further divided into calibration (80) and prediction (80) sets for 10 times. Similarly, 210 female samples were divided into modeling (140 samples) and validation (70) sets, and the 140 modeling samples were further divided into calibration (70) and prediction (70) sets for 10 times. By combining corresponding samples, the division of mixed sample set were determined as follows: 450 samples were divided into modeling (300) and validation (150) sets, and the 300 modeling samples were divided into calibration (150) and prediction (150) sets for 10 times. Among them, the ten divisions of calibration and prediction sets were used for modelling and parameter selection stability, and the validation set that does not participate in the modelling was used to test the established model for obtaining objective evaluation results.

For the three datasets of men, women and mixed groups, the evaluation indicators of modeling and validation were as follows. For each division i of calibration and prediction sets, the root-mean-square error and correlation coefficient for prediction were calculated and denoted as SEP_i and $\text{R}_{\text{P},i}$, respectively, where $i = 1, 2, \cdots, 10$. The mean values (SEP_{Ave} and $\text{R}_{\text{P,Ave}}$) and standard deviations (SEP_{SD} and $\text{R}_{\text{P,SD}}$) for all the divisions were further calculated. The optimal model parameters were selected according to the minimum SEP_{Ave}. The SEP_{SD} was used as the second optimization objective to take into account the stability of the parameters. Finally, the selected models were validated using independent validation samples. The corresponding SEP, R_P and R-SEP (relative error, %), $\text{R-SEP} = \text{SEP}/\text{C}_{\text{Ave}}$, C_{Ave} was the mean of validation samples' HCT values, were further determined.

2.3 Norris Derivative Filter

In spectral preprocessing, appropriate smoothing and derivative can effectively eliminate noises and improve spectral information quality. The famous Norris derivative filter (NDF) is an effective spectral pretreatment method, which is an algorithm group with various parameters [3, 4]. NDF includes two steps: the moving average smoothing and differential derivation and uses three parameters: the derivative order (*D*), the number of smoothing points (*S*, odd), and the number of differential gaps (*G*). The appropriate Norris parameters should be chosen according to the analytical object. It is necessary to make large-scale optimization selection for Norris parameters (*D*, *S*, *G*) according to the modeling discrimination effect. The multi-parameter optimization platform about PLS combined with NDF (Norris-PLS) was constructed to optimize PLS model. The loop parameters were set to $D = 0, 1, 2$; $S = 1, 3, \cdots, 31$; $G = 1, 2, \cdots, 30$.

2.4 EC-PLS

The EC-PLS method used the cycle parameters of initial wavelength (*I*), number of wavelengths (*N*), and number of wavelength gaps (*G*) to perform wide range of wavelength selection for improving prediction accuracy [1, 2]. The optimal EC-PLS model was determined according to the minimum SEP_{Ave}. Corresponding to the wavelength-screening range of 400–2498 nm, the parameters *I*, *N*, *G*, and the number of PLS latent variables (LV) were set as $I \in \{400, 402, \cdots, 2498\}$, $N \in \{1, 2, \cdots, 200\}$, $G \in \{1, 2, \cdots, 20\}$, and $LV \in \{1, 2, \cdots, 15\}$, respectively. Notably, when $G = 1$, the equidistant wavelength combination was corresponded to a continuous waveband. Therefore, the wavelength selection method of equidistant combination covers the well-known moving-window waveband selection method.

3 Results and Discussion

3.1 Norris-PLS Models

NDF was used for the preprocessing of the Vis-NIR spectra of the human peripheral blood samples. The optimal Norris-PLS models in full scanning spectral region (400–2498 nm) of HCT analysis for the male, female and mixed groups were selected, which outperformed the direct PLS models. The Norris parameters and modeling effects are summarized in Table 1. The Norris derivative spectra in of the male and female sample groups are shown in Fig. 1.

3.2 Optimal EC-PLS Models

The On the basis of the NDF spectra, further wavelength model optimization was performed using the EC-PLS. The optimal parameters and modelling effects were summarized in Table 2. Compared to the full-spectrum models ($N = 1050$), the optimal EC-PLS models were greatly simplified ($N = 53, 48, 15$) and modeling effect better.

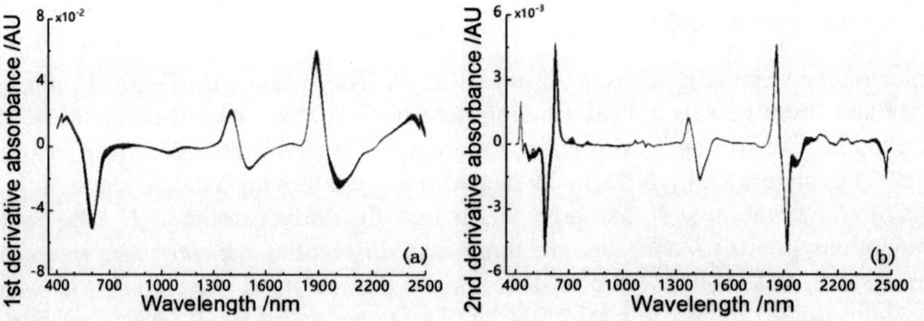

Fig. 1. Norris derivative spectra of the male and female sample groups of peripheral blood samples: (a) male; (b) female

Table 1. Modeling effects for optimal Norris-PLS models of male, female and mixed groups

Group	*D*	*S*	*G*	LV	SEP_{Ave}	SEP_{SD}	$R_{P,Ave}$	$R_{P,SD}$
Male	1	27	16	3	1.67	0.14	0.81	0.04
Female	2	29	3	6	1.82	0.11	0.81	0.03
Mixed	2	11	22	6	2.08	0.03	0.83	0.01

Table 2. Modeling effects for the optimal EC-PLS models of male, female and mixed groups

Group	*I*	*N*	*G*	LV	SEP_{Ave}	SEP_{SD}	$R_{P,Ave}$	$R_{P,SD}$
Male	698	53	15	5	1.37	0.13	0.88	0.02
Female	726	48	7	5	1.07	0.08	0.94	0.01
Mixed	1218	15	14	9	1.79	0.06	0.86	0.01

3.3 Independent Validation

A total of 150 validation samples (male 80 and female 70) not involved in modeling were used to validate the optimal EC-PLS models of the male, female and mixed groups. See Table 3 for a comparison of the effects of hybrid modeling and grouping modeling. The results showed that the effect of grouping modeling of each sample group was significantly better than that of hybrid modeling. Among them, the prediction errors (SEP's) of male, female and mixed sample groups were decreased by 12.7%, 32.4% and 20.4%, respectively. Based on grouping modeling, the relationship between the predicted and clinical actual values of the all validation samples is shown in Fig. 2, and high correlation ($R_P = 0.93$) and low prediction error (SEP = 1.21%, R-SEP = 2.7%) were observed.

Table 3. Validation effects for the optimal EC-PLS models of male, female and mixed groups

Group	Modeling	SEP (%)	R_P	R-SEP (%)
Male	Male	1.37	0.85	2.9
	Hybrid	1.57	0.79	3.4
Female	Female	1.00	0.93	2.4
	Hybrid	1.48	0.82	3.6
Mixed	**Grouping**	**1.21**	**0.93**	**2.7**
	Hybrid	1.52	0.89	3.4

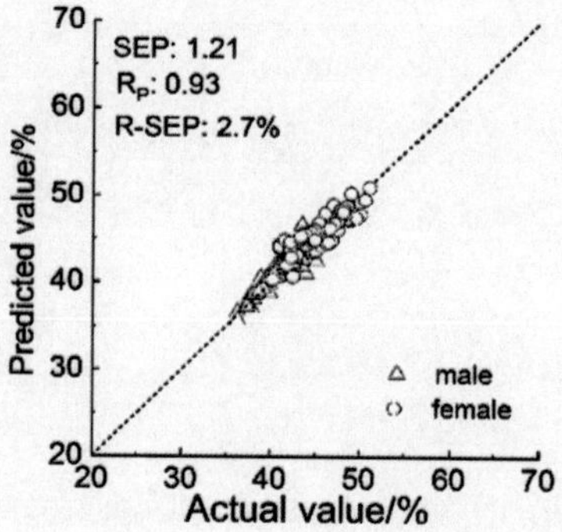

Fig. 2. Relationship between the predicted and clinical actual values of the all validation samples based on grouping modeling.

4 Conclusion

The Vis-NIR spectroscopy combined with EC-PLS method was used to establish the rapid and reagent-free analysis models of blood HCT. The multi-parameter optimization platform Norris-PLS was constructed to select appropriate spectral preprocessing. Multi-partition modeling for calibration-prediction samples made parameter selection stable. The optimal EC-PLS models were very simplified and modeling effect better than the full-spectrum models. The effects of grouping modeling of male, female and mixed groups were significantly better them of hybrid modeling. The results showed that the predicted and clinical actual values of the validation samples have high correlation and low prediction bias, and thus have potential for clinical application. The proposed method framework and grouping modeling strategy are also expected to be applied to other fields.

Acknowledgments. This work was supported by National Natural Science Foundation of China (No. 61078040) and Guangdong Province Project of China (No. 2014A020213016, No. 2014A020212445).

References

1. Pan, T., Li, M.M., Chen, J.M.: Selection method of quasi-continuous wavelength combination with applications to the near-infrared spectroscopic analysis of soil organic matter. Appl. Spectrosc. **68**, 263–271 (2014)
2. Han, Y., Chen, J.M., Pan, T., Liu, G.S.: Determination of glycated hemoglobin using near-infrared spectroscopy combined with equidistant combination partial least squares. Chemometr. Intell. Lab. **145**, 84–92 (2015)
3. Norris, K.H.: Applying Norris derivatives understanding and correcting the factors which affect diffuse transmittance spectra. NIR News. **12**, 6–9 (2001)
4. Pan, T., Zhang, J., Shi, X.W.: Flexible vitality of near-infrared spectroscopy–talking about Norris derivative filter. NIR News. **31**, 24–27 (2020)

Instrument, Accessory and Experimental Technology

The AS7265x Chipset as an Alternative Low-Cost Multispectral Sensor for Agriculture Applications Based on NDVI

A. Ducanchez[1(✉)], S. Moinard[1], G. Brunel[1], R. Bendoula[2], D. Héran[2], and B. Tisseyre[1]

[1] ITAP, University Montpellier, INRAE, Institute Agro Montpellier, 34060 Montpellier, France
arnaud.ducanchez@supagro.fr

[2] ITAP, University Montpellier, INRAE, 34000 Montpellier, France

Abstract. Recently, new low-cost multispectral sensors have been commercialized, paving the way for a large number of new agricultural applications (fertilization, grass cover, etc.), particularly for small farms. However, such sensors have never been tested for agricultural applications taking into account practical constraints (external environment, etc.). This study proposes to investigate the potentialities of the "AS7265x" chipset that presents a real interest for a wide range of applications in agriculture due to 18 spectral bands available. The first study involved the testing of three sensors in laboratory to assess the accuracy of the different spectral bands as well as the reproducibility of the measurements from one sensor to another but not presented here. In a second step, the work aimed at testing the potential of the sensor on real fields with two applications in proxy-detection to estimate the percentage of weeds over soil and the vine vigor. These field experiments focused on NDVI that is a vegetation index widely used in precision agriculture for proxy-sensing. Results show that although accurate, the sensors present some different bias for each wave bands and each sensor. These drawbacks require each sensor to be specifically calibrated before use which may limit their dissemination in agriculture. Once the sensor measurements are normalized, the NDVI values are consistent compared to the reference values given by the Greenseeker ($R^2 = 0.87$ for NDVI < 0.75). Hence, the accuracy obtained was sufficient to differentiate the levels of grass cover and the differences in vegetative expression of the vine induced by local environmental effects.

Keywords: Sensor · Multispectral · Low-cost · Agriculture

1 Introduction

Spectroscopy techniques (ST) are widely used in agriculture sector [1]. In particular, spectral data from visible and near-infrared (VIS-NIR) spectrometry is a powerful, fast and non-destructive method for measuring a large number of chemical and physical properties of agricultural products [2]. Indeed, ST offers a great diversity for agricultural

X. Chu et al. (Eds.): ICNIR 2021, *Sense the Real Change: Proceedings of the 20th International Conference on Near Infrared Spectroscopy*, pp. 201–206, 2022.
https://doi.org/10.1007/978-981-19-4884-8_21

applications in terms of remote sensing [3] or proximal sensing [4], depending on the measurement scale. They are currently used for soil monitoring to determine composition and soil properties as well as organic matter [5]. They are also widely used for plant monitoring to obtain crop characteristics and to detect biotic and abiotic stresses in plants as well as water stress [6], diseases in plant tissues [7] or management of nitrogen status [8]. The majority of these applications for crop health monitoring is currently based on multispectral vegetation indices like Normalized Difference Vegetation Index (NDVI) [9]. In that way, application of these spectrometry techniques and methods contribute to the implementation of precision agriculture, allowing farmers to adapt their practices to the real needs of the crops. Some of these measurement tools, and associated services, are currently commercialized. However, in agricultural regions dominated by small farms, the cost of access to these techniques remains an issue which partly explains low adoption in these specific contexts [10]. This question justifies research efforts to design low-cost remote sensing tools [11] and low-cost handheld multispectral devices [12] in an attempt to popularize these technologies. Recently, very low-cost multispectral sensors have been launched by AMS manufacturer (Premstätten, Austria). In particular, the "AS7265x" distributed by SparkFun (Boulder, US) is a new multispectral sensor component available at 65 dollars. The AS7265x can be used for spectral measurements in a range from visible to near-infrared. This sensor returns 18 spectral bands, from which different vegetation indices such as NDVI can be easily obtained to highlight particular properties of a plant, soil with great potential of application in precision agriculture. As this sensor is new, there is no study, to our knowledge, demonstrating its potential for real agronomic applications. Despite its great interest and as it is a multispectral sensor at very low cost, the question of the quality of the measurements carried out with such a sensor need to be studied before promoting its use for agricultural applications. Although the characteristics of the AS7265x low-cost multispectral sensor in controlled condition in terms of accuracy, repeatability, reproducibility and stability were tested (but not presented here), this paper focuses on the possibilities of this sensor in terms of proxy-detection applications. In particular, the spectral bands around the red and infrared to measure NDVI in order to validate the interest of this sensor as a tool for quantifying vine vigor and biomass in the field in comparison with reference methods.

2 Materials and Methods

2.1 Acquisition System

In order to perform hand-held measurements with the AS7265x optical sensor, this latter has been associated with different elements. In the rest of the document the term VG-sensor will be used to refer to the whole device. The VG-sensor is presented in Fig. 1. It includes i) an Arduino Uno Rev3 microcontroller platform (Arduino, Italy), ii) an AS7265x optical sensor (AMS, Austria) delivering 18 spectral bands from 410 nm to 940 presented in Fig. 1a, iii) a 5700 k white LED, a 405 nm UV LED, and a 875 nm IR LED to collect measurements with active lighting in a wide spectral range, iv) a VMA203 LCD display (Velleman, Germany) for data visualization, v) a SD shield (Catalex, US) for data storage, and vi) 3D-printed protective shells allowing the VG-sensor to be used on the field for proxidetection (Fig. 1b). The total price of the VG-sensor does not exceed

100 dollars. The software is programmed in the C+ + language. It allows to collect and store i) an ID of the measurement and ii) the 18 reflectance values from the sensor.

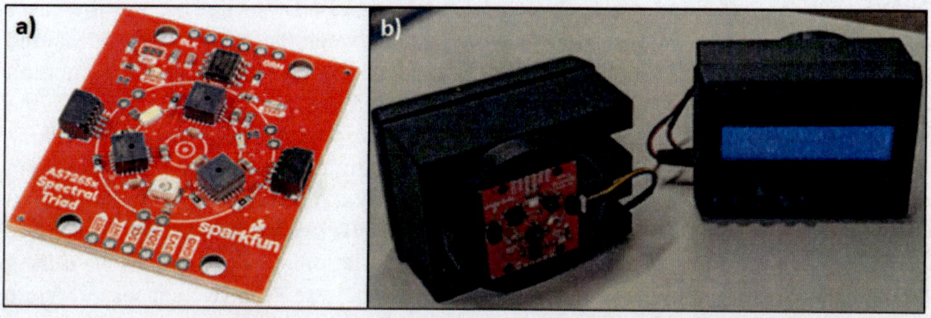

Fig. 1. a) the AS7265x sensor by AMS, b) the developed hand-held system in its 3D printed shell for use in proxy-detection mode: on the left, bottom view with the AS7265x sensor in red; on the right, front view with the LCD display.

2.2 Field NDVI Experiments

To evaluate the potential of the VG-sensors for agricultural applications, two experiments focused on NDVI measurement by proxy-detection were performed. The first study aimed at testing the ability of the VG-sensor to differentiate levels of grass cover using reference values estimated by Greenseeker (Trimble, US). The study area was a vine field in a vineyard located in the city of Villeneuve-les-Maguelone (France, 43.5323, 3.8642, WGS84). 200 measurement sites with different levels of grass cover were defined within the field. During experimentation, measurement sites were delimited by a wooden frame of 60 cm × 60 cm in order to delineate the spatial footprint of the measurement. For each sites, NDVI value was measured with two portable devices: the VG sensor and a Greenseeker. The second study aimed at testing the ability of the VG-sensor to evaluate the newly planted vine vigor and local effect on different parcels carried out by a technical institute. 40 measures of NDVI were made with the VG-sensor on 10 plots, from June to August (during the vegetative growth of the newly planted vines), every 2 weeks. The 10 plots were placed in two different vine fields A and B. The first field A was a sandy soil vine field in the Camargue region (43.5633, 4.3214, WGS84) and is known to be susceptible to the disease with low vigor. The second field B was a silty-clay soil vine field inland (43.7513, 4.2896, WGS84) and is known for a high potential of vigor. Both plots were not grassed. NDVI measurements with the VG-sensor were performed from the row at a distance of 50 cm in the direction of the canopy of each vine plant.

3 Results and Discussion

Figure 2 shows the NDVI values measured with the VG-sensor (NDVI-VG) and the Greenseeker (NDVI-Greenseeker) on the same plot with different grassing levels. NDVI-VG shows a strong linear correlation with NDVI-Greenseeker ($R^2 = 0.84$). However, the VG-sensor i) overestimates the NDVI values compared to the Greenseeker, and ii) saturates faster than the Greenseeker for NDVI values higher than 0.75. With the exception of the NDVI values from which saturation of the VG-sensor is observed (NDVI > 0.75), the correlation between NDVI-VG and NDVI-Greenseeker increases ($R^2 = 0.87$), showing a very good correspondence between the NDVI values measured with these two sensors. The unexplained variance (13%) may be related to the acquisition conditions. Indeed, both sensors were compared in real situation, on grass cover in the field, considering that the spatial footprint of both sensors are not rigorously identical, slight variations at the time of acquisition may explain that the measured surfaces are not rigorously identical between the two sensors. In addition, the GreenSeeker is an active sensor that emits its own radiation while the VG-sensor was used as a passive sensor in this experiment. As a result, even small variations in light conditions during the experiment can explain the observed variability between the two sensors. Despite these issues, the NDVI-VG values remain consistent and allow the ranking of the grass cover level of the different plots. The experiment also highlights some limitations of the VG-sensor. Indeed, NDVI-VG values remain biased compared to the values provided by a reference sensor such as the Greenseeker. Considering the Greenseeker sensor as a reference can naturally be a subject of discussion. However, it is a sensor already adopted for which professionals have already established references.

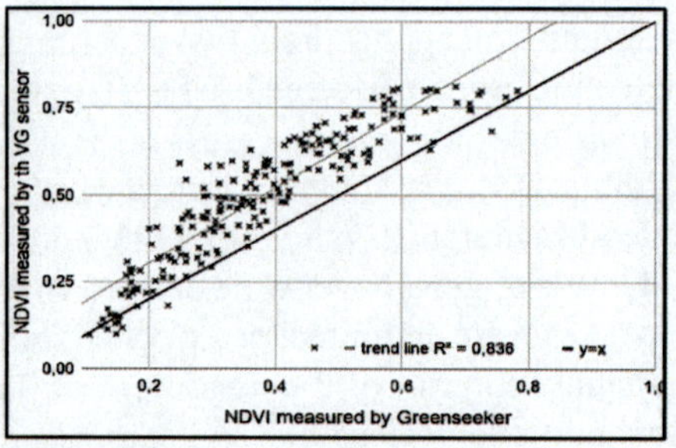

Fig. 2. NDVI measured by the VG-sensor on soils with different levels of grass covers compared to the NDVI measured by the Greenseeker.

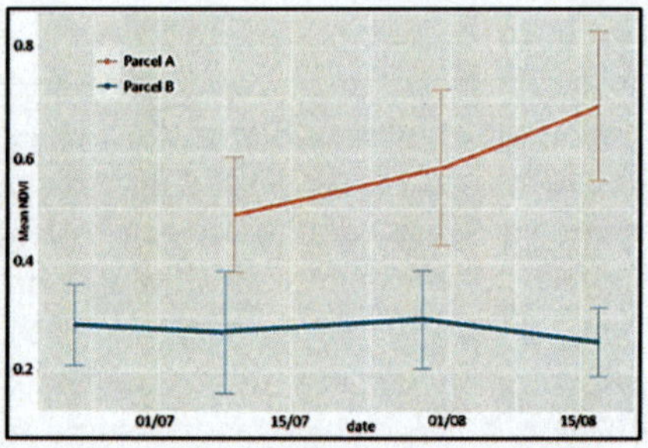

Fig. 3. Mean NDVI of newly planted vine plants in parcel A known for its low vigor (green line) and in parcel B known for its high vigor (red line).

Concerning vine vigor experimentation, NDVI-VG measurements taken at regular intervals throughout the vegetative cycle of the vine clearly show the significant increase in biomass in field B (over the month of July) while the biomass remains low throughout the season for vines in field A (Fig. 3). In spite of an important variability explained by the intra- and inter-plant variance, the sensor allows to discriminate significantly the field effect ($p < 0.05$) from mid-July onwards. This result confirms the interest of the VG-sensor to monitor a dynamic phenomenon corresponding to the vine growth (increase in biomass) but also, in the framework of an experiment to highlight the potential effect of environmental factors on plant vigor.

4 Conclusion

The study made at the earliest stages under controlled conditions and not presented here showed that low-cost multispectral sensors like the AS7265x present limitations if accurate reflectance measurements are to be produced. However, its practical characteristics within NDVI experimentations reported in this article seem sufficient to meet many needs in agriculture, including high data repeatability. They represent an opportunity for small-scale agriculture to have access to simple and robust decision support tools and precision agriculture. Two limitations are now identified before recommendations can be made. The first one concerns the calibration of the reflectance of the sensors in order to be able to compare values from different sensors. This calibration process is still a strong limitation despite the low price of the sensor. The second limitation is the need to include the optical sensor in a portable device like the VG-sensor for field measurements. This limitation is also a strength because it is possible to adapt the device to perform contact detection or proxidetection at a lower cost. These encouraging results of low-cost multispectral sensors in an agricultural context open the way to future professional uses for decision support, but also in research with the possibility of dissemination of this kind of low-cost sensors for larger scale analysis. These aspects are currently being developed in the framework of a co-construction project with a group of farmers and advisors in connection with a Fab Lab (manufacturing laboratory).

References

1. Yeong, T.J., Pin Jern, K., Yao, L.K., Hannan, M.A., Hoon, S.: Applications of photonics in agriculture sector: a review. Molecules **24**(10), 2025 (2019)
2. Cortés, V., Blasco, J., Aleixos, N., Cubero, S., Talens, P.: Monitoring strategies for quality control of agricultural products using visible and near-infrared spectroscopy: a review. Trends Food Sci. Technol. **85**, 138–148 (2019)
3. Sishodia, R.P., Ray, R.L., Singh, S.K.: Applications of remote sensing in precision agriculture: a review. Remote Sens. **12**, 3136 (2020)
4. Gholizadeh, A., Kopačková, V.: Detecting vegetation stress as a soil contamination proxy: a review of optical proximal and remote sensing techniques. Int. J. Environ. Sci. Technol. **16**(5), 2511–2524 (2019). https://doi.org/10.1007/s13762-019-02310-w
5. Nocita, M., et al.: Soil spectroscopy: An alternative to wet chemistry for soil monitoring. Adv. Agron. **132**, 139–159 (2015)
6. Gerhards, M., Rock, G., Schlerf, M., Udelhoven, T.: Water stress detection in potato plants using leaf temperature, emissivity, and reflectance. Int. J. Appl. Earth Obs. Geoinf. **53**, 27–39 (2016)
7. Farber, C., Mahnke, M., Sanchez, L., Kurouski, D.: Advanced spectroscopic techniques for plant disease diagnostics. a review. TrAC Trends Anal. Chem. **118**, 43–49 (2019)
8. Padilla, F.M., Farneselli, M., Gianquinto, G., Tei, F., Thompson, R.B.: Monitoring nitrogen status of vegetable crops and soils for optimal nitrogen management. Agric. Water Manag. **241**, 106356 (2020)
9. Xue, J., Su, B.: Significant remote sensing vegetation indices: a review of developments and applications. J. Sens. **10**(1155), 1353691 (2017)
10. Lachia, N., Pichon, L., Tisseyre, B.: A collective framework to assess the adoption of precision agriculture in France: description and preliminary results after two years. In: Precision Agriculture 2019. ED. John v. Stafford, Ampthill, UK, Wageningen Academic Publisher, pp. 851–857 (2019)

11. Cucho-Padin, G., Loayza, H., Palacios, S., Balcazar, M., Carbajal, M., Quiroz, R.: Development of low-cost remote sensing tools and methods for supporting smallholder agriculture. Appl. Geomatics **12**(3), 247–263 (2019). https://doi.org/10.1007/s12518-019-00292-5
12. Kitić, G., et al.: A new low-cost portable multispectral optical device for precise plant status assessment. Comput. Electron. Agric. **162**, 300–308 (2019)

PAT and Imaging

Application of On-line Near Infrared Spectroscopy in the Production of Traditional Chinese Medicine

Jun Wang[1], Yerui Li[1], Jiapeng Huang[1], Xiaoxue Zhang[1], Jingnan Wu[1], and Xuesong Liu[2](✉)

[1] Suzhou ZeDaXingBang Pharmaceutical Co., Ltd., Suzhou 215000, China
[2] College of Pharmaceutical Sciences, Zhejiang University, Hangzhou 310058, China
liuxs@sino-essence.com

Abstract. The production process of Chinese medicine is characterized by complex processes, tedious steps and complicated influences, as all aspects of formulation production affect the final quality of a Chinese medicine product. Online NIR spectroscopy has the advantages of rapid and non-destructive, and can be used as an analytical technique for rapid evaluation of critical quality properties in the production process of traditional Chinese medicine. This paper systematically described the analysis and control methods of online NIR spectroscopy in the production process of traditional Chinese medicine from the application perspective of enterprises, and takes the online NIR spectroscopy analysis platform for traditional Chinese medicine built by Zeda Xingbang Pharmaceutical Technology Co., Ltd. as an example, to elucidate the feasibility of the application of online NIR spectroscopy in traditional Chinese medicine more comprehensively, discussed the economic benefits of the application of online NIR detection technology, and provided insights into the future The feasibility of the application of online NIR spectroscopy for Chinese medicine production is more comprehensively elucidated. The economic benefits of the application of on-line NIR detection technology were discussed and an outlook was also made for the future application of NIR technology in the field of Chinese medicine.

Keywords: Manufacture process of Chinese materia medica (CMM) · Near-infrared spectroscopy · Process quality control · Continuous manufacturing

1 Current Status of Traditional Chinese Medicine Production

As a distinctive and advantageous industry in China, traditional Chinese medicine has an important strategic position in the field of biomedicine and has become one of the most important pillars to Chinese pharmaceutical industry after years of development. The industrialized production process of traditional Chinese medicine, which integrates

X. Chu et al. (Eds.): ICNIR 2021, *Sense the Real Change: Proceedings of the 20th International Conference on Near Infrared Spectroscopy*, pp. 209–219, 2022.
https://doi.org/10.1007/978-981-19-4884-8_22

multiple processes such as raw material control, production control, and quality inspection, is characterized with complex process, cumbersome steps, multiple influential factors, nonlinearity and significant interactions [1]. In terms of quality control of Chinese medicine, most domestic enterprises only focus on raw materials and finished products, while neglecting the quality control of the production process and its intermediate steps. The quality of intermediate and final products have been evaluated by manual sampling and analysis and offline testing for a long term, which is time-consuming and subjective, and the testing results are lagging behind the production process, resulting in the inability to adjust the production process timely on the basis of real-time quality fluctuations [2, 3]. In recent years, it frequently occurs that the intermediate or final products have to be reworked or discarded due to quality problems.

2 The Development of NIR in the Production of Chinese Medicine

In recent years, technologies such as online testing, process analysis technology (PAT), and quality control systems have been applied into the production process to enhance the understanding of the process and eliminate the uncertainties and risks in the process through rational process design, analysis and control, so as to guarantee the quality of the final products. At present, the commonly used process analysis technologies include online near infrared spectroscopy online Raman spectroscopy, online UV, etc., among which NIR spectroscopy is superior due to its rapidity, efficiency, multi-component simultaneous measurement and absence of sample pretreatment [4, 5]. Without sample pretreatment and can be transmitted through optical fiber, NIR spectroscopy analysis technology is a good choice for the rapid analysis of the quality of raw materials of complex Chinese medicines and online inspection of the production process, which covers the identification of the origin of the Chinese herbs, the determination of the active ingredient content and the online detection and monitoring of the pharmaceutical processes [6]. Since the "13th Five-Year Plan", Zeda Xingbang Pharmaceutical Technology Co., Ltd. has joined forces with several "top 100" enterprises in the pharmaceutical industry and successfully implemented several cases in the field of traditional Chinese medicine production, the cases are shown in Table 1 (Fig. 1).

Table 1. Implementation examples of PAT in traditional Chinese medicine production monitoring process. (Zeda Xingbang)

Customer	Type	Description
Yangtze River Pharmaceutical Group	Lanqin Oral Liquid	Offline, Online
Shanghai Xingling Technology Pharmaceutical Co., Ltd.	Ginkgo ketone ester	Offline, Online
Jiuzhitang Co., Ltd.	Liuwei dihuang wan, Lvjiao buxue granules	Online, Offline
Jiangsu Kangyuan Group Co., Ltd.	Reduning, Guizhi Fuling	Offline, Online
Huarun Sanyao (Benxi) Pharmaceutical Co., Ltd.	Qizhi weitong granules	Offline,Online
Huarun Sanjiu (Zaozhuang) Pharmaceutical Co., Ltd.	Ganmaoling granules	Offline,Online
Shandong Green Leaf Pharmaceutical Co., Ltd.	Rotigotine	Offline, Online
Chongqing Tai Chi Group Co., Ltd.	Huoxiang zhengqi oral liquid	Offline, Online
Beijing Weixin Biotechnology Co., Ltd.	Xuezhikang	Offline, Online
Guangdong Zhongsheng Pharmaceutical Co., Ltd.	Fufang naoshuantong	Offline, Online
Xiangyu Pharmaceutical Co., Ltd.	Fufang Hongyibuxue oral liquid	Offline, Online
……		

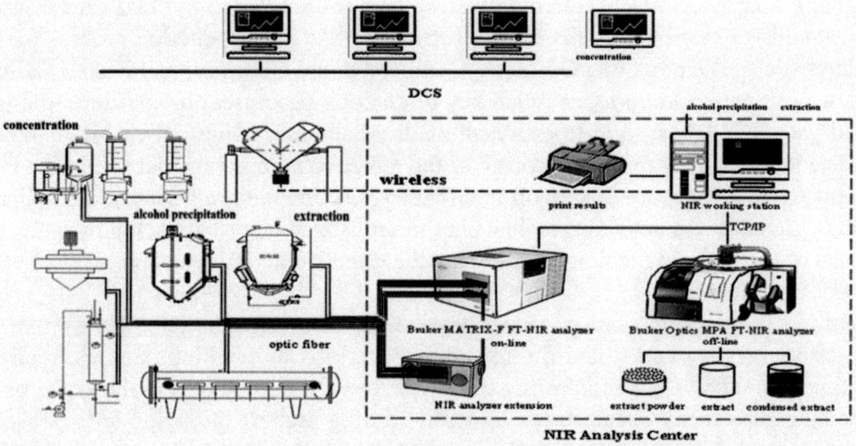

Fig. 1. Near-infrared online detection system in Chinese medicine production process

3 Application Examples of Near Infrared in the Production of Chinese Medicine

3.1 Huarun Sanjiu Ganmaoling Granules: Concentration and Total Mixing Section

The Ganmaoling Granules is used to expel the wind-cold evil, clear heat and relieve pain. It is composed of Sanchaku, wild chrysanthemum, chlorpheniramine maleate, caffeine, etc. and is widely used for headaches, fever, nasal congestion, running nose and sore throat caused by colds. The active ingredients such as the anthocyanin in the wild chrysanthemum stand as the important indicators for the quality of Ganmaoling granules. It is complicated to produce, so ensuring the stable product quality in each process is an effective guarantee for the final product. However, the current analysis method is disadvantageous since it is time-consuming and information lagging, which seriously affect the product quality and production costs, so it is urgent to develop a fast and accurate detection technology.

At present, the near-infrared spectroscopy detection technology has developed from offline experiments or small-scale simulation experiments to online monitoring of large-scale production processes gradually. Compared with the former, the near-infrared online monitoring technology is of practical guiding significance in that it can monitor the quality of the production process while ensuring that the indicators of the object can be applied to establish accurate quantitative models. In the new intelligent manufacturing model application project of the Ministry of Industry and Information Technology, Zeda Xingbang Pharmaceutical Technology Co., Ltd has established the fast detection and online quality detection systems for key technical links and production process for key products such as Ganmaoling Granules, Ganmao Qingre Granules, and Xiaoer Ganmao granules of Sanjiu Group to establish key production processes. In addition, it is integrated with SCADA system to establish quality database. Among them, it covers the fast detection and real-time monitoring of the effective ingredients and solid content of the midstream extracts of Ganmaoling Granules, Ganmaoqingre Granules and Xiaoerganmao granules, the active ingredients in the semi-finished products, the moisture and extracts of the original medicinal materials, the effective ingredients and extracts of the concentrated liquid.

During the implementation of the project, the near-infrared detection system has been effectively applied to the production process of Ganmaoling granules, realizing real-time dynamic online monitoring of the quality of intermediates in the key process, reducing the volatility of intermediate quality during process operation to improve the quality control during finished medicine production. The Fig. 2 shows the combined application of near-infrared technology and the total mixing process of Ganmaoling granules. Taking the semi-finished product as an example, the model established for the content of montanoside, acetaminophen, caffeine, and chlorpheniramine maleate is satisfactory in the prediction results. The correlation coefficient R is 0.9757, 09523, 0.9705, 0.9803, and the RMSEP is 0.0115, 0.219, 0.202, 0.126, which can meet the accuracy requirements of real-time analysis of semi-finished products of Ganmaoling granules (Table 2).

Fig. 2. The effect of online detection of concentrated solid content of Xiaoer ganmaoling granules

Table 2. On-line detection results of concentrated solid content in Xiaoerganmao Granules

File name	Method	Component	Prediction (%)	True value (%)	Mahalanobis distance	Range	Component value density
2008003-1.0	Concentrated solid content model of Xiaoerganmao Granules. q2	Solid content	20.67	19.73	0.11	0.47	0.65
2008003-2.1	Concentrated solid content model of Xiaoerganmao Granules. q2	Solid content	30.01	30.46	0.011	0.47	1.62
2008003-3.1	Concentrated solid content model of Xiaoerganmao Granules. q2	Solid content	34.18	34.02	0.03	0.47	0.76
2008004-1.0	Concentrated solid content model of Xiaoerganmao Granules. q2	Solid content	29.67	29.26	0.051	0.47	1.67

(*continued*)

Table 2. (*continued*)

File name	Method	Component	Prediction (%)	True value (%)	Mahalanobis distance	Range	Component value density
2008004-2.0	Concentrated solid content model of Xiaoerganmao Granules. q2	Solid content	30.30	30.21	0.0094	0.47	3
2008004-3.1	Concentrated solid content model of Xiaoerganmao Granules. q2	Solid content	31.46	31.54	0.028	0.47	3.72
2008004-4.0	Concentrated solid content model of Xiaoerganmao Granules. q2	Solid content	32.60	32.63	0.028	0.47	20.12
2008004-5.0	Concentrated solid content model of Xiaoerganmao Granules. q2	Solid content	32.79	33.13	0.04	0.47	19.87
2009002-1.0	Concentrated solid content model of Xiaoerganmao Granules. q2	Solid content	26.64	27.03	0.022	0.47	0.67
2009002-2.0	Concentrated solid content model of Xiaoerganmao Granules. q2	Solid content	26.74	27.25	0.024	0.47	0.78
2009002-3.1	Concentrated solid content model of Xiaoerganmao Granules. q2	Solid content	27.87	27.50	0.025	0.47	0.73
2009002-4 1	Concentrated solid content model of Xiaoerganmao Granules. q2	Solid content	27.87	27.72	0.025	0.47	0.73
2009004-5.0	Concentrated solid content model of Xiaoerganmao Granules. q2	Solid content	28.14	28.05	0.024	0.47	0.78

3.2 Shanghai Xingling Technology Pharmaceutical Co., Ltd.—Ginkgo Ketone Ester: Column Chromatography Section

As an extract of Ginkgo biloba leaves, Ginkgo ketone ester is a brown-yellow to yellow-brown powder. The main active substances are flavonol glycosides and terpene lactones and is mainly used clinically for blood stasis typed chest pain, angina pectoris due to coronary heart disease and dizziness due to blood stasis typed mild cerebral arteriosclerosis. It can increase cerebral blood flow, reduce cerebrovascular resistance, improve cerebrovascular circulation, protect brain cells, stabilize cell membranes, and spare brain cells from damage caused by ischemia. It can also expand the coronary arteries, increase the blood flow of the coronary arteries, improve the blood supply to the heart, and prevent the angina and myocardial infarction from forming. Nevertheless, there is a wide range of sources and varieties of its raw materials. The quality of the same medicinal materials varies due to their different growth conditions, harvest seasons, processing methods and storage conditions, resulting in a certain quality difference in the finished products of traditional Chinese medicine. The traditional quality evaluation methods are cumbersome and time-consuming, which is not conducive to large-scale rapid quality testing. Therefore, the choice of a simple analysis technology realizing fast analysis, non-destructive is helpful to greatly reducing the quality inspection time and labor costs in the production process, reducing the waiting time for product release.

In order to realize the intelligent monitoring for the production of ginkgo ketone esters, Zeda Xingbang Pharmaceutical Technology Co., Ltd. joined hands with Shanghai Xingling Technology Pharmaceutical Co., Ltd. on the ginkgo ketone ester PAT project, during which they developed online and offline rapid detection methods for medicinal materials, intermediates (extracting solution, concentrated solution, alcohol precipitant solution, chromatography solution, dry matter) and quality indicators of finished product, achieved rapid quality detection and control throughout the life cycle, and solved the problems in the existing detection mode, such as lagging results, long analysis time and low efficiency, etc. Taking the chromatography process of Ginkgo ketone ester as an example, the chromatography process is combined with online detection technology to realize the real-time and rapid detection of the liquid quality indicators in the chromatography process. During the production process, it can be used to collect quality data of the liquid in real time. Figure 3 shows the online detection installation diagram of the chromatography process as well as the online monitoring results of the chromatography process. It provides a data source for the construction of process and quality databases when combined with the process data collected by the DCS system while laying a technical foundation for process and quality data mining in the production line later on (Fig. 4).

Fig. 3. Online inspection and installation diagram of chromatography section

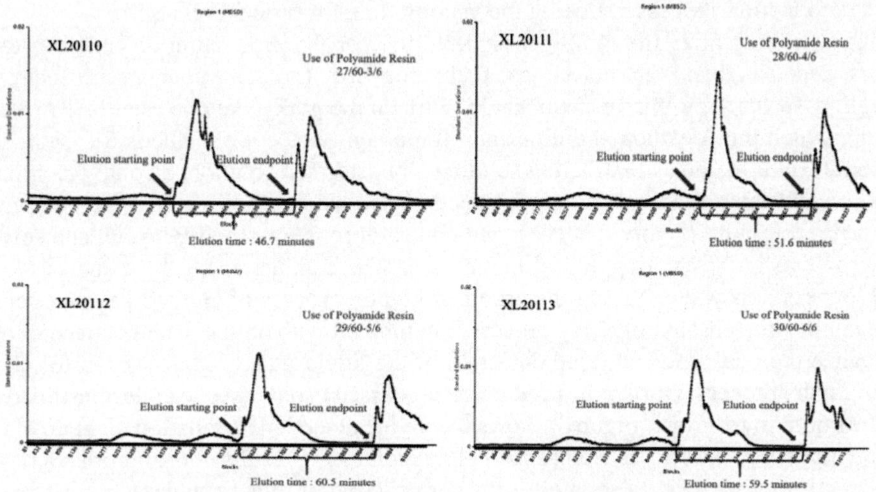

Fig. 4. Online monitoring results of chromatography section

4 Economic Benefits

The near-infrared online detection technology is of broad application prospects in that it can reduce the position setting and labor intensity of the inspectors, improve the processing volume and accuracy of data, guide the production operation in real time, reduce the energy consumption of processing and production and shorten the production of traditional Chinese medicine, bringing good economic benefits to enterprises. Taking the ginkgo ketone ester above as an example, in the pharmaceutical process of traditional

Chinese medicine, it often encounters the problem of the end-point judgment of alcohol precipitation and column chromatography. The traditional end-point judgment method is subjective and without no actual theoretical basis. Establishing the MBSD qualitative model for the ginkgo ketone ester chromatography section to track different production batches can obtain a real-time prediction map of the elution process of the ginkgo ketone ester chromatography section. The model can be divided into a static section, a water washing section, an elution stage, and an ethanol recovery stage when combined with the process. The starting point and end point of the elution section can be seen clearly from the figure, indicating that the model can be used to judge the elution starting point and end point. Using near-infrared spectroscopy technology to determine the end point of the traditional Chinese medicine production process is helpful to identifying the end point of the process in a timely and accurate manner which reduces collection time, greatly reduces energy consumption, improves raw material utilization, and ensures uniform and stable product quality, laying a theoretical foundation for improving the quality of Ginkgo ketone ester.

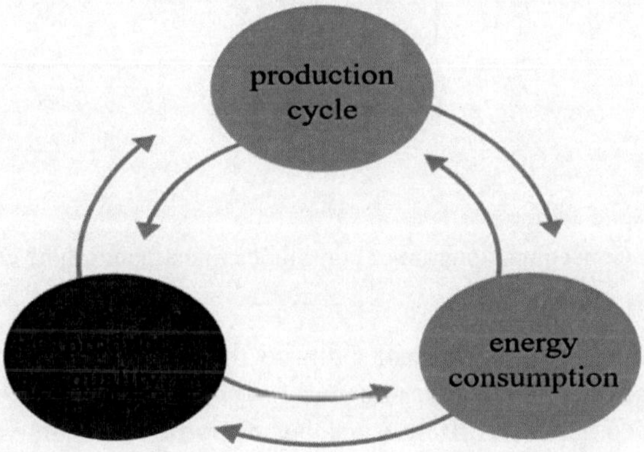

5 Prospect

For traditional Chinese medicine production, near-infrared spectroscopy technology still faces some limitations. As an analysis technology, NIR relies more on the established model. The reliability of the model varies between production batches and production time. Therefore, it is a crucial issue to update the model and transmit between different NIR devices [7]. At the same time, many types of chemical substances are involved in the pharmaceutical process, and the raw materials vary to a greater extent. It is often required to monitor multiple CPPs or CQAs which is difficult to monitor, with complicated process control and more uncontrollable factors. In addition, the current near-infrared detection of raw materials for Chinese medicine requires powdering the raw materials, and it is a challenge to develop near-infrared online detection without pretreatment.

As the development trend of pharmaceutical manufacturing in the future, drug developers and manufacturers have shown great interest in continuous manufacturing. The

Fig. 5 shows the concept of continuous manufacturing of traditional Chinese medicine granules. The continuous ingredients, continuous soft materials, continuous granulation, continuous drying, continuous total mixing processes allow the uninterrupted passage of materials or products between each operation of the unit operation through the equipment and control system designed. On the basis of real-time monitoring and control, the water, acetaminophen, chlorpheniramine maleate, and caffeine are measured for soft material particles, dry particles, and total mixed particles to form a real-time linkage feedback control system, and after combining the physical and chemical properties of the materials, a data model of simulated release is established to perform real-time release checks on the packaged formulations.

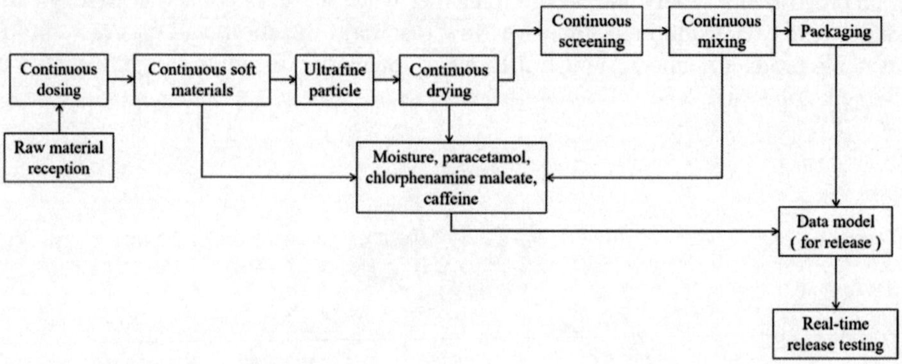

Fig. 5. Conceptual diagram of continuous manufacturing of granules

The original product of traditional Chinese medicine is characterized with large fluctuations in quality while comparing with western medicine. To a certain extent, the quality difference between different batches of traditional Chinese medicine affects the stable performance of the clinical efficacy of traditional Chinese medicine. The proposal of guiding principle of "homogenization" aims to batch charge the qualified prescription medicines of different batches in an appropriate proportion and reach the expected quality target. The integration of data intelligence and NIR nodes is the trend with the development of data technology and network technology. The near-infrared online monitoring technology serves as an extra choice for online real-time monitoring of key quality attribute of drugs in the continuous manufacturing process, supporting the traditional Chinese medicine manufacturing develops toward continuous manufacturing.

References

1. Pan, C.N.: Study on quality risk and countermeasure in the production of traditional Chinese medicine preparations. Pharm. Chem. **45**, 235–265 (2019)
2. Xiong, H.S., Tian, G., Liu, P., et al.: Research progress on key technologies of quality control in traditional Chinese medicine production process. Chin. Trad. Herb Drug. **51**, 4331–4337 (2020)

3. Cheng, Y.Y., Qu, H.B., Zhang, B.L.: Innovation guidelines and strategies for pharmaceutical engineering of Chinese medicine and their industrial translation. China J. Chin. Mater. Med. **38**, 3–5 (2013)
4. Velasco, L., Becher, H.C.: Estimating the fatty acid composition of oil in intact-seed rapeseed by near-infrared reflectance spectroscopy. Euphytica **101**, 221–230 (1998)
5. Liu, S.H., Zhang, X.G., Zhou, Q., Sun, S.Q.: Determination of geographical origins of Chinese medical herbs by NIR and pattern recognition. Spectrosc. Spect. Anal. **34**, 171–174 (2006)
6. Li, W.L., Cheng, Z.W., Wang, Y.F., et al.: Quality control of Lonicerae Japonicae Flos using near infrared spectroscopy and chemometrics. J. Pharm. Biomed. **72**, 33–39 (2013)
7. Lu, W.H., He, S.G., Cao, J.G., et al.: Analysis and application on technique of near infrared spectroscopy. Eucal. Sci. Tech. **29**, 49–54 (2012)

Coating Control on a Functional Digestion Tablet by Portable Near-Infrared Spectroscopy

Yewei Zhu[1,2], Yizhi Shi[1], Rui Chen[1], Shuai Wang[2], Zhijian Zhong[2], and Yue Huang[1(✉)]

[1] College of Food Science and Nutritional Engineering, China Agricultural University, Beijing 100083, People's Republic of China
huangyue@cau.edu.cn
[2] Beijing Great Tech Technology Co., Ltd., 100142 Beijing, People's Republic of China

Abstract. Process analysis can effectively stabilize pharmaceutical quality and optimize the control of production process. This study attempted to use a portable near-infrared spectroscopy for rapid detection of a Chinese medicine tablets from production line. First, PLS regression models were established for coating film at twelve different locations of the tablet section, and the results showed that the correlation coefficients of training and validation sets were all over 0.80. Subsequently, the twelve locations were divided into six groups to further establish regressions. After chemometrics optimization, the optimal of six group models were generally better than single location models, with Rc^2 and Rv^2 all above 0.85, and RMSEV values all below 2.0. The proposed approach can successfully realize on-site and online pharmaceutical monitoring and has a promising practical value.

Keywords: Functional digestion tablet · Chinese medicine · Coating · Near-infrared spectroscopy · Chemometrics

1 Introduction

The thickness and uniformity of the coating film are important indicators of the quality of tablets [1]. During tablet production, quality monitoring of coating thickness is usually performed by measuring the thickness of the cross-section of the tablet using a visible microscope [2]. However, this measurement is often time-consuming and laborious, which is also not convenient for quality control of large batches of samples. With development of spectroscopic techniques, the application of non-visible wavelengths to investigate tablet coatings has become an exploratory endeavor. In terms of detection mechanism, various spectroscopic techniques basically use the cavity, defect or foreign matter in the tablet coating to change the length of electromagnetic wave irradiation time [3], or intensity of reflection and refraction, finally to obtain the physicochemical information such as thickness consistency and component distribution of the tablet coating film.

X. Chu et al. (Eds.): ICNIR 2021, *Sense the Real Change: Proceedings of the 20th International Conference on Near Infrared Spectroscopy*, pp. 220–226, 2022.
https://doi.org/10.1007/978-981-19-4884-8_23

The functional digestion tablet involved in this study is a typical Chinese medicine with homology of medicine and food, which consists of radix ginseng, tangerine peel, yam, fried malt, and hawthorn as the active ingredients. It can increase the secretion of gastric juice and total acidity, improve the activity of pepsin, promote digestion and absorption [4]. In the production process, the use of spectroscopy to rapidly obtain the quality information of the coating film of the tablets in time is very critical for production management. Actually, this requirement is often reserved for laboratory-type testing instruments such as electron microscopes, spectral microscopes, etc. In this case, mobility and portability become the advantages of miniaturized spectroscopies in online control.

Currently, despite much inspiring research results have been achieved by the bench-top spectroscopies or microscopes. However, for manufacturers, quality evaluation of samples from production line to the central laboratory often results in data that lags behind the production reality. Such data would be relatively stationary and inefficient for quality management. Hence, the miniaturization and portability of spectrometers like dispersive near-infrared sensors for quality control of instant production batches of tablets is of considerable research and development worth. NIR spectroscopy often needs to be combined with chemometric methods to establish prediction models for coating film thickness. In particular, for low-energy NIR spectra obtained by non-Fourier transform methods, the fitting of target information and spectra requires more optimal selection of spectral variables. Therefore, in this study, a portable NIR spectrometer was used to rapidly evaluate the coating film uniformity of a Chinese functional digestion tablet.

2 Materials and Methods

2.1 Sample Preparation

Samples for this study were obtained directly from the production line of the CR Jiangzhong pharmaceutical workshop. The sampling was mainly divided into four batches, each batch was sampled in 9–11 groups, and 2 tablets were randomly sampled in each group (that is, about 20 samples per batch), and finally about 80 samples were selected. Spectral scans were performed immediately after sample selection, followed by coating film thickness measurements.

2.2 Instruments

Tablet coating machine is from Harbin Nano Pharmaceutical Machinery Group, China. Manual rotary microtome is from Leica HistoCore BIOCUT, Germany. Optical stereo microscope is purchased from Nikon SMZ745T, Japan. NIR spectra were acquired from VIAVI Micro-NIR PAT-W, U.S., equipped with a sample funnel and reference whiteboard.

2.3 Measurement of Coating Film

In order to accurately measure the coating thickness of samples, the tablet was longitudinally sectioned according to the position in Fig. 1, then the plan view of section was observed by an optical microscope, and the film thickness at different specified locations was measured by optical calculation. Eventually, the coating film thickness data of 12 specified locations in the longitudinal section were recorded.

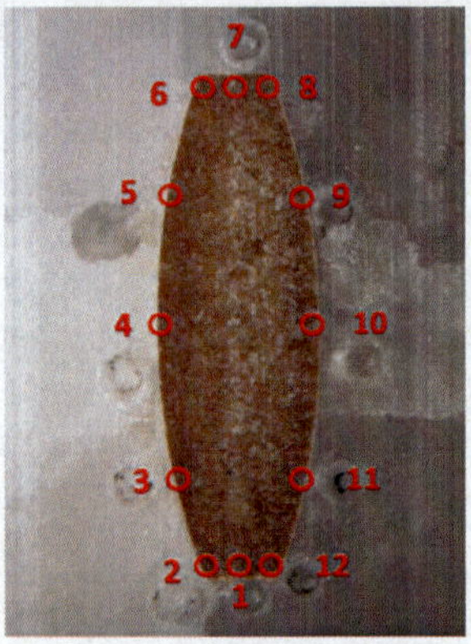

Fig. 1. Sampling position on 12 locations from tablet section

2.4 Spectra Collection and Data Processing

Group samples in each batch were picked out according to the sampling time in experimental design, from which then two tablets were randomly selected. First one tablet was put into the sample funnel, covered with a reference whiteboard, two spectra scanned, then turn the tablet over, measured the other side, and another two spectra were scanned. The same scanning was done to the other one tablet. The spectral acquisition was diffuse reflection mode, the spectral wavelength range was 900–1700 nm, the integration time was 13.5 ms, and the number of scans for each measurement was 100 times. Figure 2 shows the original NIR spectra obtained by the portable dispersive spectroscopy. Next, in order to investigate the correlation between coating film thickness data and spectra at different locations, models on microscopy calculation data and spectra were established respectively. Different spectral preprocessing methods and wavelength screening methods for each tablet were combined to establish different models. Regression mainly adopted PLS2 linear modeling. Finally, the optimal calibration model would be picked from all the comparing models, and samples of the external prediction set would be predicted to survey the relative error of the average prediction.

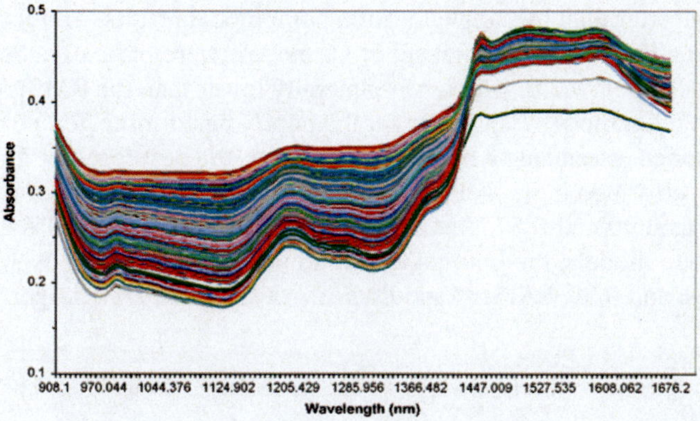

Fig. 2. Near infrared spectra of functional digestion tablets from portable spectrometer.

3 Results and Discussion

3.1 Modeling of Different Locations

First, the regression models between coating film and spectra were established for 12 specified locations on the tablets. Three models were calculated and compared for each location. The optimal PLS model at certain location was finally obtained by the different treatments of spectral pre-processing, selection of principal components, and the optimization of wavelength ranges. From the results, there was an acceptable correlation between the coating film thickness and the NIR spectra. Except for location 7 and 9, the correlation coefficients of calibration and validation set in all models basically were over 0.80, indicating that it is feasible to use portable NIRS to detect the coating film thickness of this kind of functional digestion tablets. However, it was also observed that, models established on the different locations had moderately large deviations in the prediction results of each other with RMSEV values ranged from 1.82 to 3.47. It was revealed that the certain single location was not sufficient to represent the coating level of the entire tablet.

3.2 Modeling of Different Groups

In order to enhance the representativeness of the coating film thickness, the 12 locations on the tablet cross-section were divided into 6 groups according to their relative distributions. Among which, group A contained location 1 and 7; group B contained locations 2, 6, 8, and 12; group C consisted of location 3, 5, 9, and 11; group D were locations 4 and 10; group E included locations 3, 4, 5, 9, 10 and 11; group F included all 12 locations. Herein, when modeling for each location group, the film thickness value after averaging different locations in the group was taken as the coating film value of the group. Afterward, the regression was calculated by correlating this thickness values with the spectra of the tablet, respectively (Fig. 3). Group modeling results are presented in Table 1. It can be found that, after grouping, the optimal results of the 6 groups were generally better than the single location point model results. The correlation coefficients of calibration and validation set of all models were basically above 0.85, with the RMSEV value below 2.0, which was generally lower than the RMSEV value of the single location point model. Especially for the poorly performing 7th location model in single point model, after being a member of group A, the best result of group A, Model A-2, actually performed quite well, with the correlation coefficients of calibration and validation set as 0.90 and 0.87, RMSEC and RMSEV of 1.68 and 1.84, respectively. Among all group models, the optimal calibration came from Model E-3, with the Rc^2 and Rc^2 of 0.94 and 0.93, RMSEC and RMSEV of 0.99 and 1.17, respectively.

Table 1. PLS regression models established on six different location groups on functional digestion tablets

Groups	Models	Pretreatments	PCs	Wavelength ranges (nm)	Rc^2	Rv^2	RMSEC	RMSEV
A	A–1	SNV	5	957–1484	0.85	0.81	2.11	2.39
	A–2	SNV + FD + SGS	3	951–1539	0.90	0.87	1.68	1.84
	A–3	Baseline + Detrend	3	932–1552	0.82	0.82	2.21	2.37
B	B–1	SNV	1	939–1100, 1211–1657	0.84	0.83	1.92	1.99
	B–2	SNV + FD + SGS	4	951–1254, 1285–1589	0.91	0.89	1.51	1.76
	B–3	Baseline + Detrend	5	908–982, 1050–1242, 1372–1676	0.89	0.87	1.69	1.88
C	C–1	SNV + Detrend + SGS + FD	4	939–1663	0.88	0.86	1.41	1.54
	C–2	SNV + Detrend + SGS + FD	5	1056–1651	0.88	0.85	1.66	1.84
D	D–1	SNV + Detrend + SGS + FD	7	939–1663	0.89	0.85	1.81	2.11
	D–2	SNV + Detrend + SGS + FD	7	1056–1651	0.86	0.82	1.65	1.91
E	E–1	SNV + Detrend + FD	5	1056–1651	0.90	0.88	1.50	1.65
	E–2	SNV + Detrend + SGS + FD	5	939–1663	0.90	0.88	1.55	1.75
	E–3	SNV + Detrend + SGS	7	939–1663	0.94	0.93	0.99	1.17
	E–4	SNV	7	1056–1651	0.90	0.87	1.51	1.73

(*continued*)

Table 1. (*continued*)

Groups	Models	Pretreatments	PCs	Wavelength ranges (nm)	*Rc*^{2}	*Rv*^{2}	RMSEC	RMSEV
F	F–1	SNV + Detrend + SGS + FD	5	939–1663	0.93	0.91	1.24	1.45
	F–2	SNV + Detrend + SGS + FD	7	908–1676	0.94	0.91	1.20	1.41
	F–3	SNV + Detrend + SGS + FD	6	1056–1651	0.93	0.91	1.21	1.36
	F–4	SNV + SGS + FD	6	939–1663	0.92	0.90	1.32	1.49
	F–5	SNV + SGS + FD	7	1056–1651	0.93	0.91	1.20	1.39
	F–6	SGS + FD	6	939–1663	0.91	0.89	1.35	1.49
	F–7	SGS + FD	6	1056–1651	0.90	0.88	1.41	1.56

*FD the first derivative, SNV standard normal variate, SGS Savitzky-Golay smoothing, Rc^2 correlation coefficient of calibration, Rv^2 correlation coefficient of validation, RMSEC root mean square error of calibration, RMSEV root mean square error of validation set

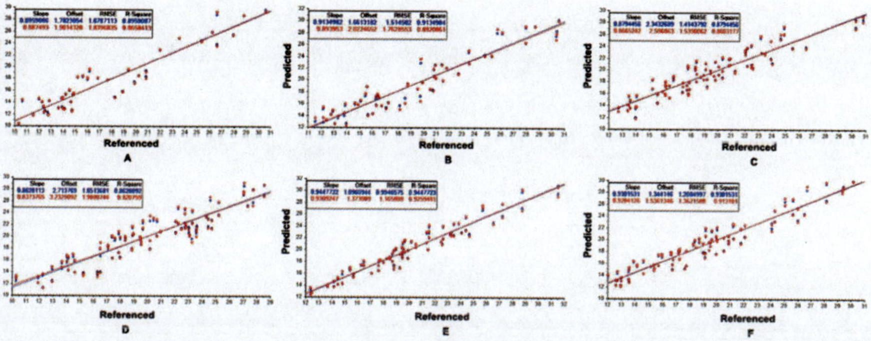

Fig. 3. PLS regression of spectral predicted versus reference values for functional digestion tablets

4 Conclusion

This study conducted a process analysis for Chinese functional digestion tablet by using the portable NIR spectroscopy. Modeling of coating film at twelve different locations on tablet slices proved that there was a good correlation between coating film and NIR spectra. Further, to improve the representativeness of the tablet coating film, the twelve locations on the tablet section were divided into different groups according to their characteristics to establish a more comprehensive model. Results showed that the location grouping models can bring the better prediction. Overall data indicates that the portable NIR spectroscopy has a good application to quantitatively detect functional digestion tablet coating film.

Acknowledgments. This research is financially supported by the project from the health food industry research institute (Xinghua, China Agricultural University (No.201905)).

References

1. Xie, C.H., You, Y., Ma, H.Q., Zhao, Y.Z.: Mechanism of inter-tablet coating variability: investigation about the motion behavior of ellipsoidal tablets in a pan coater. Powder Tech. **379**, 345–361 (2021)
2. Bikiaris, D., Koutri, I., Alexiadis, D., Damtsios, A., Karagiannis, G.: Real time and non-destructive analysis of tablet coating thickness using acoustic microscopy and infrared diffuse reflectance spectroscopy. Int. J. Pharm. **438**, 33–44 (2012)
3. Haaser, M., et al.: Evaluating the effect of coating equipment on tablet film quality using terahertz pulsed imaging. Eur. J. Pharm. Biopharm. **85**, 1095–1102 (2013)
4. Kambayashi, A., Sako, K., Kondo, H.: Scintigraphic evaluation of the in vivo performance of dry-coated delayed-release tablets in humans. Eur. J. Pharm. Biopharm. **152**, 116–122 (2020)

Rapid Screening of Industrial Hemp Based on Handheld Near Infrared Spectrometer

P. P. Zhang[1], W. J. Shi[2(✉)], G. Z. Ji[1], and Y. X. Cheng[3]

[1] Chenguang Biotech Group Co., Ltd., Handan 057250, China
[2] Hebei Chenguang Testing Technology Service Co., Ltd., Handan 057153, China
shiwenjiekey@126.com
[3] Hebei Province Natural Pigment Industry Technology Research Institute, Handan 057250, China

Abstract. To realize the rapid screening of industrial hemp for procurement, a method for evaluating the content of total cannabidiol (CBD) and total tetrahydrocannabinol (THC) of industrial hemp was established based on near infrared (NIR) reflectance spectroscopy. Both smashed un-decarboxylation industrial hemp samples and smashed decarboxylation industrial hemp samples were scanned. The spectral information was optimized by combining spectral pretreatment. These quantitative models were established based on partial least squares (PLS) and leave one cross validation. The content of total CBD and total THC models of industrial hemp were established on the condition of first derivative, standard normalization and de-trend pretreatment methods and the band range of 950–1650 nm. Through model comparison, it was found that sample pretreatment affects model accuracy. The calibration correlation coefficient (R^2(c)), the cross validation root mean square error (RMSECV) and the prediction root mean square error (RMSEP) of the best total CBD model was 0.9803, 0.2888, 0.2425, respectively. The R^2(c), the RMSECV and the RMSEP of the best total THC model was 0.9726, 0.0486, 0.0285, respectively. In the practice aplication, it could identify samples with high content CBD and avoid samples with high content THC (more than 0.3%). The handheld spectrometer could be used for the rapid determination of industrial hemp content.

Keywords: Industrial hemp · Cannabidiol · Tetrahydrocannabinol · Near infrared spectrometer

1 Introduction

Industrial hemp refers to hemp with the content of tetrahydrocannabinol (THC) less than 0.3% [1–3]. It is an annual herb of cannabinaceae genus which was called hemp in China. Studies showed that the cannabidiol (CBD) is not addictive and plays important roles in anti-inflammatory, anti-epileptic, anti-convulsive, neuroprotective, anti-cancer, anti-vomiting and so on [4–6]. At present, Industrial hemp has been widely used in textile, cosmetic, food and pharmaceutical industries. CBD and THC of industrial hemp

X. Chu et al. (Eds.): ICNIR 2021, *Sense the Real Change: Proceedings of the 20th International Conference on Near Infrared Spectroscopy*, pp. 227–232, 2022.
https://doi.org/10.1007/978-981-19-4884-8_24

mainly exist in the form of cannabidiol acid (CBDA) and tetrahydrocannabinol acid (THCA) respectively. CBDA can be converted to CBD under the condition of lighting, heating, alkalization and other conditions [7, 8].

Until now, the main method for detecting the CBD, CBDA, THC and THCA of industrial hemp was high performance liquid chromatography. The pre-processing of traditional detection methods was time-consuming and not suitable for rapid detection [9–12]. At present, there were methods for rapid detection of THC and CBD content in cannabis by near infrared spectroscopy [13]. However, conversion of phenolic acids to phenolic substances was involved in sample production and chemical testing. Therefore, the selection of NIR pretreatment was closely related to the conversion rate. In this method, two pretreatment ways were used to establish the model and verified the effect of decarboxylation rate on the near infrared model. When the decarboxylation rate was up to 90%, the prediction accuracy of the total CBD and total THC models established by the two pretreatment methods was similar. When the decarboxylation rate was lower than 90%, the prediction accuracy of the models may be poor.

2 Materials and Methods

100 batches of mature industrial hemp were from yunnan cannabis supplier, Yunma 1–7 series, Gushi Kui ma, Lu 'an cold ma, Yu County big white PI, Yanchi hemp, Wuchang 40, Lingyuan hemp, Linjiang hemp. Handheld near infrared spectrometer was German (AUAR) ZEISS. The data were acquired in the 950−1650 nm wavelength range and four spectra were acquired for each sample. The Ucal software and Origin8 were used for data treatment and presented.

Sample pretreatment of chemical method: The original industrial hemp samples were separated by quartering method. Each part was around 200 g. One part was smashed by a pulverizer and 80% of them could pass 40-mesh screen. Another part after smashed and then put into the oven at 130 °C for 60 min to finish the process of decarboxylation. Decarboxylation rate can be as high as 90%.

HPLC system coupled to UV detector (Waters) was used as a reference technique to quantify THC and CBD in the samples in order to calibrate and validate the model.

NIR sample pretreatment: (1) smashed un-decarboxylation samples: The original industrial hemp samples were separated by quartering method. Each part was around 200 g. One part was smashed by a pulverizer. (2) smashed decarboxylation samples: Another part after smashed and then put into the oven at 130 °C for 60 min to finish the process of decarboxylation.

NIR method: Both of smashed un-decarboxylation samples and smashed decarboxylation samples were scanned by handheld near infrared spectrometer. Each sample was scanned 4 times under 20–25 °C with the wavelength 950–1650 nm. In order to improve the stability and accuracy of the model, the sample thickness was kept consistent and the 4 times average spectrum were calculated to build models.

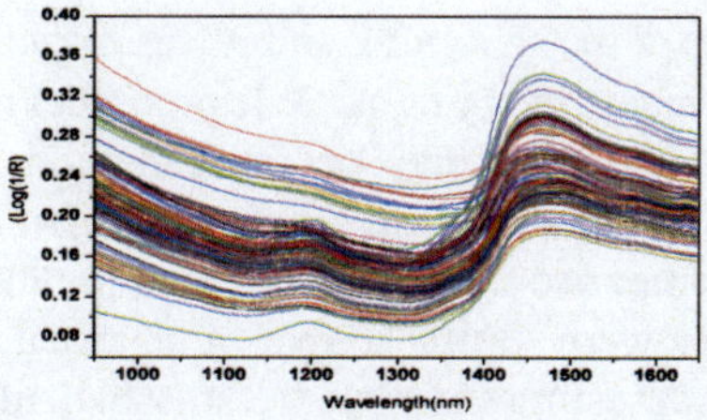

Fig. 1. The near infrared spectroscopy of smashed un-decarboxylation industrial hemp

3 Results and Discussion

3.1 Near Infrared Spectroscopy

As shown in Fig. 1, this was the near infrared spectroscopy of smashed un-decarboxylation industrial hemp. The results showed that the NIR absorption of 100 samples was not more than 0.5 in the wavelength range of 960–1650 nm. The trend of the NIR spectra of each sample was almost the same, while the absorbance at the same wavelength was different, indicating that the composition of the samples was similar, but the content of each component was different.

3.2 Spectral Preprocessing Method

Table 1. The optimal spectral preprocessing selection for the models

Sample pretreatment	Model	Pretreatment method	R^2(c)	RMSECV	RMSEC	RMSEP	Factors
Smashed un-decarboxylation samples	CBD	SG + 1st + SNV	0.9803	0.2888	0.2311	0.2425	4
	THC	SG + 1st + SNV	0.9726	0.0486	0.026	0.0285	7
Smashed decarboxylation samples	CBD	SG + 1st + SNV	0.9812	0.2845	0.2208	0.2418	5
	THC	SG + 1st + SNV	0.9782	0.0426	0.0281	0.0282	6

The difference of sample particle size directly affects the absorption and scattering of near-infrared, leading to spectral variation [14]. Therefore, the original spectrum must be preprocessed. The common pretreatment method was first derivative (1st), second derivative (2nd), multiple scattering correction (MSC), standard normalization (SNV), de-trending (DTD), Savitzky-Golay (SG), etc. [15] However, in practice, there were many kinds of interferences in the spectrum, and it was difficult to get the desired result only by one kind of pretreatment. Selecting suitable preprocessing method could improve the effect of model. Different spectral preprocessing methods were adopted to compare the effects of each model. The best spectral preprocessing of each model was shown in Table 1. Through model comparison, it was found that the modeling effect of smashed un-decarboxylation samples were similar to that of smashed decarboxylation samples. In view of the convenience of testing samples, the smashed un-decarboxylation samples were considered as the best pretreatment way. It also proved that the prediction accuracy of the total CBD and total THC models established by the two pretreatment methods was similar, when the decarboxylation rate was up to 90%.

3.3 Model Analysis and Validation

These models of un-decarboxylation crushed sample were established by Ucal software and figures were made by origin software. As shown in Figs. 2, 3, Due to the low content and narrow distribution range of THC, the linearity and stable of THC model was not so good as CBD model. However, both models were helpful to be applied in sample screening. As shown in Fig. 4, the validation correlations of CBD and THC model were all greater than 0.8, indicating that the models had a good effect.

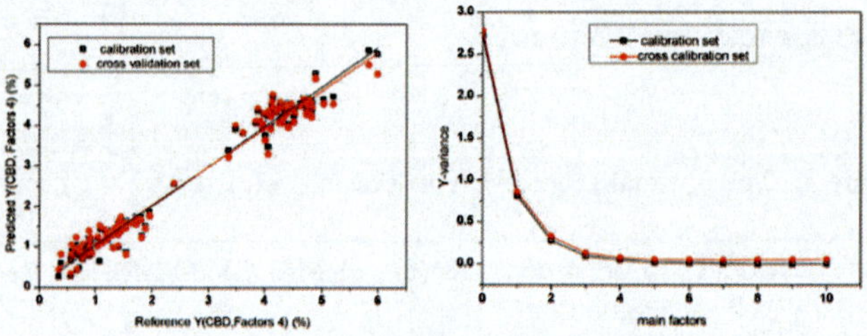

Fig. 2. The parameters of CBD model

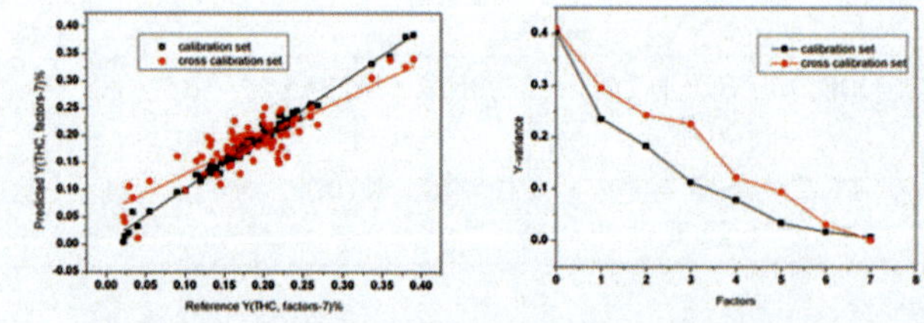

Fig. 3. The parameters of THC model

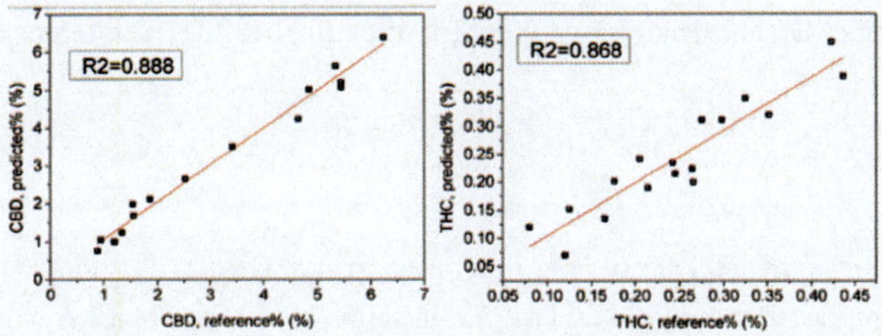

Fig. 4. The validation results of CBD and THC model

3.4 Practical Applications

These models of smashed un-decarboxylation industry hemp were put to screen raw materials for procurement in Yunnan of China. As shown in Fig. 5, during the forty samples, 3 of them THC content exceed 0.3%. Most of the samples between 0.15%–0.22%, and the lowest comes to 0.02%. The content of CBD ranges from 0.3%–5.2% with varieties. After using handheld NIR spectrometer, the CBD, THC and moisture values can be determined immediately. Furthermore, the CBD content of various varieties of industry hemp could be obtained quickly in the mature stage, taking the lead in the market competition.

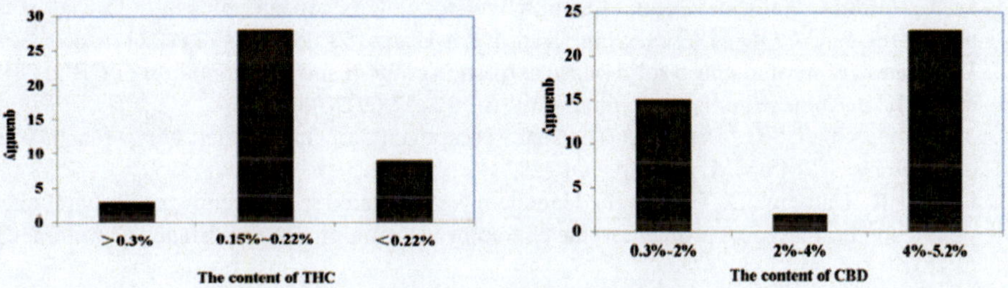

Fig. 5. The distribution of THC and CBD content

4 Conclusion

The total CBD and total THC models were established based on NIR reflectance spectroscopy. Through model comparison, it was found that the modeling effect of smashed un-decarboxylation samples were similar to that of smashed decarboxylation samples. In view of the convenience of testing samples, the smashed un-decarboxylation sample model was considered as the best pretreatment way. It also proved that the prediction accuracy of the total CBD and total THC models established by the two pretreatment methods was similar, when the decarboxylation rate was up to 90%. The CBD and THC models were helpful to be applied in material screening, which helped the company select more high CBD material and avoid material that THC exceed 0.3%, taking the lead in the market competition.

Acknowledgments. Great thanks for your contribution to NIR 2021 Conference.

References

1. Zhang, X.Y., Cao, K., Han, C.W.: Study on growth and development characteristics of three imported hemp varieties in light and moderate saline soil. J. Northeast Agri. Sci. 1–7 (2021)
2. Zhang, Q.Y., Guo, R., Xu, Y.P., Chen, X., Lv, P.: Effects of different types of plastic mulch on growth and yield of hemp. Plant Fiber Sci. China **42**, 239–243 (2020)
3. Sun, Z., Wang, J.E., Qiao, Y.G.: Study on identification method of male and female hemp plants in early development of industrial hemp. Agri. Tech. **41**, 70–71 (2021)
4. Huang, T.F., Yan, C.: Study on the antitumor effect and mechanism of cannabidiol in glioma. Chin. J. Pharm. Toxico. **35**, 725–726 (2021)
5. Yu, Y.Z., Jiang, W.: Mechanism of cannabidiol against febrile convulsion, **35**, 675 (2021)
6. Liu, J., Yu, S.Y., Zhai, W.L., et al.: Study on anti-inflammatory and antibacterial activity of cannabidiol. China Surfactent Deterg. Cosmet. **51**, 655–661 (2021)
7. Yu, X.J., Liu, C.Y., Yang, L.R., et al.: Research progress of cannabidiol in industrial hemp. Chin. Trad. Patent Med. **43**, 1275–1279 (2021)
8. Wang, Y.N., Zeng, L.B., Wang, H.Y., et al.: Effects of temperature on growth and cannabidiol content of industrial hemp. Hunan Agri. Sci. 27–31 (2021)
9. Liu, S.G., Ma, H.Y., Li, Z.G., et al.: Determination of CBD and THC in hemp Mosaic leaves by HPLC, Yunnan. Chem. Tech. **47**, 62–64 (2020)
10. Li, J., Mi, Y.L., Wang, S.J., et al.: Quantitative study of six cannabinoids in industrial hemp based on UPLC-QQQ-MS/MS. Chin. Trad. Herb. Drugs. **53**, 1163–1172 (2021)
11. Preparation of normal phase solid phase extraction column and determination of CBD, CBN and THC in Chinese hemp, J. Qiqihar Univ. **36**, 13–17 (2020)
12. Yan, Z.E., Guan, J.L., Ding, B.: The main components of interplanted hemp plants were detected by GC/MS/SIM, pp. 115–121 (2021)
13. Deidda, R., Damergi, D., Coppey, F.: Handheld Near Infrared spectroscopy for cannabis analysis: from the analytical problem to the chemometric solution. In: Conference Chimiometrie (2020)
14. Wang, D.M., Ji, J.M., Gao, H.Z.: Influence of multiple scattering correction on calibration model of near infrared spectral analysis. Spectrosc. Spect. Anal. **34**, 2387–2390 (2014)
15. Wang, L., Meng, Q.X., Ren, L.P., Yang, J.S.: Near infrared spectroscopy rapid analysis technology and its application in animal feed and product quality inspection. Spectrosc. Spect. Anal. **30**, 1482–1487 (2010)

Pharmaceutical and Chemistry

Embedded NIR Spectroscopy for Rotary Tablet Press

Yves Roggo[1](✉), Laurent Pellegatti[1], Anna Novikova[2], Alexander Evers[2], Simon Ensslin[1], and Markus Krumme[1]

[1] Novartis Pharma AG, Technical Research and Development, Continuous Manufacturing, Basel, Switzerland
yves.roggo@novartis.com
[2] Fette Compacting, Schwarzenbek, Germany

Abstract. The Near Infrared Spectroscopy (NIRS) was employed for control and monitoring of the tableting process during a continuous manufacturing process. The tableting production is key step in the production of solid dosage forms. Two spectrometers were embedded in the press and controlled by the press automation without any additional computer in order to have a fast and robust measurement. 72 batches were produced to calibrate and to validate the NIRS models at the BU and the CU position. A precise calibration (API content in %) have been obtained and a full speed acquisition was possible. The embedded spectroscopy for the control and the monitoring of the tableting process was demonstrated. The proposed strategy is an adequate Process Analytical Technology tool for continuous manufacturing and will enable opportunities for Real Time Release.

Keywords: Process Analytical Technology · Embedded spectroscopy · Tablet press · Pharmaceutic

1 Introduction

Continuous Manufacturing (CM) of pharmaceutical drug products is a new approach within the pharmaceutical industry, opposing traditional batch manufacturing process based on its potential to increase manufacturing flexibility and efficiency. In CM, all process units are directly connected to each other. Process Analytical Technology is a key element of the control strategy of the CM production. The use of Near Infrared Spectroscopy (NIRS) technique was employed for the control and monitoring of the tableting step during a continuous manufacturing process. The tableting process is key step in the production of solid dosage forms. The final product quality depends on this step. It is important to verify the blend uniformity prior the tableting and the content uniformity of the tablet.

X. Chu et al. (Eds.): ICNIR 2021, *Sense the Real Change: Proceedings of the 20th International Conference on Near Infrared Spectroscopy*, pp. 235–239, 2022.
https://doi.org/10.1007/978-981-19-4884-8_25

2 Materials and Methods

2.1 Materials

Two NIRS instruments were installed in a high throughput rotary tablet press (FE55, Fette Compacting) in order to in-line control the input of the press and the final product. The first instrument was installed in the tablet press feed frame (Blend Uniformity - BU - probe), to ensure the absence of segregation issues that could occur after milling and transfer (Fig. 1). The second instrument was installed in the tablet press to measure every single tablet right before being ejected from the press (Content Uniformity - CU - probe). The novelty of this work first lies in the acquisition speed of NIR spectra covering the whole speed range of a production press. The embedded system can acquire spectra with the frequency of 120 Hz. Moreover, the two spectrometers are fully embedded in the press automation. The automation of the press controls the spectrometer and performed the online prediction. A single rejection based on the NIR prediction out of specification tablet is done by the tablet press via compressed air (fast gate ejection).

2.2 Methods

Design of Experiments: A DoE was prepared to assess the impact of the Active Pharmaceutical Ingredient (API) content, the turret speed and the feed frame speed. Spectra were measured in reflection mode using one scan of 0.004 s and a spectral range of 1000–2000 nm. 72 batches were prepared: 9 levels with different API contents and 8 levels with different speeds (4 levels of turret speed (2.4 k, 45 k, 60 k, 90 k Tablets per hours corresponding to 1, 19.2, 25.6 and 38.4 rpm) at constant feed frame speed - and 4 levels of feed frame speed (10, 30, 45 and 60 rpm) at constant turret speed of xx rpm).

Chemometrics: The spectral preprocessing has been selected (Savitzky Golay second derivative - filter window length = 31 and polynomial order = 3). Principal Component Analysis (PCA) is used to visualize the impact of the main process parameters and Partial Least Squares (PLS) regression is applied for the quantitative prediction of the API content.

Fig. 1. FETTE FE55 press with two NIR probes: blend uniformity of powder and content uniformity of tablet

3 Results and Discussion

It was possible to obtain high spectral quality for the two NIR sensors (CU and BU - Fig. 2). The Principal Component Analysis is used as a clustering method in order to detect outliers and to assess the main sources of spectral variations (*i.e.* process understanding). The API content has a clear impact on the NIR spectra for CU and BU (Fig. 3). The feed frame speed has only a small influence on the BU spectra. The speed of the turret has a clear influence of the CU spectra. This variability has to be taken into account for the model calibration.

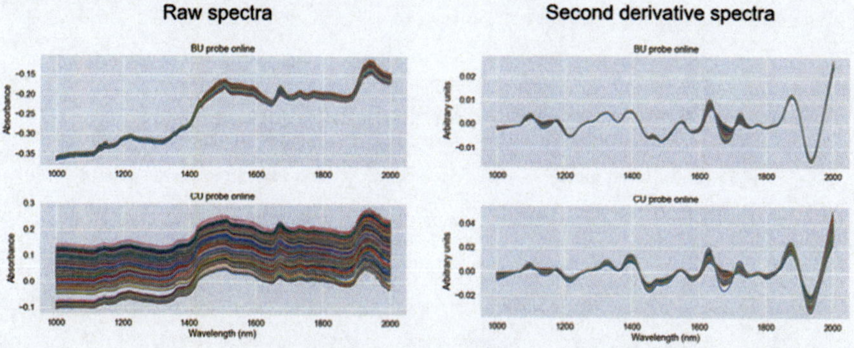

Fig. 2. NIRS spectra acquired online or in the laboratory

PLS calibration for the BU was developed with online spectra. Excellent accuracy was obtained with this sensor. Concerning the CU sensor, a first calibration was created in a laboratory and transferred to the production line. The direct calibration transfer was not possible (Fig. 4). A robust calibration developed with spectra of the lab equipment and the production system was computed in order to solve the calibration transfer issue (Fig. 5). A high correlation coefficient (close to one) and a low error of prediction (less than 3.0%) were obtained for the two NIRS probes. The quantitative prediction of the API content is used to evaluate the uniformity of mixing and the content uniformity of tablets. The content of API for BU and CU has to be within predefined limits. Therefore, NIRS enables the online quality control of the product.

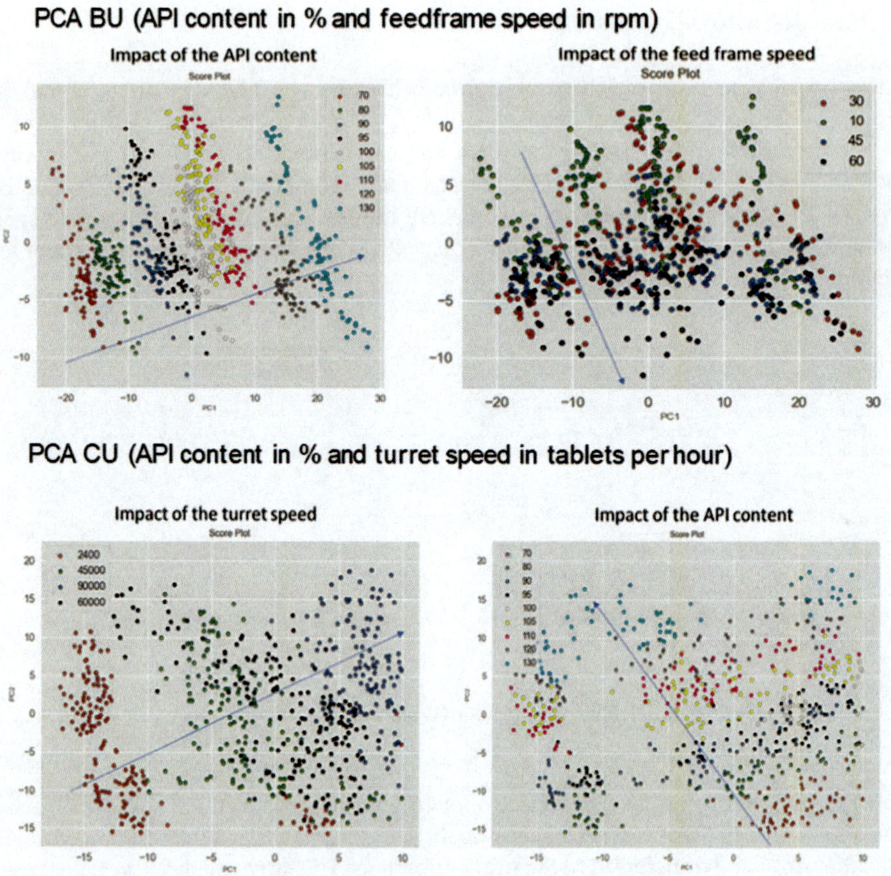

Fig. 3. Principal component analysis for BU probe

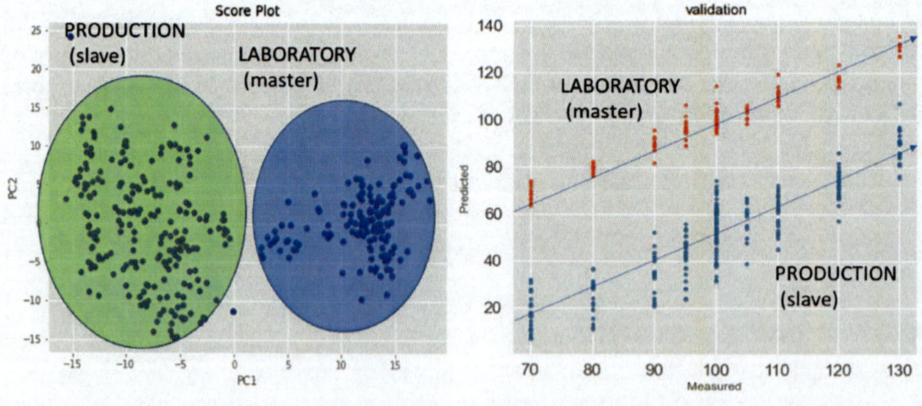

Fig. 4. NIRS spectra acquired online or in the laboratory and PLS calibrations.

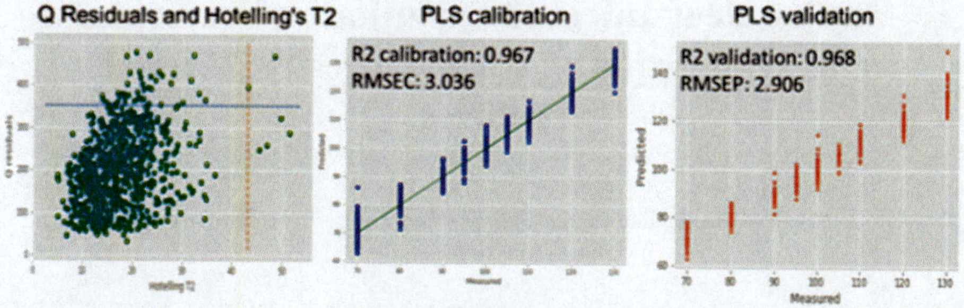

Fig. 5. Robust calibration

4 Conclusion

The results presented in this study demonstrate that the tableting step in a continuous manufacturing process can be monitored through the combination of two near infrared probes. Quantitative regression models were developed for this application. The first probe analyzed the dried granules in the tablet press feed frame and verifies their uniformity in API content to avoid segregation issues that could occur after milling and transfer. The second probe allowed 100% control of tablet content uniformity in real-time. Due to the fact that the PAT is integrated into the tablet press control system, the PAT outcome can be linked to the sorting functionality of the press, which enables the opportunity to reject out-of-specification tablets. The proposed strategy is a performant Process Analytical Technology tool for continuous manufacturing and will enable Real Time Release.

Acknowledgments. Great thanks to Fette, Development team in Mechelem (BE).

Reference

1. Roggo, Y., et al.: A review of near infrared spectroscopy and chemometrics in pharmaceutical technologies. J. Pharm. Biomed. Anal. **44**, 683–700 (2007)

On-line Near-Infrared Quantitative Prediction and Verification of Waste Polyester Blended Fabrics

Yue Wang[1], Wenqian Du[1], Peng Jiang[2], Wenxia Li[1](✉), Zhengdong Liu[3], and Huaping Wang[4]

[1] College of Materials Design and Engineering, Beijing Institute of Fashion and Technology, Beijing 100029, China
liwenxia307@163.com
[2] Beijing Wool, Jute & Silk Fabrics Quality Supervision Laboratory, Beijing 100085, China
[3] School of Fashion, Beijing Institute of Fashion and Technology, Beijing 100029, China
[4] College of Materials Science and Engineering, Donghua University, Shanghai 201620, China

Abstract. Polyester is an important textile material and the main part of waste textiles. Predicting the content of waste polyester fibers is the key to realizing the classification and recycling of waste textiles. In this paper, a total of 273 samples of polyester/nylon, polyester/wool, and polyester/cotton were used as the research targets, and quantitative analysis models of three types polyester blends were established by using near-infrared online analysis technology combined with partial least squares (PLS). The selection of preprocessing methods and the optimization of evaluation factors in the modeling process were also discussed. When the preprocessing method is Savitzky-Golay Derivative + Vector Normalization + Multiplicative Scatter Correction + Mean Centering, the quantitative analysis models of polyester/nylon and polyester/cotton blended fabrics predicted the best results. Among them, the number of evaluation factors for the polyester/nylon model is 9, and the evaluation factor for the polyester/cotton model is 5. When the Savitzky-Golay Derivative + Mean Centering was selected, and the number of evaluation factors is 6, the prediction effect of the polyester/wool model is the best. In this experiment, the built model was internally verified, and the accuracy of the model is higher than 95% under a tolerance of 3%. The model was tested externally using 90 polyester samples that were not involved in the modeling, and the overall prediction accuracy of the model was 94.4%. The established models can be applied to the quantitative prediction of three polyester blended fabrics.

Keywords: On-line near-infrared · Quantitative analysis models · Classification and recycling

1 Introduction

Polyester fiber has become the main raw material of textile industry because of its excellent performance and low cost. Under the current "fast fashion" consumption mode,

X. Chu et al. (Eds.): ICNIR 2021, *Sense the Real Change: Proceedings of the 20th International Conference on Near Infrared Spectroscopy*, pp. 240–250, 2022.
https://doi.org/10.1007/978-981-19-4884-8_26

the production and consumption of global textile and apparel products are increasing year by year, of which polyester textiles account for about 70% of all textile products. As a major producer and consumer of polyester fibers in the world, China has steadily increased its polyester production capacity. In 2019, China's polyester fiber production exceeded 43 million tons, a year-on-year increase of 14.82% [1]. With the improvement of people's level of consumption and the rapid change of fashion trends, the use cycle of textiles is shortened, and the number of waste textiles is increasing, of which waste polyester textiles occupied a considerable proportion. Polyester fibers are difficult to biodegrade under natural conditions. A large number of discarded polyester textiles not only cause waste of resources, but also have a great impact on the environment. In 2020, the national standard for the classification and code of waste textiles has been promulgated and implemented [2], which provides technical specifications for the recycling and sorting of waste textiles, but quickly and non-destructively carry out online identification and content prediction of polyester blended fabrics is an unsolved technical problem. the "14th Five-Year" Development Outline of the textile industry also clearly stated that one of the key projects for China's textile industry to realize resource recycling is to break through the industrialization and large-scale technology of chemically recycled polyester, and promote the construction of waste textile recycling, sorting, dismantling and standardized processing bases [3]. Therefore, it is of great practical significance for China's textile industry to focus on the recycling and reuse of polyester waste textiles and to overcome the key technologies of component identification and content prediction.

Online Near Infrared Spectroscopy (Online NIR) analysis technology is a new type of analysis and detection technology. In practical applications, it has the advantages of fast, non-destructive, low-cost, zero pollution, etc. It has been successfully applied to the analysis of agricultural, food, pharmaceutical and other products [4–6]. The information carried by the near-infrared spectrum is mainly the multiple frequency and combined frequency information of the stretching vibration of hydrogen-containing groups (C-H, N-H, O-H). In terms of textile detection, the characteristic spectral regions corresponding to different components are different, and the component content has a direct impact on the characteristic peak intensity, so it can be applied to the composition identification and content prediction of waste textiles [7, 8].

In this paper, online NIR analysis technology is used to study a total of 273 samples in three categories: polyester/nylon, polyester/wool, and polyester/cotton. The quantitative analysis models of three types of polyester fabrics were established by using the chemometrics software combined with the PLS method. And by choosing different pretreatment methods and evaluation factors, the optimal modeling conditions are discussed to realize the prediction of polyester content in polyester blended fabrics, and provide a basic model for subsequent online sorting.

2 Experimental Section

2.1 Experimental Device and Test Conditions

In this experiment, the self-developed "BIFT NIRMagic 6701 efficient identification and automatic sorting device of fiber products", referred to as "sorting device", was used in the near-infrared spectrum acquisition of fabric samples, with a built-in array NIR spectrometer. The spectrum collection method is diffuse reflection, and the wavelength range is 900–2500 nm. The device mainly includes four parts: sample delivery system, NIR detection system, intelligent identification system and purge sorting system. It is equipped with textile online master control program (TOCP) and chemometrics software (ChemoStudio2019, CS2019), the sorting speed is better than 30 pieces per minute. At the beginning of the experiment, identification of sample components was performed, using the Fourier Transform Mid-Infrared (FT-MIR) spectrometer of Thermo Fisher Scientific, USA, supplemented by Smart Orbit accessory, the wavenumber range is 400–4000 cm^{-1}, the resolution is 8 cm^{-1}, and its number of scans is 32.

When collecting near-infrared spectra, the optimal test conditions explored by the research group in the early stage were used, the speed of the conveyor belt is 0.282 $m\ s^{-1}$, and the measurement integration time of the sample is 10 ms. Since the NIR detector quickly obtains the fabric information by scanning the sample on the conveyor belt, it is difficult to capture the information of the knitted sample with too loose structure. And the test thickness of sample is too thin, which will shift up from the spectral baseline and weaken the characteristic signal [9]. Therefore, the test thickness of sample in the experiment for the tight fabric is above 1 mm, and the loose fabric is above 2.5 mm.

2.2 Sample Selection and Content Determination

The samples used in this experiment were provided by the National Wool Textile Quality Inspection Center, Ningbo Entry-Exit Inspection and Quarantine Bureau, Beijing Textile Fiber Inspection Institute, Wujiang Yuehua Weaving Co., Ltd., Chuangyi (Fujian) Textile Technology Co., Ltd., etc. There are 66 polyester/nylon samples, 97 polyester/wool samples, and 110 polyester/cotton samples, for a total of 273 samples.

First, according to "FZ/T 01057.8-2012 Test Method for identification of textile fibers Part 8: Infrared Spectroscopy" [10], using Fourier transform mid-infrared spectrometer and attenuated total reflection sampling accessory, collecting the spectra of the front and back sides of waste textiles and the warp and weft yarns respectively to determine fabric composition. For fibers that are difficult to distinguish by the mid-infrared, according to "FZ/T 01057.3-2007 Test Method for identification of textile fibers Part 3: Microscopy" [11], the fiber's microscope image is used for identification.

The content of each component of the modeling sample constitutes the basic data of the quantitative model and plays a crucial role in the accuracy of the model prediction results. Therefore, the three types of fabrics in the experiment were determined according to the chemical dissolution method of the national standard. Among them, the polyester/nylon blended fabric sample adopts GB/T 2910.7-2009 "Quantitative Chemical Analysis of Textiles Part 7: Polyamide Fiber and Certain Other Fiber Mixtures (formic acid method)"; the polyester/wool blended fabric sample adopts GB/T 2910.4-2009 "Quantitative Chemical Analysis of Textiles Part 4: Mixtures of Certain Protein Fibers and Certain Other Fibers (Hypochlorite Method)"; Polyester/cotton blended fabric samples adopt GB/T 2910.11-2009 "Quantitative Chemical Analysis of Textiles Part 11: Mixtures of Cellulose Fibers and Polyester Fibers (Sulfuric Acid Method)" [12–14].

2.3 Online NIR Spectrum Acquisition

The samples were scanned one by one using the "sorting device" to obtain their online raw near-infrared spectra. The front and back sides of each fabric sample were scanned more than 3 times, and the spectra with good repeatability and consistent spectra on the front and back sides were selected and stored. The components and their contents of the fabric samples selected in this experiment were evenly distributed.

2.4 Establishment of Quantitative Analysis Model

Three types of on-line near-infrared quantitative analysis models of waste polyester textiles were established with polyester content as a reference, there are polyester/nylon, polyester/wool and polyester/cotton quantitative analysis models. The modeling steps are as follows:

(1) The original online NIR spectra of the three types of blended fabric samples were added to the CS2019 software, and the polyester content (T%) was used as the model parameter property. In order to make the model have better universality, the content of polyester is as evenly distributed as possible between 0–100%;
(2) Select the K-S automatic classification method to divide the data set into the calibration set and the validation set;
(3) Select the modeled spectral band according to the correlation coefficient figure. The closer the correlation coefficient of the selected spectral band is to 1, the better the correlation, and should contain the main characteristic peaks of this type of fabric.
(4) Use the PLS method to establish a quantitative model. Mean Centering is the default preprocessing method. For other preprocessing methods, such as Savitzky-Golay Smoothing (S-G Smoothing), Savitzky-Golay Derivative (S-G Derivative), Multiplicative Scatter Correction (MSC), Standard Normal Variate (SNV), Maximum Minimum Normalization (MMN), Vector Normalization (VN), etc. To select and apply them in combination. Eliminate outlier samples and select the best evaluation factor, establish different quantitative models and compare model parameters.
(5) Use the quantitative model under the optimal preprocessing method to predict the polyester content of the samples in the validation set to determine the prediction accuracy of the model.

3 Results and Discussion

3.1 Basis for NIR Quantitative Analysis of Polyester Blended Fabrics

Analyze the online NIR spectra of each component of the modeling sample. As shown in Fig. 1, the main characteristic peaks of polyester are around 1665 nm and 2250 nm, nylon around 1730 nm and 2300 nm, wool around 1500 nm and 1950 nm, and cotton around 1470 nm and 2090 nm [15].

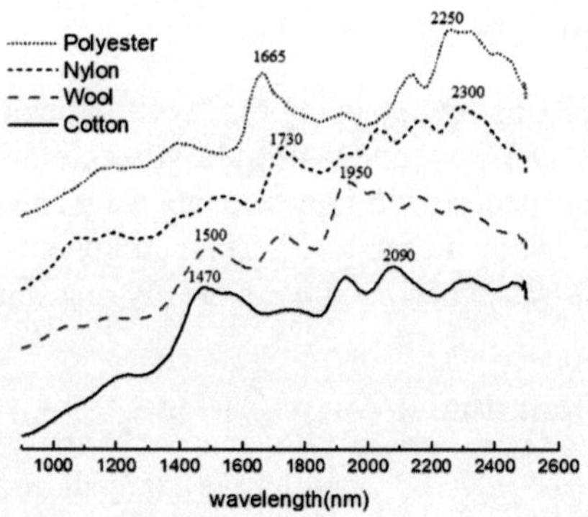

Fig. 1. Online raw NIR spectra of each component of the modeled sample

Different polyester content in the fabric corresponds to different characteristic peak intensities, and the position of the peak also appears slightly shifted. Taking polyester/nylon as an example, with the decrease of polyester content, the characteristic peak intensity of polyester at 1665 nm and 2265 nm gradually weakened, while the characteristic peak intensity of nylon at 1725 nm and 2300 nm gradually increased. The changes of the online NIR spectra of polyester/nylon with different contents are shown in Fig. 2. The change of the component content of the fabric is positively correlated with the change of the corresponding characteristic peak intensity, which lays the foundation for the establishment of the quantitative model of the polyester/nylon blended fabrics. Therefore, The online NIR spectroscopy can be used to predict the component content of polyester blended fabrics.

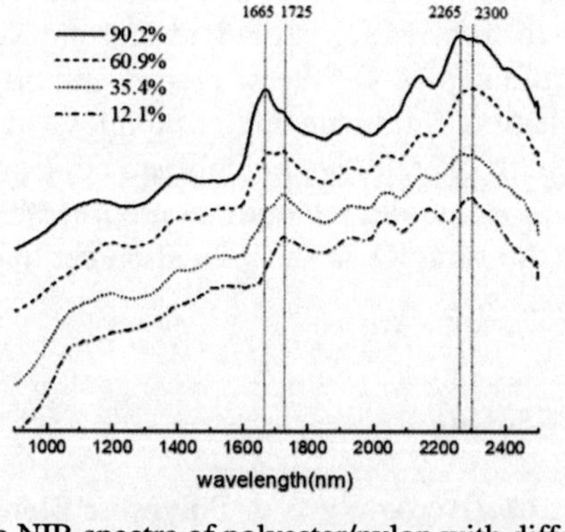

Fig. 2. Online NIR spectra of polyester/nylon with different contents

3.2 Establishment of Online NIR Quantitative Model and Internal Inspection

3.2.1 Selection of Different Preprocessing Methods and Modeling Parameters

Compared with the mid-infrared (MIR) spectrum, the absorption band observed in the NIR region is wider, and there are overlapping bands. When the online NIR test is performed on the fabric, due to the short scanning time and high scanning speed, the original spectrum signal is relatively small and weak, and low intensity [16]. In order to filter spectral noise, reduce the drift of spectral baseline, and reduce background interference, in the modeling process, the original online NIR spectrum must be pre-processed to enhance the spectral feature information and improve the robustness of the model [17]. For the three polyester blended fabrics, under the default selection of mean centering, other different preprocessing methods and their combinations were selected respectively, and the parameters of each model were compared to select the optimal method combination.

When choosing modeling parameters, the setting of evaluation factors is crucial. If the number of evaluation factors is too low, the equation will be insufficiently fitted and useful information will be lost; if the number of evaluation factors is too high, it will lead to overfitting and increase the model prediction error [18]. On the basis of the selected preprocessing method, observe the model curve by adjusting the evaluation factors of the model, and select the number of factors when the curve tends to be flat. Some parameters are used as evaluation indicators to compare different preprocessing methods and evaluation factors, such as Root Mean Square Error of Calibration (RMSEC), Root Mean Square Error of Validation (RMSEV), Relation Coefficient of Calibration (RC), Relation Coefficient of Validation (RV), Relative Prediction Deviation of Calibration (RPDC), etc. The smaller the RMSEC, the closer the correlation coefficient R is to 1, and the larger the RPD, indicating that the higher the accuracy of the model and the better the prediction performance.

The evaluation parameters of the polyester/nylon quantitative model are shown in Table 1. The evaluation factor of the optimal model is set to 9, the optimal preprocessing method is S-G Derivative + VN + MSC, and the Mean Centering is selected. Under this condition, the minimum RMSEC of the model is 0.98, the maximum RPDV is 12.68, the ratio of RMSEV to RMSEC is 0.98, less than 1, and the model is robust.

The model evaluation parameters of polyester/wool blended fabrics under different pretreatment methods and evaluation factors are shown in Table 2. After selecting Mean Centering and S-G Derivative preprocessing, the minimum RMSEC is 1.21, the maximum RPDV is 8.05, and the standard deviation ratio between the calibration set and the validation set is 0.95, which is less than 1, and the model is robust. And when the number of evaluation factors is 6, the prediction accuracy of the model is the highest.

Table 1. Evaluation parameters of polyester/nylon quantitative model under different preprocessing methods

Preprocessing	Factor	Calibration set			Validation set		
		RMSEC	RC	RPDC	RMSEV	RV	RPDV
Raw	4	2.15	0.98	4.98	1.7	0.98	3.79
S-G Derivative	7	1.7	0.99	6.97	1.44	0.99	5.72
S-G Derivative +S-G Smoothing	7	1.68	0.99	7.23	1.40	0.99	5.98
S-G Derivative +MSC	3	1.34	1.00	11.39	1.04	1.00	10.77
S-G Derivative +SNV	6	1.12	1.00	14.91	1.14	0.99	8.96
S-G Derivative +MMN	6	1.45	0.99	9.20	1.36	0.99	6.32
S-G Derivative +VN	7	2.45	1.95	1.00	2.72	1.53	0.99
S-G Derivative +VN+MSC	**9**	**0.98**	**1.00**	**18.30**	**0.96**	**1.00**	**12.68**
S-G Derivative +VN+SNV	6	1.4	1.00	10.58	1.24	0.99	7.68
S-G Derivative +MMN+SNV	6	1.4	1.00	10.58	1.24	0.99	7.68
S-G Derivative +MMN+MSC	8	1.38	1.00	11.03	1.08	1.00	10.20

Table 2. Evaluation parameters of polyester/wool quantitative model under different pretreatment methods

Preprocessing	Factor	Calibration set			Validation set		
		RMSEC	RC	RPDC	RMSEV	RV	RPDV
S-G Derivative	**6**	**1.21**	**1.00**	**11.19**	**1.27**	**0.99**	**8.05**
S-G Derivative +MSC	7	1.37	0.99	8.79	1.5	0.99	5.81
S-G Derivative +SNV	3	1.45	0.99	7.86	1.34	0.99	7.28
S-G Derivative + MMN	6	1.38	0.99	8.62	1.35	0.99	7.16
S-G Derivative +VN	3	1.44	0.99	7.80	1.36	0.99	7.01
S-G Derivative +VN+MSC	8	1.27	1.00	10.09	1.46	0.99	6.16
S-G Derivative +VN+SNV	3	1.45	0.99	7.86	1.34	0.99	7.28
S-G Derivative +MMN+SNV	3	1.45	0.99	7.86	1.34	0.99	7.28
S-G Derivative +MMN+MSC	8	1.27	1.00	10.10	1.46	0.99	6.16

The model evaluation parameters of polyester/cotton fabrics under different pretreatment methods are shown in Table 3. After selecting Mean Centering and preprocessing by S-G Derivative + VN + MSC, the minimum RMSEC is 1.28, and the maximum RPDV is 13.71. And when the number of evaluation factors is set to 5, the prediction accuracy of the model is the highest.

Table 3. Evaluation parameters of polyester/cotton quantitative model under different pretreatment methods

Preprocessing	Factor	Calibration set			Validation set		
		RMSEC	RC	RPDC	RMSEV	RV	RPDV
S-G Derivative	4	1.35	1.00	10.85	1.13	1.00	13.32
S-G Derivative +MSC	4	1.29	1.00	11.00	1.17	1.00	12.45
S-G Derivative +SNV	4	1.35	1.00	10.87	1.24	1.00	11.16
S-G Derivative +VN	4	1.34	1.00	11.16	1.25	1.00	11.01
S-G Derivative +VN+MSC	**5**	**1.28**	**1.00**	**12.37**	**1.12**	**1.00**	**13.71**
S-G Derivative +VN+SNV	4	1.35	1.00	10.87	1.24	1.00	11.16

3.2.2 Internal Testing of the Model

Under the premise that the tolerance is within 3%, the absolute error between the actual value of a certain component content in the blended fabric and the predicted value of the quantitative model is used as the evaluation standard. Three optimal models were internally tested respectively, using 20 polyester/nylon samples, 21 polyester/wool samples, and 33 polyester/cotton samples from the validation set. Among them, the prediction accuracy of the polyester/nylon model is 100%, and the absolute error range is 0.04–2.43%. The fitting curve of the model is shown in Fig. 3. The prediction accuracy of the polyester/wool model is 95.2%, and the absolute error range is 0.04–3.77%. The fitting curve of the model is shown in Fig. 4. The prediction accuracy of the polyester/cotton model is 97%, and the absolute error range is 0.01–3.3%. The fitting curve of the model is shown in Fig. 5.

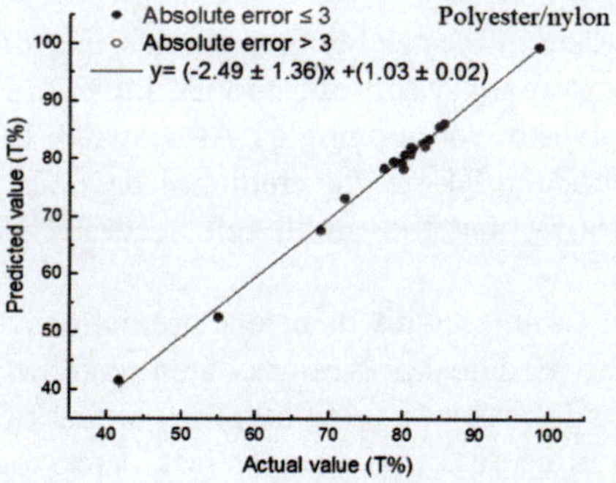

Fig. 3. Internal prediction fitting curve of polyester/nylon quantitative model

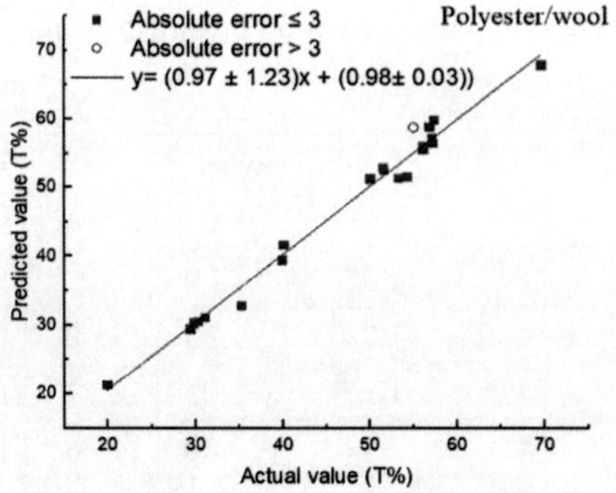

Fig. 4. Internal prediction fitting curve of polyester/wool quantitative model

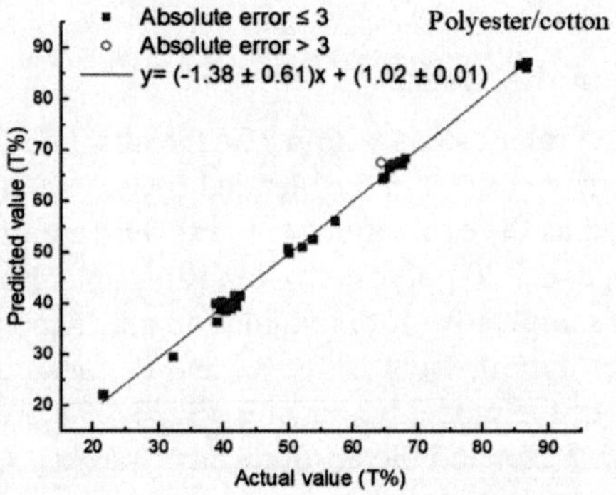

Fig. 5. Internal prediction fitting curve of polyester/cotton quantitative model

3.3 External Validation and Results Analysis of Quantitative Models

In this experiment, 90 waste polyester blended fabrics that did not participate in the modeling were used to externally verify the model, including 30 polyester/nylon, 30 polyester/wool, and 30 polyester/cotton samples. As shown in Table 4, the overall prediction accuracy rate of the model is 94.4%, the prediction accuracy rate for polyester/nylon and polyester/wool is 100%, and the prediction accuracy rate for polyester/cotton samples is lower, is 83.3%.

Five polyester/cotton samples with incorrect predictions were analyzed, and the absolute error between the predicted polyester content and the true value was between 5.30% and 8.09%. The main reason may be that the content of one of the components is less than 5%, and the online detection speed is fast, cause the characteristic peaks of low-content components are inconspicuous. On the other hand, there may be errors in the content of the samples used for verification, and the content of these five samples will be checked by chemical dissolution method in the future.

This experiment will continue to optimize and improve the established polyester quantitative analysis model, so that the predicted value of polyester content is as close to the real value as possible, providing a fast and accurate online quantitative prediction

Table 4. Online NIR content prediction accuracy of waste polyester textiles

Textile category	Quantity of samples	Correct quantity of online prediction	Prediction accuracy (%)
Polyester/nylon	30	30	100
Polyester/wool	30	30	100
Polyester/cotton	30	25	83.3
Total	90	85	94.4

technology for waste polyester textiles, promoting the grading and subsequent reuse of waste polyester blended fabrics in the recycling process.

4 Conclusions

In this paper, using the self-developed "efficient identification and automatic sorting device of fiber products", based on the online near-infrared analysis technology, combined with the PLS method, the quantitative analysis models of three polyester waste textiles were established, And the optimal modeling conditions were discussed. The best pretreatment methods for the quantitative models of polyester/nylon and polyester/cotton is S-G Derivative + VN + MSC + Mean Centering, and the best pretreatment methods for the quantitative models of polyester/wool is S-G Derivative + Mean Centering. 90 waste polyester blend samples that were not involved in the modeling were selected for online prediction of polyester content, and the prediction accuracy is 94.4%. The disadvantage of the model is that the prediction results of the two-component fabrics that the content of a certain component is less than 5% are not accurate enough, and there are still errors higher than 3%. The model will continue to be optimized in the later stage. This study can predict the content of waste polyester textiles covered by the model online, and the prediction time of each sample is less than 2 s, which provides a technical reference for the grading recycling and high-value utilization of waste polyester blended fabrics.

Acknowledgments. Great thanks for National Key R&D Program Project (2016YFB0302900) and ICNIRS 2021 Committee.

References

1. Weiran, Q., Pinghua, X., Laili, W.: Review on polyester fiber recycling and progress of its environmental impact assessment. Adv. Text. Technol. **29**(01), 22–26 (2021)
2. GB/T 38923-2020. Classification and code of textile waste
3. "14th Five-Year" Development outline of textile industry, Text. Sci. Res. **07**, 40–49 (2021)
4. Samantha, J.N., Henry, J., Bach Knudsen, K.E.: Prediction of protein and amino acid composition and digestibility in individual feedstuffs and mixed diets for pigs using near-infrared spectroscopy. Anim. Nutr. **7**(4) (2021)

5. Jing, Z., Yang, X., Yanwu, J., et al.: Recent advances in application of near-infrared spectroscopy for quality detections of grapes and grape products. Spectrosc. Spect. Anal. **41**(12), 3653–3659 (2021)
6. Yanlong, T., Yi, W., Xiao, W., et al.: Advances in detection of microorganisms using Near-infrared spectroscopy. Spectrosc. Spect. Anal. **42**(01), 9–14 (2022)
7. Mengting, Z., You, L., Yi, C., et al.: Fast determination of lipid and protein content in green coffee beans from different origins using NIR spectroscopy and chemometrics. J. Food Compos. Anal. (2021)
8. Jiaojiao, Z., Xiaoyang, W., Juan, Y., et al.: Rapid determination of the textural properties of silver carp (Hypophthalmichthys molitrix) using near-infrared reflectance spectroscopy and chemometrics. LWT **129** (2020)
9. Cura, K., Rintala, N., Kamppuri, T., et al.: Textile recognition and sorting for recycling at an automated line using near infrared spectroscopy. Recycling **6**(1) (2021)
10. FZ/T 01057.8-2012. Test method for identification of textile fibers part 8: Infrared Spectroscopy
11. FZ/T 01057.3-2007. Test method for identification of textile fibers part 3: Microscopy
12. GB/T 2910.7-2009. Quantitative chemical analysis of textiles part 7: Polyamide fiber and certain other fiber mixtures (formic acid method)
13. GB/T 2910.4-2009. Quantitative chemical analysis of textiles part 4: Mixtures of certain protein fibers and certain other fibers (hypochlorite method)
14. GB/T 2910.11-2009. Quantitative chemical analysis of textiles part 11: Mixtures of cellulose fibers and polyester fibers (sulfuric acid method)
15. Zihan, W., Wenxia, L., Huaping, W., et al.: Establishment of quantitative near-infrared analysis models and their application for prediction of common textiles contents. J. Beijing Inst. Fash. Technol. (Nat. Sci. Ed.) **39**(02), 30–37 (2019)
16. Krzysztof, B., Grabska, J., Badzoka, J., et al.: Spectra-structure correlations in NIR region of polymers from quantum chemical calculations. The cases of aromatic ring, C=O, C≡N and C-Cl functionalities. Spectrochim. Acta A Mol. Biomol. Spectrosc. **262** (2021)
17. Zhushanying, Z., Hanwen, G., Kaiwe, X., et al.: Pretreatment and combined method based on near infrared spectroscopy. Laser Optoelectron. **58**(16), 472–479 (2021)
18. Xuexue, M., Ying, M., Haoru, G., et al.: NIR spectroscopy coupled with chemometric algorithms for the prediction of cadmium content in rice samples. Spectrochim. Acta A Mol. Biomol. Spectrosc. **257** (2021)

Spectroscopy Theory and Chemometrics

Theoretical Simulation of Near-Infrared Spectrum of Piperine. Insight into Band Origins and the Features of Regression Models from Different Spectrometers

Justyna Grabska(✉), Krzysztof B. Bec, and Christian W. Huck

Institute of Analytical Chemistry and Radiochemistry, University of Innsbruck, Innrain 80-82, 6020 Innsbruck, Austria
Justyna.Grabska@uibk.ac.at

Abstract. Strong convolution of numerous overtones and combination bands makes NIR spectra difficult to interpret. Recent advances in anharmonic simulations decisively improved comprehension of NIR bands. Still, computational cost of accurate simulation remains very high, which hinders its wide use by non-specialized laboratories. In this proceedings article of NIR-2021 Conference, we discuss effective approaches to this problem, with optimizations for less time-consuming computations. Taking the example of piperine, the most popular spice ingredient in world trade, we directly compared two time-efficient approaches to this problem. The simulated NIR spectrum reveals an inherently complex structure with a large number of convoluted bands, mostly binary combinations, in particular in the 5500–4000 cm^{-1} range. The detailed assignments of NIR bands of piperine allowed to interpret the characteristics of the PLS regression models of the piperine in black pepper. Two models were compared, developed for spectral data sets obtained with the benchtop instrument (NIRFlex N-500) and a miniaturized spectrometer (microPHAZIR). These two spectrometers use different optical principles (benchtop: FT-NIR with a polarization interferometer and microPHAZIR: a programmable MEMS Hadamard mask), leading to profound instrumental differences. However, both are able to capture the most significant NIR absorption of piperine. In conclusion, the sensitivity of the two instruments to certain types of piperine NIR vibrations is different, with the stationary spectrometer being much more selective. This difference in capturing chemical information from the sample results in the difference in performance between the laboratory FT-NIR spectrometer and narrow-waveband miniaturized spectrometer in analyzing the piperine content in black pepper.

Keywords: Anharmonic calculation of NIR spectra · Overtones · Combination bands · Instrumental difference · Interpretation of PLS regression

X. Chu et al. (Eds.): ICNIR 2021, *Sense the Real Change: Proceedings of the 20th International Conference on Near Infrared Spectroscopy*, pp. 253–261, 2022.
https://doi.org/10.1007/978-981-19-4884-8_27

1 Introduction

Near-infrared (NIR) spectroscopy is a major analytical tool with marked importance in various applications [1–3]. In recent years a dynamic development of handheld, miniaturized NIR spectrometers enhanced the potential of this spectroscopic technique by enabling analysis directly on-site and enabled a new spectrum of applications of this technique in science and industry [4, 5]. The advantages of miniaturized NIR spectroscopy are particularly exposed in several field of natural products, because of their chemical diversity that can vary depending on the medicinal plant cultivation conditions, geographical origin or harvest time [6]. The ability to perform direct in-field scanning of medicinal plants, as well as agricultural crops, provides critical advantage not offered by conventional laboratory-based spectroscopy.

Handheld spectrometers are becoming essential in controlling the cultivation and production process, as the suitability of the final retail product is defined at the pre-harvest time. Furthermore, the determination of the optimal harvest moment is equally important, and decides, for example, about the concentration of the therapeutic ingredient in the harvested plant. However, the sensor miniaturization required implementing a number of distinct engineering solutions, and these sensors differ by the key elements used for their construction, e.g. wavelength selectors and detector configurations [4, 5]. Different design philosophies affect the performance and applicability of various miniaturized NIR spectrometers, as the consequence of the operating spectral region, resolution and sensitivity. Furthermore, analysis of natural products, which feature a complex chemical matrix typical for plant material, is often challenging [2, 7–11].

Recent advancements in basic research in NIR spectroscopy have brought feasibility of theoretical in silico simulation of NIR spectra [12–14]. Narrow spectral working regions of these instruments limit their ability to measure part of meaningful vibrational bands, which changes from one to another analyzed chemical. This effect was investigated in our recent studies [15]. Importantly, there appears opportunity to employ spectra simulation to gain understanding of calibration models in chemical sense. This breakthrough opens the possibility to interpret the results of multivariate analysis performed on the spectral sets acquired by different NIR spectrometers.

The location of meaningful variables can be associated with specific molecular vibrations, and the sensitivity and specificity of a spectrometer can be analyzed. In other words, it becomes feasible to assess the ability of how a certain NIR spectrometer discriminates narrow bands of analyzed chemical ingredient in a complex NIR lineshape of a multi-constituent natural sample [6, 16].

2 Materials and Methods

2.1 NIR Spectra Measurements – Benchtop and Miniaturized Spectrometers

Piperine (98.5%) and carbon tetrachloride (anhydrous, ≥99.5%) were purchased from Sigma Aldrich. Carbon tetrachloride was distilled and stored over molecular sieves (5 Å, Sigma Aldrich). Piperine was used with no further purification. The NIR spectra were measured using a Fourier transform Büchi NIRFlex N-500 spectrometer equipped with polarization interferometer and the accessory for measurement of liquid samples in transmittance mode. The samples were deposited in a quartz cell of 10 mm thickness (Hellma QX). The spectra were acquired in the region of 10 000 to 4000 cm^{-1} with the spectral resolution of 8 cm^{-1}, interpolated to 4 cm^{-1}, resulting in 1501 data points per spectrum. For each spectrum, a total number of 64 averaged scans were collected. The interpretation of the quantitative performance in analyzing piperine content in black pepper by NIRFlex N-500 was compared with the same results obtained by microPHAZIR.

2.2 Chemometrics

The construction of PLS regression (i.e. PLSR) models, as well as the preceding spectral pretreatments was performed with the use of The Unscrambler X v.10.5 (CAMO Software) software. The measured spectra were transformed from reflectance R to log(1/R), i.e., into intensity scale common with absorbance. Replicate spectra measured for each sample were averaged to one spectrum per sample. Afterwards, numerical second-order differentiation of the spectra using the Savitzky–Golay algorithm was carried out. PLSR was performed for mean-centered spectral sets, with NIPALS algorithm and full cross-validation using a leave-one-out scheme.

PLSR models with the best prediction performance were calculated independently for all spectrometers and ultimately compared regarding their observed differences. In addition, a quantum mechanical simulation of NIR spectrum of piperine (validated through a comparison with the spectra measured for an analytical standard of piperine) was performed. This step unveiled the specific NIR bands of piperine and demonstrated which chemical information on piperine content can be observed by each of the evaluated miniaturized spectrometers. This also allowed better understanding of the interplay between the chemical and physical properties of the sample that contribute to the calibration model in relation to the instrumental difference between various spectrometers.

To examine the relevance of the accessibility of a spectrometer to the specific chemical information present in NIR spectrum, the case study is based on the microPHAZIR (Thermo Fisher Scientific) handheld spectrometer. This instrument is a miniaturized Hadamard spectrometer with a programmable micro–electro–mechanical system mask, which makes it distinctly different in the used technology from a generic design of a benchtop FT-NIR spectrometer. The microPHAZIR is particularly interesting for this study, as being one of the first broadly available portable NIR spectrometers it is a fairly popular instrument in various analyses. However, it operates over a particularly narrow spectral range of 6266–4173 cm^{-1}, which for most molecules only allows to capture the region of strongly overlapping combination bands.

2.3 In Silico (Quantum Chemical) Simulation of NIR Spectra

On the example of piperine, we directly compared DVPT2//PM6 and DVPT2//ONIOM[PM6:B3LYP/6-311þþG(2df,2pd)] approaches for highly efficient simulation of NIR spectra and assessment of the resulting accuracy trade-off. The former offers very high efficiency and the accuracy level suitable for the needs of fast, approximate NIR band assignments. ONIOM is a hybrid scheme, in which the electronic structure of the system may be determined by applying different methods to distinct atomic centers. In this case, to describe the alkene fragment, we employed a lower-level method, a semi-empirical parametric method 6 (PM6). The remaining atoms were treated at density functional theory (DFT) level, with single-hybrid B3LYP functional combined with 6-311++G(2df,2pd) basis set. The following anharmonic vibrational analysis was performed using the deperturbed vibrational second-order perturbation theory (DVPT2). The calculations included up to two quanta transitions, i.e., the first overtones and binary combination; this is sufficient to accurately reconstruct an NIR spectrum [17, 18]. All quantum mechanical calculations were performed with Gaussian 09 Revision E01 software [19]. The modeling of the spectral lineshape was carried out through parameterized band broadening. Lorentz–Gauss (Cauchy–Gauss) product function was used as the band shape model [20].

3 Results and Discussion

3.1 Band Assignment

The assessment of performance vs. accuracy of NIR spectra simulation performed the two ways (described in *Materials and methods*), indicated that ONIOM scheme is preferable for acceptable accuracy. While full semi-empirical approach offers further gains in efficiency, these were concluded not favorable in light of the much less accurate predictions of the positions of NIR bands associated with stretching modes.

To present the NIR assignments in an exhaustive way and dissect the intricate band overlapping (Fig. 1), we illustrate the NIR assignments of piperine in the form of a heatmap, where the significance of each vibrational mode of interest is represented in a false-color scale (Fig. 2). The vibrational contributions to the NIR spectrum of piperine unveiled this way show the dominance in the upper region (6200–5500 cm^{-1}) of the first overtones of C–H and CH_2 stretching as well as the binary combinations involving either of these modes (Fig. 2). Noteworthy, the νCH bands appear at the higher wavenumbers than the νCH_2 ones, reflecting a similar order between the respective calculated fundamental bands appearing in IR region of piperine. Combinations including stretching and deformation vibrations of νCH and νCH_2 groups, as well as with ring deformation modes, mostly populate the lower NIR region of piperine (4500–4000 cm^{-1}). Note, the relative contributions of C = O stretching and C–N stretching are relatively minor. The contribution from C = O stretching can be seen in the region free from very strong bands, i.e. in the region of 4900–4600 cm^{-1}, where its combination bands with C–H stretching are observed.

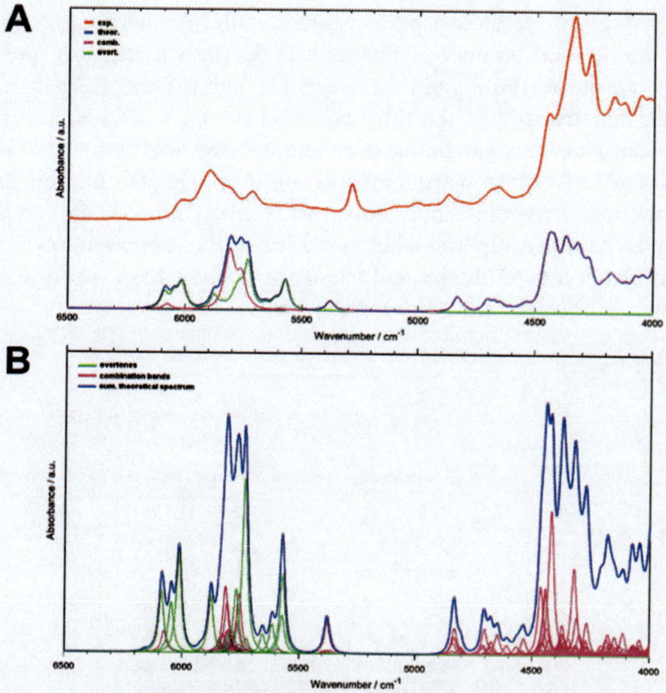

Fig. 1. NIR spectra of piperine. (A) The experimental and the calculated spectrum; summarized contributions from the first overtones and binary combination bands are presented. (B) The individual simulated first overtones and binary combination bands (narrower simulated bands presented for a better view of detail). Reproduced with permission from Ref. [8].

3.2 Vibrational Interpretation of the PLS Regression Factors Corresponding to Piperine Content in Black Pepper

The detailed band assignments of piperine offer insights into the correspondence of the PLS factors in the models describing piperine content in black pepper samples. The NIR vibrational contributions of piperine can be roughly established. The analysis by a benchtop spectrometer NIRFlex N-500 (Fig. 3A) and a handheld microPHAZIR (Fig. 3B) is dissected. The narrow spectral region in which microPHAZIR operates (6266–4173 cm^{-1}) is just enough to acquire the most meaningful absorption of piperine with only weak second overtones and ternary combination bands populating the spectrum above 6150 cm^{-1}. Firstly, as expected, the structure of the loadings plots for all factors indicates a clear correspondence with the absorption features of piperine. Interestingly, exclusion of the region between 5550 and 4950 cm^{-1} improved the performance of the prediction for microPHAZIR. Comparing this information with the determined vibrational assignments, one may conclude that the contributions from weak νCH combination bands of piperine to the NIR spectrum of black pepper are not acquired well enough by microPHAZIR. The model constructed for microPHAZIR required 4 factors to obtain the maximum predictive performance, while for the dataset from benchtop spectrometer the optimal number of factors was 3. The structures of the first factors for both cases are quite similar, and these seem to capture the most intense bands of piperine. Note, the structure of the third factor in the case of benchtop spectrometer clearly stands out from the remaining ones, e.g., above 6000 cm^{-1} and in the region of 5300–4900 cm^{-1}. At these wavenumbers, one may see, standing out from the rest as well, contributions from the combination bands involving νCH and, to lesser extent, ring deformation bands of piperine. The presence of the combinations involving ring deformation is also viable in the second factor for the benchtop spectrometer, where a distinct structure is observed near 4750–4500 cm^{-1}.

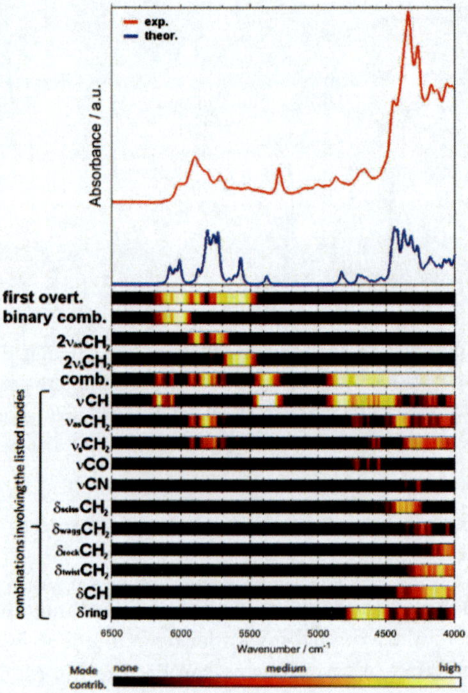

Fig. 2. Vibrational contributions to NIR spectrum of piperine. Reproduced with permission from Ref. [15].

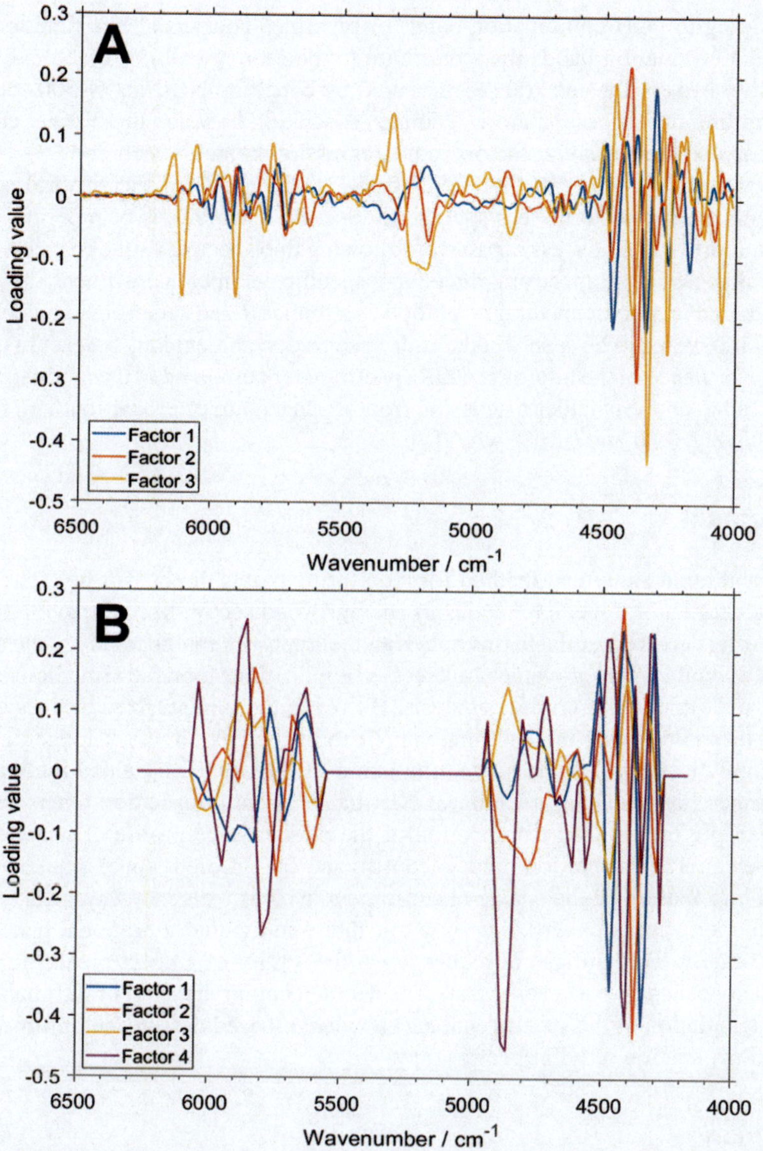

Fig. 3. Loadings plots for the PLSR model of piperine content in black pepper developed for the NIR spectral sets measured with (A) benchtop Buchi NIRFlex N-500; (B) miniaturized microPHAZIR spectrometer. Reproduced with permission from Ref. [15].

4 Conclusion

In short, the conclusions drawn from this study indicate that the laboratory spectrometer appears to be more sensitive to the specific vibrations of piperine in black pepper. As shown, for example, by the clearly distinct to the others (i.e. F-1 and F-2) factor 3 corresponding to νCH combination bands of piperine. There is also a significant part of the ring deformation bands that contribute to the second and third factors in PLSR model calibrated for the spectral set measured by benchtop NIRFlex N-500 spectrometer. Therefore, in this case, a more accurate association between the distinct chemical information and the particular factors in the regression model is recorded.

In contrast, the factors in the PLSR model constructed for the spectral set from microPHAZIR, appear to be less specific to individual vibrations of piperine. Consequently, microPHAZIR, is less capable of following fine spectral variations representing intensity changes in the spectrum related to a specific chemical constituent.

It is likely that poorer spectral resolution of a miniaturized spectrometer also plays a role here. It remains to be seen whether this observation can explain, at least in part, the poorer performance of the microPHAZIR spectrometer compared to the laboratory spectrometer in the analysis under discussion (root mean square of prediction, i.e., RMSEP of, respectively, 0.30 and 0.18% w/w) [9].

5 Summary

Flexible and rapid analytical method for monitoring plants directly in-field has critical importance for natural medicine industry and agri-food sector. Miniaturized NIR spectroscopy offers great potential in this role. Handheld sensors enable rapid on-site analysis and mark a significant improvement in practical applications for natural medicines industry, food and agricultural product analysis. However, the miniaturized devices differ in the used technology and their ultimate applicability varies largely, depending on the specific analytical problem. In silico simulation of NIR spectra yields highly detailed and accurate chemical interpretation of NIR bands. This information opens new possibilities to perform a deep examination of the performance profile of handheld NIR spectrometers. The calibration models constructed for different spectrometers capture chemical information on the analyzed constituent in clearly distinct way, with benchtop high-resolution spectrometer being able to capture individual vibrational bands much more accurately. This brings consequences to the ability of a spectrometer to acquire fine intensity changes in a specific task. The detailed comprehension of NIR bands from accurate simulation of the spectra enables knowledge-based design and optimization of analytical application of NIR spectroscopy.

References

1. Ozaki, Y., Huck, C.W., Beć, K.B.: Near-IR spectroscopy and its applications. In: Gupta, V.P. (ed.) Molecular and Laser Spectroscopy. Advances and Applications, pp. 11–38. Elsevier, San Diego (2018)
2. Huck, C.W., Beć, K.B., Grabska, J.: Near infrared spectroscopy in natural product research. In: Meyers, R.A. (ed.) Encyclopedia of Analytical Chemistry: Applications, Theory and Instrumentation, pp. 1–29. Wiley (2020)
3. Czarnecki, M.A., Beć, K.B., Grabska, J., Hofer, T.S., Ozaki, Y.: Overview of application of NIR spectroscopy to physical chemistry. In: Ozaki, Y., Huck, C., Tsuchikawa, S., Engelsen, S.B. (eds.) Near-Infrared spectroscopy, pp. 297–330. Springer, Singapore (2021). https://doi.org/10.1007/978-981-15-8648-4_13

4. Beć, K.B., Grabska, J., Huck, C.W.: Principles and applications of miniaturized near-infrared (NIR) spectrometers. Chem. – A Eur. J. **27**(5), 1514–1532 (2021). https://doi.org/10.1002/chem.202002838
5. Beć, K.B., Grabska, J., Siesler, H.W., Huck, C.W.: Handheld near-infrared spectrometers: where are we heading? NIR News **31**, 28–35 (2020)
6. Beć, K.B., Grabska, J., Huck, C.W.: NIR spectral analysis of natural medicines supported by novel instrumentation, methods of data analysis and interpretation. J. Pharm. Biomed. Anal. **193**, 113686 (2020)
7. Mayr, S., et al.: Challenging handheld NIR spectrometers with moisture analysis in plant matrices: performance of PLSR vs. GPR vs. ANN modelling. Spectrochim. Acta A **249**, 119342 (2021)
8. Mayr, S., et al.: Quantification of Silymarin in Silybi mariani fructus: challenging the analytical performance of benchtop vs. handheld NIR spectrometers on whole seeds. Planta. Med. **87**, 1–13 (2021)
9. Mayr, S., Beć, K.B., Grabska, J., Schneckenreiter, E., Huck, C.W.: Near-infrared spectroscopy in quality control of Piper nigrum: a Comparison of performance of benchtop and handheld spectrometers. Talanta **223**, 121809 (2021)
10. Mayr, S., et al.: Theae nigrae folium: comparing the analytical performance of benchtop and handheld near-infrared spectrometers. Talanta **221**, 121165 (2021)
11. Beć, K.B., Grabska, J., Huck, C.W.: Miniaturized near-infrared spectroscopy in natural product analysis. Current and future directions. In: Gupta, V.P. (ed.) Molecular and Laser Spectroscopy - Advances and Applications, vol. 3. Elsevier (2022)
12. Beć, K.B., Huck, C.W.: Breakthrough potential in near-infrared spectroscopy: spectra simulation. A review of recent developments. Front. Chem. **7**, 48 (2019)
13. Beć, K.B., Grabska, J., Huck, C.W.: Current and future research directions in computer-aided near-infrared spectroscopy: a perspective. Spectrochim. Acta A **254**, 119625 (2021)
14. Ozaki, Y., et al.: Advances, challenges and perspectives of quantum chemical approaches in molecular spectroscopy of the condensed phase. Chem. Soc. Rev. **50**, 10917–10954 (2021)
15. Grabska, J., Beć, K.B., Mayr, S., Huck, C.W.: Theoretical simulation of near-infrared spectrum of piperine. Insight into band origins and the features of regression models. App. Spectr. **75**, 1022–1032 (2021)
16. Beć, K.B., Grabska, J., Huck, C.W.: Near-infrared spectroscopy in bio-applications. Molecules **25**, 2948 (2020)
17. Grabska, J., Czarnecki, M.A., Beć, K.B., Ozaki, Y.: Spectroscopic and quantum mechanical calculation study of the effect of isotopic substitution on NIR spectra of methanol. J. Phys. Chem. A **121**, 7925–7936 (2017)
18. Grabska, J., Beć, K.B., Kirchler, C.G., Ozaki, Y., Huck, C.W.: Distinct difference in sensitivity of NIR vs. IR bands of melamine to inter-molecular interactions with impact on analytical spectroscopy explained by anharmonic quantum mechanical study. Molecules **24**, 1402 (2019)
19. Frisch, M.J., Trucks, G.W., Schlegel, H.B., Scuseria, G.E., et al.: Gaussian 09, Revision E.01. Gaussian, Inc., Wallingford (2009)
20. Grabska, J., Ishigaki, M., Beć, K.B., Wójcik, M.J., Ozaki, Y.: Correlations between structure and near-infrared spectra of saturated and unsaturated carboxylic acids. Insight from anharmonic density functional theory calculations. J. Phys. Chem. A **121**, 3437–3451 (2017)

Vis-NIR Spectroscopy Combined with Bayes Classifier Applied to Wine Multi-brand Identification

Xianghui Chen[1], Jiaqi Li[2], Nailiang Chang[2], Jiemei Chen[1], Lifang Fang[3], and Tao Pan[2](✉)

[1] Department of Biological Engineering, Jinan University, Guangzhou 510632, China
[2] Department of Optoelectronic Engineering, Jinan University, Guangzhou 510632, China
tpan@jnu.edu.cn, 466945939@qq.com
[3] Guangdong Langtao Information Technology Co., Ltd., Dongguan, China

Abstract. The multi-brand identification technology of wine has important application prospects. Since the main components of wine are roughly the same, and the characteristic components that can distinguish wine brands are usually trace amounts. The conventional quantitative detection method for brand identification is complicated and difficult. The naive Bayes classifier is an algorithm based on probability distribution, which is simple and particularly suitable for multi-class discriminant analysis. However, the absorbance probability between spectral wavelengths is not necessarily strongly independent, which limits the application of Bayes method in spectral pattern recognition. In this paper, a Bayes classifier algorithm based on wavelength optimization was proposed. First, a large-scale wavelength screening for equidistant combination (EC) was performed, and then wavelength step-by-step phase-out (WSP) was carried out to reduce the correlation between wavelengths and improve the accuracy of Bayes discrimination. The proposed EC-WSP-Bayes method was applied to the 5-category discriminant analysis of wine brand with visible and near-infrared (Vis-NIR) spectroscopy. The wavelength combination of the optimal EC-WSP-Bayes model was 412, 510, 1098, 1980, 2274, 2372 nm located in the visible, short-NIR and combination frequency regions. In modelling and independent validation, the total recognition accuracy rate (RAR_{Total}) reached 97.6% and 98.7%, respectively. The technology is quick, easy, and has potential application in market. The proposed model of less-wavelength and high-efficiency ($N = 6$) can provide a valuable reference for small specialized spectrometer design. The proposed integrated EC-WSP-Bayes method can reduce the correlation between wavelengths, improve the recognition accuracy and applicability of Bayes method.

Keywords: Wine · Multi-brand identification · Bayes classifier · Equidistant combination wavelength screening · Wavelength step-by-step phase-out

1 Introduction

The identification of high-quality wine brands can avoid adulteration and fraud, and protect the rights and interests of producers and consumers. The compositions of wine

X. Chu et al. (Eds.): ICNIR 2021, *Sense the Real Change: Proceedings of the 20th International Conference on Near Infrared Spectroscopy*, pp. 262–268, 2022.
https://doi.org/10.1007/978-981-19-4884-8_28

are similar, and the quantitative detection method for brand identification is complicated and difficult. Vis-NIR spectral discriminant analysis uses the similarity of the same type of spectra and the heterogeneity of different type of spectra for pattern recognition.

Naïve Bayes classification algorithm [1] is based on probability distribution for multi-classification, which is simple and fast. However, the absorbance probability between spectral wavelengths is not necessarily strongly independent, which limits the application of Bayes method in spectral pattern recognition. This paper proposed a Bayes classification algorithm based on wavelength optimization. First, a large-scale wavelength screening for equidistant combination (EC) [2, 3] was performed, and then wavelength step-by-step phase-out (WSP) [4, 5] was carried out to reduce the dependence between wavelengths and improve the accuracy of Bayes discrimination. It has been applied to the 5-category discriminant analysis of wine brand identification.

2 Materials and Methods

2.1 Samples and Measurement

Four types of wine brands were collected from regular sales channels as identification brands (not in order as I, II, III, and IV), as follows: Great Wall, Chile Aoyo, Dynasty and Changyu (20 bottles each, 5 samples/bottle, a total of 100 samples for each category). The fifth type of samples collected was regarded as the interference brands (denoted as V, 111 samples in total), which includes 21 other commercial brand wines (one bottle each brand, 3 samples/bottle, 63 samples in total) as well as home-brewed wines from different sources (48 bottles, 1 sample/bottle, 48 samples in total). In total, 511 samples were used for spectral measurements.

The XDS Rapid Content™ Liquid Grating Spectrometer (FOSS, Denmark) and a transmission accessory with 1 mm cuvette were used for spectral measurement. Spectral scope ranged as 400–2498 nm with a 2-nm wavelength interval. Wavebands of 400–1100 nm and 1100–2498 nm were used for Si and PbS detection, respectively. Each sample was measured three times, and a total of 1533 spectra (I, II, III, IV: 300 each, V: 333) were obtained. The experimental temperature and humidity were 25 ± 1 °C and 45 ± 1%, respectively.

2.2 Calibration-Prediction-Validation Framework and Evaluation Indicators

A sample-independent experimental design in calibration-prediction-validation was adopted. The calibration and prediction sets were used for modeling and parameter optimization; and the selected model was validated using the independent validation samples that were excluded in the modeling, thus objective evaluation was obtained. Each identification brand of wine (20 bottles, 100 samples, 300 spectra) was randomly divided into calibration (8 bottles, 40 samples, 120 spectra), prediction (6 bottles, 30 samples, 90 spectra) and validation (6 bottles, 30 samples, 90 spectra) sets. The fifth type of samples (V, interference brands) were divided calibration (39 samples, 117 spectra), prediction (36 samples, 108 spectra) and validation (36 samples, 108 spectra) sets.

Evaluation indicators were set as recognition accuracy rate of each type sample ($RAR_i, i = 1, 2, \cdots, 5$), their standard deviation (RAR_{SD}), and total recognition accuracy rate (RAR_{Total}) of all samples, as follows:

$$RAR_i = \frac{\tilde{M_i}}{M_i}, i = 1, 2, \cdots, 5 \tag{1}$$

$$RAR_{Total} = \frac{\sum_{i=1}^{5} \tilde{M_i}}{\sum_{i=1}^{5} M_i} \tag{2}$$

where $M_i (i = 1, 2, \cdots, 5)$ was the number of samples of i-th category of the prediction (or validation), and $\tilde{M_i}$ was the number of accurately identified samples in i-th category samples of the prediction (or validation). In the modeling process, in order to take into account balance, the wavelength models were preferred according to a comprehensive indicator ($RAR_{Total} - RAR_{SD}$).

2.3 Bayes Classification Algorithm

The probability for the sample to be judged as the kth category according to the spectrum was as follows:

$$P(\text{Class} = k|\text{Spectrum}) = \frac{P(\text{Sepctrum}|\text{Class} = k)P(\text{Class} = k)}{\sum_{i=1}^{5} P(\text{Spectrum}|\text{Class} = i)P(\text{Class} = i)}, k = 1, \cdots, 5 \tag{3}$$

Based on the assumption that absorbance of a single wavelength obeys the normal distribution and probability independence, and the logarithm is used to avoid calculation overflow, the following calculation method was proposed:

$$\ln(P(\text{Spectrum}|\text{Class} = k)) = \sum_{j=1}^{N} \ln(P(\text{Spectrum}_j|\text{Class} = k)) \tag{4}$$

where N was the total number of wavelengths; j corresponded to the j-th wavelength; the probability $P(\text{Spectrum}_j|\text{Class} = k)$ was calculated according to the mathematical expectation and standard deviation of j wavelength's absorbance of k-th category sample in calibration set. Finally, the unknown sample was judged as the category corresponding to the maximum value of $\ln(P(\text{Spectrum}|\text{Class} = k))$.

2.4 Bayes Classifier Based on Wavelength Model Optimization

EC-Bayes Equidistant combination wavelength selection combined with Bayes: Use the initial wavelength (I), the number of wavelengths (N) and the number of wavelength gaps (G) to perform a wide range of wavelength screening. The search parameters are $I \in \{400, 402, \cdots, 2498\}$; $N \in \{1, 2, \cdots, 1050\}$; $G \in \{1, 2, \cdots, 50\}$.

EC-WSP-Bayes Based on EC-Bayes, the wavelength step-by-step phase-out was carried out to further optimize the model: first, each time eliminated the wavelength, whose removing resulted in the best recognition accuracy, until only one wavelength remained; then, the optimal model was selected from the above-mentioned process of wavelengths elimination by step-by-step phase-out mode. The specific algorithm was as follows.

Step 1. The combination of N wavelengths that required optimization was denoted as follows:

$$\Omega^{(N)} = \{\lambda_1^{(N)}, \lambda_2^{(N)}, \cdots, \lambda_N^{(N)}\} \tag{5}$$

Step 2. By arbitrarily eliminating one wavelength, N combinations of (N-1) wavelengths were as follows:

$$\Omega_i^{(N)} = \Omega^{(N)} - \{\lambda_i^{(N)}\}, i = 1, 2, \cdots, N \tag{6}$$

They were used to build Bayes models, and the corresponding $\text{RAR}_{\text{Total}}$ was denoted as $\text{RAR}_{\text{Total},i}^{(N)}$, $i = 1, 2, \cdots, N$. The optimal wavelength combination was selected by $\max\limits_{1 \le i \le N} RAR_{Total,i}^{(N)}$, $i = 1, 2, \cdots, N$ and additionally denoted as follows:

$$\Omega^{(N-1)} = \{\lambda_1^{(N-1)}, \lambda_2^{(N-1)}, \cdots, \lambda_{N-1}^{(N-1)}\} \tag{7}$$

Step 3. The method of *Step 2* was used to eliminate wavelengths step by step until the number of wavelengths of the combination was 1. For each step, the optimal wavelength combination and the $\text{RAR}_{\text{Total}}$ were denoted as follows: Ω_k, $RAR_{Total}^{(N)}$, $k = 1, 2, \cdots, N$. The global optimal wavelength combination was further selected according to $\max\limits_{1 \le k \le N} RAR_{Total}^{(k)}$ and denoted as follows:

$$\Omega* = \{\lambda_1^*, \lambda_2^*, \cdots, \lambda_{N*}^*\} \tag{8}$$

where N^* was the number of wavelengths in this combination, and $1 \le N^* \le N$.

3 Results and Discussion

3.1 EC-Bayes Models

Based on the entire scanning region (400–2498 nm, $N = 1050$), the direct Bayes model was established first. The RAR_{Total} of modelling was 95.1%, and the RAR_i of the five types were 92.2%, 100%, 94.4%, 94.4%, and 94.4%, respectively.

The EC-Bayes method was used for wavelength model optimization based on the selection of multi-parameter combination (I, N, G). The parameters of the optimal model were $I = 412$ nm, $N = 21$, $G = 49$. The RAR_{Total} of modelling reached 96.6%, and the RAR_i of the five types were 94.4%, 100%, 95.6%, 95.6% and 97.2%, respectively.

3.2 EC-WSP-Bayes Models

Since the equidistant wavelength combination obtained by the EC-Bayes was likely to still contain redundant wavelengths, the WSP-Bayes method was further used to improve the selected Top 10 EC-Bayes models. The wavelength combination of the optimal EC-WSP-Bayes model was 412, 510, 1098, 1980, 2274, 2372 nm. The RAR_{Total} of modelling was 97.6%, and the RAR_i of the five types were 92.2%, 100%, 100%, 95.6%, and 100%, respectively. The results showed that the effect of EC-WSP-Bayes method was good and the wavelength model was simple ($N = 6$). The modeling effects of the Top10 EC-Bayes models and corresponding EC-WSP-Bayes models were shown in Fig. 1. It illustrated that for all of the Top 10 EC-Bayes models, after the process of wavelength step-by-step phase-out, the numbers of wavelengths were all greatly reduced and the discrimination effects were all improved.

3.3 Independent Validation

A total of 468 spectra of the validation samples (90 for each of I, II, III and IV, 108 for V) not involved in modeling, were used to validate the effect of the selected optimal EC-WSP-Bayes model. Using the mathematical expectation and standard deviation of the spectral absorbance in the calibration set, the conditional probability of the spectra in validation set were calculated and the type of validation samples were determined. In validation, the RAR_i of the five types were 94.4%, 100%, 98.9%, 100%, and 100% respectively and the RAR_{Total} was 98.7%. The results showed that the optimal EC-WSP-Bayes model also achieved a good performance in validation.

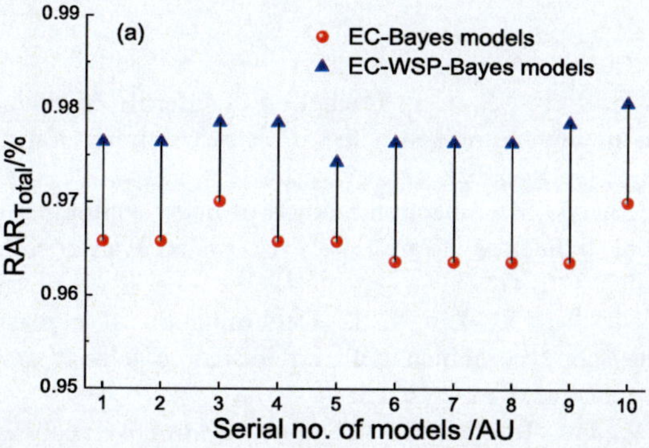

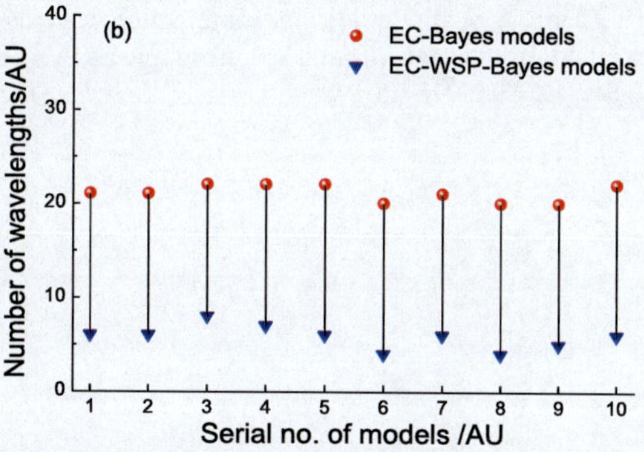

Fig. 1. Comparison of modeling effects of the Top 10 EC-Bayes models and corresponding EC-WSP-Bayes models: (a) RAR_{Total}; (b) number of wavelengths.

4 Conclusion

The application of Naïve Bayes algorithm in spectral classification is limited by the strong independence of the absorbance probability between wavelengths. In this paper, the equidistant combination wavelength screening and wavelength step-by-step phase-out were combined to reduce the correlation between wavelengths, improve the recognition accuracy of Bayes model, and reduce the complexity of the model. It was successfully applied to the five-category discriminant analysis of wine brands. The selected optimal EC-WSP-Bayes model used only six wavelengths to achieve high-precision discrimination effect in independent validation ($RAR_{Total} = 98.7\%$). The proposed method is also expected to be applied to spectral discriminant analysis in other fields.

Acknowledgments. This work was supported by National Natural Science Foundation of China (No. 61078040) and Guangdong Province Project of China (No. 2014A020213016, No. 2014A020212445).

References

1. Dong, J., Dong, X.G., Li, Y.L., et al.: Prediction of infertile chicken eggs before hatching by the Naive-Bayes method combined with visible near infrared transmission spectroscopy. Spectrosc. Lett. **53**, 327–336 (2020)
2. Pan, T., Li, M.M., Chen, J.M.: Selection method of quasi-continuous wavelength combination with applications to the near-infrared spectroscopic analysis of soil organic matter. Appl. Spectrosc. **68**, 263–271 (2014)
3. Han, Y., Chen, J.M., Pan, T., Liu, G.S.: Determination of glycated hemoglobin using near-infrared spectroscopy combined with equidistant combination partial least squares. Chemometr. Intell. Lab. **145**, 84–92 (2015)
4. Chen, J.M., Li, M.M., Pan, T., et al.: Rapid and non-destructive analysis for the identification of multi-grain rice seeds with near-infrared spectroscopy. Spectrochim. Acta. A. **219**, 179–185 (2019)
5. Yang, Y.H., Lei, F.F., Zhang, J., et al.: Equidistant combination wavelength screening and step-by-step phase-out method for the near-infrared spectroscopic analysis of serum urea nitrogen. J. Innov. Opt. Heal. Sci. **12**, 1950018 (2019)

Outlier Detection in Calibration Transfer for Near Infrared Spectra

Kaiyi Zheng(✉), Ye Shen, Wen Zhang, Xiaowei Huang, Zhihua Li, Di Zhang, Jiyong Shi, and Xiaobo Zou(✉)

Key Laboratory of Modern Agriculture Equipment and Technology, School of Food and Biological Engineering, Jiangsu University, Zhenjiang 212013, China
kaiyizhengjsu@126.com, zou_xiaobo@ujs.edu.cn

Abstract. Based on model population analysis (MPA), the ensemble refinement (ER) has been proposed for outlier detection in calibration transfer. The ER first constructs many subsets of transfer set, and then computes the validation errors of each subset. After that, for each sample, the average error for subsets which include the one sample can be obtained. Finally, the samples with large average errors can be identified as outliers. The simulated dataset has been used to testing this method. The results showed that for calibration transfer methods such as CCA-ICE, DS and SST, ER can all identify outliers.

Keywords: Calibration transfer · Model population analysis · Near infrared spectra · Outlier detection

1 Introduction

With advantages of low measurement cost, fast analysis period and easy operation, near infrared spectra have been widely used in pharmaceutical, environmental and agricultural areas. However, as an indirect analytical method, the feasible model of near infrared spectra should be calibrated in advance, such as partial least square (PLS). Usually, the model calibrated in one condition may be invalid for the spectra scanned under another condition. Thus, under this circumstance, recalibrating a new model can solve this problem. But recalibrating model may be expensive and laborious. Thus, calibration transfer can be another solution to this problem. For the batch of spectra in calibration transfer, the samples constructing model are called primary spectra, while the samples not being calibrated but only using the model of primary spectra are called secondary spectra.

In recent years, many calibration transfer methods have been proposed, such as: direct standardization (DS), piecewise direct standardization (PDS), canonical correlation analysis (CCA), canonical correlation analysis combined with informative component extraction (CCA-ICE) and spectral space transformation (SST). In addition to calibration transfer, the methods of selecting samples for transfer set are also important, such as Kennard–Stone (KS) method.

X. Chu et al. (Eds.): ICNIR 2021, *Sense the Real Change: Proceedings of the 20th International Conference on Near Infrared Spectroscopy*, pp. 269–273, 2022.
https://doi.org/10.1007/978-981-19-4884-8_29

However, the transfer set can only be selected by the distance of samples from calibration set. In our presume, similar as calibration set, there may still exist outliers in calibration transfer which can enlarge prediction errors. Thus, the samples in transfer set need to be further refined. In recent years, the model population analysis (MPA) has become more and more widely used in chemical and biochemical data analysis, such as sample selection in multivariate calibration [1]. Thus, in this paper, an outlier detection method for calibration transfer called ensemble refinement (ER) was proposed which uses the ideology of MPA.

2 Materials and Methods

2.1 Notations

The primary and secondary spectra are noted as matrix $\boldsymbol{A}$ and $\boldsymbol{B}$, respectively. The transfer and calibration sets of spectra $\boldsymbol{A}$ are respectively assigned as $\boldsymbol{A_t}$ and $\boldsymbol{A_c}$, and $\boldsymbol{A_t}$ can be selected from $\boldsymbol{A_c}$ by KS method. In addition to spectra $\boldsymbol{A}$, the transfer, validation and prediction sets of spectra $\boldsymbol{B}$ are designated as $\boldsymbol{B_t}$, $\boldsymbol{B_v}$ and $\boldsymbol{B_p}$, respectively. Meanwhile, the samples of spectra $\boldsymbol{B}$ with same concentrations of $\boldsymbol{A_t}$ are assigned as $\boldsymbol{B_t}$. And $\boldsymbol{y}$ indicates sample concentrations.

2.2 Procedure

The detailed procedure of ER can be shown as follows:

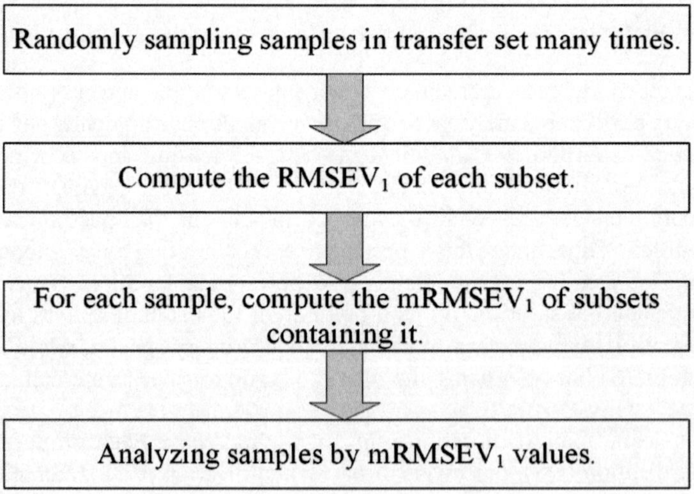

Fig. 1. The procedure of ER method

In Fig. 1, the proposed method includes four steps: (1) random sampling, (2) obtaining the RMSEV of each subset by PLS, (3) computing the average RMSEV for each sample ($mRMSEV_1$) and (4) analyzing the samples by $mRMSEV_1$.

2.3 The Simulated Dataset

Similar as calibration, there may also exist outliers in calibration transfer. In calibration transfer, the calibration model and samples are both fixed in advance in practical work. And it is supposed that the calibration model has been optimized, thus the $\boldsymbol{A_c}$ does not contain outliers. Because the $\boldsymbol{A_t}$ is generated from $\boldsymbol{A_c}$,, there are still no outliers in $\boldsymbol{A_t}$. The transfer set of secondary spectra with the same concentrations of $\boldsymbol{A_t}$ can be obtained from the spectra $\boldsymbol{B}$. The spectra $\boldsymbol{B}$ as a new batch of spectra cannot be analyzed in advance. Thus, unlike $\boldsymbol{A_t}$, there may exist outliers in $\boldsymbol{B_t}$. Based on this hypothesis, we test whether the ER method detects those abnormal samples in $\boldsymbol{B_t}$. Thus, in order to test our hypothesis, we generate the simulated dataset with outliers.

Both primary and secondary spectra contain four components and the spectrum of each component is Gaussian peaks [2]. The pure spectrum of primary spectra can be shown in Fig. 2.

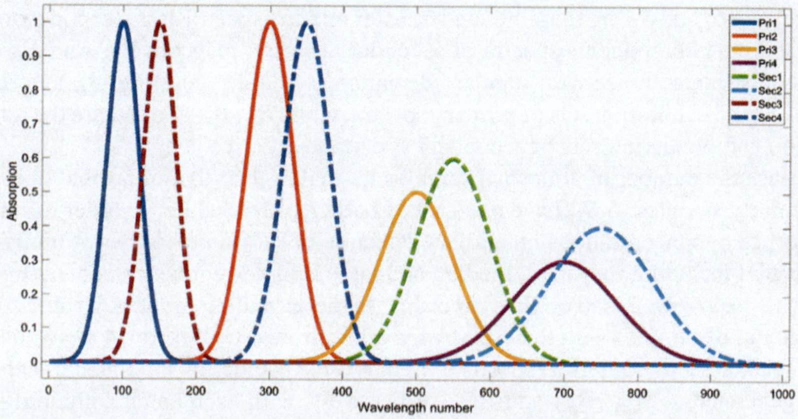

Fig. 2. Pure components spectra simulated primary (Pri1, Pri2, Pri3 and Pri4) and secondary (Sec1, Sec2, Sec3 and Sec4) spectra

In Fig. 2, it can be shown that the intensities and peak positions of pure primary spectra are different from those of secondary spectra. After obtaining the pure spectra, the concentrations of four components can also be generated by simulation. The matrix can be assigned as concentration values with size of 100 × 4 and each element is uniformly distributed from 0.2 to 0.3. Thus, the calibration set of primary spectra can be generated as follows:

$$\boldsymbol{A_{cs}} = \boldsymbol{C_s} * \boldsymbol{S_A} + \boldsymbol{E_{cs}} \tag{1}$$

In Eq. (1), $\boldsymbol{A_{cs}}$, $\boldsymbol{C_s}$ and $\boldsymbol{S_A}$ represent the simulated calibration mixture spectra, concentration matrix of calibration set and pure components spectra of primary spectra, respectively. Meanwhile, $\boldsymbol{E_{cs}}$ is the normally distributed noise with standard deviation as 0.002. Therefore, the primary spectra of calibration set can be generated and the first column of $\boldsymbol{C_s}$ can be set as $\boldsymbol{y}$ values.

Similar as Eq. (1), the sizes of concentrations in both validation and prediction sets are set as 25 × 4. And the corresponding simulated mixtures of secondary spectra can be computed as follows:

$$\boldsymbol{B_{vs}} = \boldsymbol{C_v} * \boldsymbol{S_B} + \boldsymbol{E_{vs}}, \tag{2}$$

$$\boldsymbol{B_{ps}} = \boldsymbol{C_p} * \boldsymbol{S_B} + \boldsymbol{E_{ps}}, \tag{3}$$

Moreover, in order to obtain transfer set, the $\boldsymbol{A_{cs}}$ can be sorted by KS method. Thus, according to the KS ranks, the first 30 samples can be set as the transfer set of primary spectra ($\boldsymbol{A_{ts}}$), while the concentrations of four components in the above 30 samples can be used to generate the transfer set of secondary spectra.

$$\boldsymbol{B_{ts}} = \boldsymbol{C_t} * \boldsymbol{S_B} + \boldsymbol{E_{ts}} \tag{4}$$

Here, $\boldsymbol{B_{ts}}$, $\boldsymbol{C_t}$ and $\boldsymbol{S_B}$ are the simulated transfer mixture spectra, transfer concentration matrix and pure components spectra of secondary spectra, respectively. And $\boldsymbol{E_{ts}}$ is the normally distributed noise with standard deviation as 0.002. Therefore, $\boldsymbol{A_{ts}}$ and $\boldsymbol{A_{cs}}$ are the transfer and calibration sets of primary spectra, while $\boldsymbol{B_{ts}}$, $\boldsymbol{B_{vs}}$ and $\boldsymbol{B_{ps}}$ are the transfer, validation and prediction sets of secondary spectra, respectively.

Because the number of abnormal samples is smaller than that of normal ones, only three in thirty samples in $\boldsymbol{B_{ts}}$ have been set as outliers. Meanwhile, in order to research the influence of concentrations on outliers detection of ER, three batches of outliers can be generated including the small, median and large samples with each containing three outliers, respectively. Based on this, according to the ascending order of $\boldsymbol{y}$ values of $\boldsymbol{B_{ts}}$, three batches of samples with the orders of $\boldsymbol{y}$ values from 1 to 3, from 14 to 16 and from 28 to 30 can be set as outliers, respectively. Thus, three batches of simulated transfer sets can be generated as $\boldsymbol{B_{ts1}}$, $\boldsymbol{B_{ts2}}$ and $\boldsymbol{B_{ts3}}$, respectively, with each batch containing three abnormal samples by the corresponding normal ones multiplying 1.2.

3 Results and Discussion

Based on the simulated dataset generated in Sect. 2.3, three batches of simulated datasets can be obtained with each containing three man-made outliers in transfer set of secondary spectra as $\boldsymbol{B_{ts1}}$, $\boldsymbol{B_{ts2}}$ and $\boldsymbol{B_{ts3}}$ corresponding to the small, median and large $\boldsymbol{y}$ values in transfer sets, respectively. Each one of the three batches of dataset can be tested by CCA-ICE, DS and SST, respectively. The mRMSEV_1 of each sample transferred by ER can be shown as follows:

The 1st, 2nd and 3rd columns represent the transfer sets as $\boldsymbol{B_{ts1}}$, $\boldsymbol{B_{ts2}}$ and $\boldsymbol{B_{ts3}}$, while the 1st, 2nd and 3rd rows as the results transferred by CCA-ICE, DS and SST, respectively.

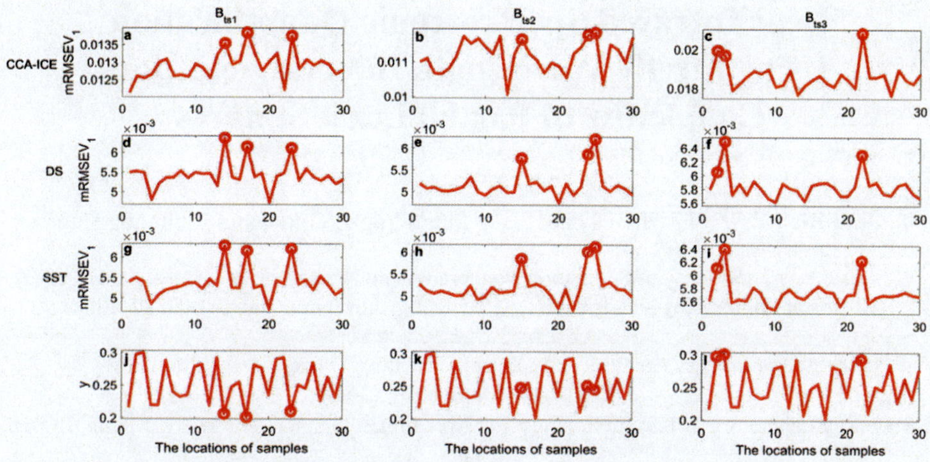

Fig. 3. The mRMSEV_1 of ER for simulated datasets

The 4th row represents the $\boldsymbol{y}$ value of each sample. In each plot, the abnormal samples have been illustrated as red circles.

In Fig. 3, it is interesting that, for transfer sets of $\boldsymbol{B}_{ts1}$, $\boldsymbol{B}_{ts2}$ and $\boldsymbol{B}_{ts3}$ with outliers corresponding to small, median and large $\boldsymbol{y}$ values, the BR can all identify the outliers as ones with high mRMSEV_1 values. Meanwhile, three calibration transfer methods including CCA-ICE, DS and SST can all isolate outliers from normal samples.

4 Conclusion

An outlier detection method called ensemble refinement (ER) has been proposed based on model population analysis (MPA) and applied in calibration transfer. The ER generates many subsets for calibration transfer at first. Then, for each sample, the average error for subsets with this sample can be obtained. Finally, samples with large errors can be identified as the outliers. The simulated spectra datasets with outliers can be used to test this method. The results showed that ER combined with different calibration transfer methods can isolate outliers obviously, including CCA-ICE, DS and SST.

Acknowledgments. The authors gratefully acknowledge the financial support provided by the National Natural Science Foundation of China (31972153), the China Postdoctoral Science Foundation (2019M661758), the Jiangsu Provincial Postdoctoral Science Foundation (2019K014) and the Foundation of Jiangsu University (19JDG010).

References

1. Deng, B., Lu, H., Tan, C., Deng, J., Yin, Y.: Model population analysis in model evaluation. Chemom. Intell. Lab. Syst. **172**, 223–228 (2018). https://doi.org/10.1016/j.chemolab.2017.11.016
2. Xie, Y., Hopke, P.K.: Calibration transfer as a data reconstruction problem. Anal. Chim. Acta. **384**, 193–205 (1999)

Near Infrared Spectroscopic Quantification Using Firefly Wavelength Interval Selection Coupled with Partial Least Squares

Xihui Bian[1,2,3(✉)], Zizhen Zhao[1], Hao Sun[1], Yugao Guo[1], and Lizhuang Hao[3]

[1] State Key Laboratory of Separation Membranes and Membrane Processes, School of Chemical Engineering and Technology, Tiangong University, Tianjin 300387, China
bianxihui@163.com

[2] Key Lab of Process Analysis and Control of Sichuan Universities, Yibin University, Yibin 644000, Sichuan, China

[3] State Key Laboratory of Plateau Ecology and Agriculture, Qinghai University, Xining 810016, China

Abstract. Firefly algorithm (FA) combined with partial least squares (PLS) are developed for near infrared (NIR) spectral interval selection and quantitative analysis of complex samples. The method firstly segments the near-infrared spectra into a number of intervals. Vectors with 1 and 0, which represent the interval selected or not, are used as the inputs of the FA. The RMSEP value predicted by PLS model is used as the fitness function of the FA. The number of spectral intervals, the population number, environmental absorbance and the constant of FA are optimized. With the optimal parameters, FA-PLS model is established and applied to predict protein, hemoglobin and cetane number in wheat, blood and diesel fuel samples, respectively. The results show that FA-PLS can significantly improve the prediction accuracy compared with full-spectrum PLS model.

Keywords: Variable selection · Firefly algorithm · Multivariate calibration · Partial least squares · Near infrared spectroscopy

1 Introduction

Analysis of complex samples is a challenging task in analytical chemistry and industries [1, 2]. In addition, traditional separation methods are difficult and time-consuming. Therefore, it is necessary to find an appropriate method to analyze complex samples. Spectral analysis provides a simple method for the analysis of complex samples due to its advantages of simplicity, rapidity and non-destructiveness [3–5]. It is widely used in agricultural commodities [6], medical [7, 8], food [9] and tobacco [10], *etc.* Nevertheless, the spectral peaks are overlapping severely in complex samples, which can reduce the prediction performance. Thus, multivariate calibration is required in the quantitative analysis of complex samples.

X. Chu et al. (Eds.): ICNIR 2021, *Sense the Real Change: Proceedings of the 20th International Conference on Near Infrared Spectroscopy*, pp. 274–282, 2022.
https://doi.org/10.1007/978-981-19-4884-8_30

The commonly used multivariate calibration methods include multiple linear regression (MLR) [11], principal component regression (PCR) [12], partial least squares (PLS) [13, 14], artificial neural network (ANN) [15] and extreme learning machine (ELM) [16], *etc.* Among these multivariate calibration methods, PLS is the most popular technique in multivariate calibration. However, with the development of modern analytical instruments, spectra data contain an enormous number of wavelength. In some cases, some wavelength consist of irrelevant information. These irrelevant wavelength usually degrade the prediction performance of model. Therefore, wavelength selection is required before multivariate calibration.

At present, many wavelength selection methods have been developed. They mainly includes individual wavelengths and spectral interval selection based on single index [17], statistics [18, 19] and swarm intelligence optimization algorithms [20, 21]. Among these methods, the swarm intelligence optimization algorithm has attracted increasing attention, especially the genetic algorithm (GA) [22, 23]. GA is inspired by natural evolution, which simulates the phenomena of crossover and mutation that occur in natural selection. This process is repeated continuously and finally obtains the optimal individual in population. However, this method has some disadvantages such as slow convergence speed and easily trapping into local optimum. Therefore, a series of new swarm intelligence optimization algorithms have been proposed.

Inspired by the flashing behavior of fireflies, Yang [24] proposed the firefly algorithm (FA). In FA, the less bright fireflies can follow the brightest firefly by attraction. The group of fireflies can gradually close to the area where the brightest firefly is located. Furthermore, the brightest individual firefly is considered as the optimal solution. The position iteration is realized by this process. Although FA has been widely used in other fields, relative few research about FA are carried out in spectral analysis fields [25, 26].

In this study, the feasibility of near infrared (NIR) spectral interval selection by FA is discussed for multivariate calibration of complex samples. The number of spectral intervals is determined firstly. Then the population number, environmental absorbance and constant of FA are optimized, respectively. With the optimal parameters, FA is used for interval selection and then PLS model is established. Compared with the full-spectrum PLS model, FA-PLS has lower root mean square error of prediction (RMSEP) and higher correlation coefficient (R) for predicting samples in prediction set.

2 Theory and Algorithm

As a swarm intelligence algorithm, FA has advantages of fast convergence speed and strong global optimization ability. The realization of this algorithm is based on the following three assumptions (i) all fireflies are unisex (ii) the brighter fireflies are more attractive than the other fireflies, *i.e.*, attractiveness is proportional to the brightness. The attractiveness and brightness can decrease with the increasing of distance, the brightest firefly can move randomly (iii) the brightness of fireflies depends on objective function. I_i is the brightness of firefly *i*, $\vec{x}_i$ is the current location of firefly *i*. The equation of firefly brightness I_i at current location is as follows.

$$I_{\mathrm{i}} = \mathrm{f}(\overrightarrow{\mathrm{x_i}}) \tag{1}$$

where the brightness of firefly is equal to the value of objective function. Fireflies are attracted to fireflies that are brighter than themselves. The equation of relative brightness between two fireflies is as follows.

$$I_{ij}(r_{ij}) = I_i e^{\gamma \times r_{ij}^2} \tag{2}$$

where I_{ij} is the light intensity. γ is the environmental absorbance. r_{ij} is the distance between firefly i and firefly j. The distance formula of the standard FA is as follows.

$$r_{ij} = \sqrt{\sum_{k=1}^{d} (x_{i,k} - x_{j,k})^2} \tag{3}$$

where d is the dimension of the solution. $x_{i,k}$ and $x_{j,k}$ are the kth dimension components of spatial coordinates x_i and x_j, respectively. The attractiveness is proportional to the brightness, the attractiveness β can be defined as

$$\beta_{ij}(r_{ij}) = \beta_0 e^{\gamma \times r_{ij}^2} \tag{4}$$

where β_0 is the maximum attraction. The distance is zero between firefly i and firefly j. The position update formula can be written as

$$\overrightarrow{x_j}(t+1) = \overrightarrow{x_j}(t) + \beta_{ij}(r_{ij})[\overrightarrow{x_i}(t) - \overrightarrow{x_j}(t)] + \alpha \overrightarrow{\varepsilon_j} \tag{5}$$

where t is the number of iterations, α is a constant and ε_j is the random number in Gaussian distribution. The schematic diagram of FA is shown in Fig. 1. The process FA is as follows. First of all, the parameters of algorithm need to be set, such as the population number, environmental absorbance and constant. The next step is to initialize

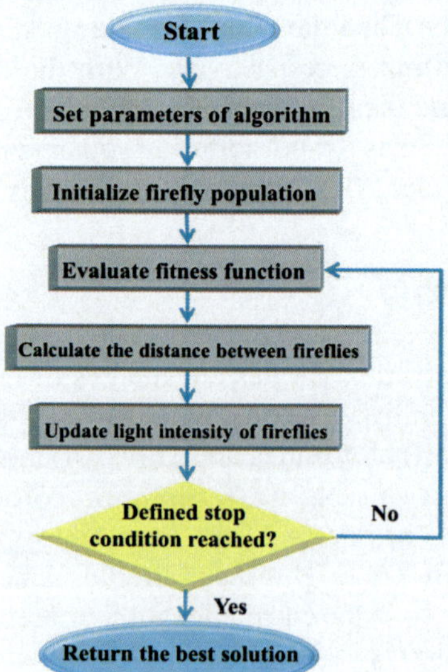

Fig. 1. The flowchart of FA algorithm.

firefly population. The third step is to evaluate fitness function. The better the fitness, the greater brightness of firefly is. The fourth step is to calculate the distance r_{ij} between fireflies. The fifth step is to update light intensity of fireflies. The last step is to output the best solution when the maximum number of iterations is reached. Otherwise, skip to the third step to continue iteration.

In the above process of FA, the population number n, environmental absorbanceγ and constant α need to be optimized. Moreover, the number of wavelength interval is also an important parameter, which needs to be determined for the input of FA.

3 Experimental

Three NIR spectral datasets were used to evaluate the predictive performance of FA-PLS. Wheat dataset was contributed by P.C. Williams, which consists of visible-NIR spectra and six properties of 884 wheat samples. [27] The Vis-NIR spectra and the protein contents are used in this study. The spectra were scanned on a Foss Model 6500 over 1050 channels recorded in the wavelength range of 400–2498 nm with the digitization interval of 2 nm. The reference values of protein contents were determined at the Grain Research Laboratory, Winnipeg. The samples Nos. 680 and 681 are two outliers and have been deleted. Figure 2(a) displays the NIR spectra of the 882 samples.

Blood dataset was provided by Norris *et al.* [28], which includes the NIR transmission and reflection spectra and four properties of 231 blood samples. The NIR reflection spectra and hemoglobin concentrations were used for this study. The spectra were scanned by model 6500 spectrometer (NIR systems, Inc., Silver Springs, USA). Each spectrum is composed of 700 variables recorded in the wavelength range 1100–2498 nm with a 2 nm interval. Figure 2(b) displays the NIR reflection spectra of the 231 samples.

Diesel fuel dataset was provided by SWRI, San Antonio, TX through Eigenvector Research, Inc. (Manson, Washington) [29]. It consists of NIR spectra and six properties of 256 fuel samples. The NIR spectra and cetane number values were investigated for this study. The spectra were measured at Southwest Research Institute (SWRI) on a project sponsored by the U.S. Army. Each spectrum is composed of 401 variables recorded in the wavelength range of 750–1550 nm. The cetane numbers were independently measured by the American Society of Testing and Materials (ASTM) standard method. Figure 2(c) displays the NIR spectra of the 256 samples.

Before calculation, the three datasets were divided into the training and prediction sets as described on the website for model building and performance validation, respectively. For wheat dataset, 775 and 107 samples were used as the training set and prediction sets. For blood dataset, 173 and 58 samples were used as the training and prediction sets. For the diesel fuel dataset, 138 and 118 samples were taken as the training and prediction sets. For PLS modeling, the optimal latent variable (LV) number is determined as 10, 11 and 9 by Monte Carlo cross-validation combined with F-test for wheat, blood and diesel fuel datasets, respectively.

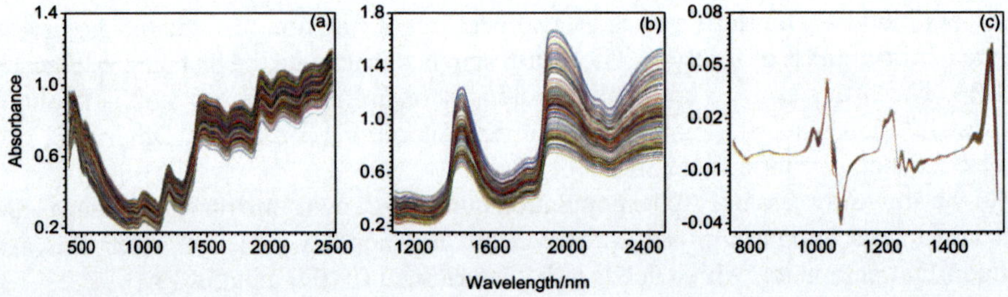

Fig. 2. NIR spectra for wheat (a), blood (b) and diesel fuel (c) datasets, respectively.

4 Results and Discussion

4.1 Determination the Interval Number

Interval number is a key parameter for FA-PLS. In order to get the optimal interval number, FA parameter takes default values, *i.e.*, the population number is 20, environmental absorbance is 1 and the constant is 0.5. The interval numbers are investigated in the range of 5–30 with an interval of 5. For each interval number, FA-PLS is performed and a RMSEP can be obtained. The variation of RMSEP with the number of intervals for wheat dataset is shown in Fig. 3.

As demonstrated in Fig. 3, with the increase of interval number, the RMSEP decreases before 10. After that, the RMSEP increases significantly. RMSEP represents predictive ability of the model, a model with a better parameter should have a lower RMSEP. Thus, the optimal interval number for wheat dataset is 10. Similarly, the optimal interval number is determined as 20 for both blood and diesel fuel datasets.

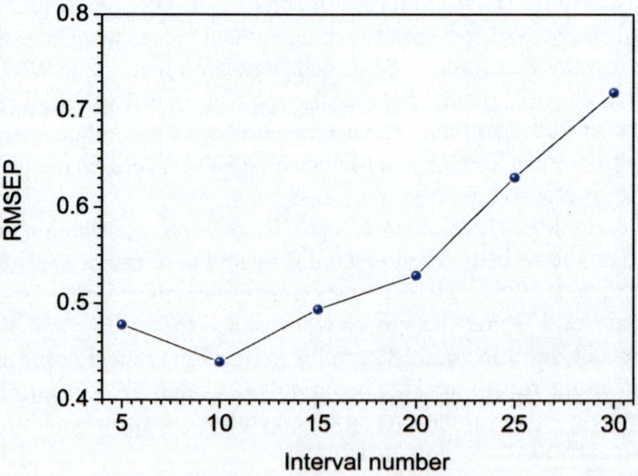

Fig. 3. Variation of RMSEP with interval number for wheat dataset.

4.2 Parameter Optimization of FA

The population number *n*, environmental absorbance γ and constant α are three important parameters for FA. The three parameters are optimized successively for each dataset. For determining the optimal population number *n*, the interval number is used the optimal value determined above. Other parameters take default values, *i.e.*, environmental absorbance is 1 and the constant is 0.5. Population number changes from 10 to 60 with an interval of 10. For each population number, FA-PLS is performed and a RMSEP can be obtained. Figure 4(a) shows the variation of RMSEP values with the population number for wheat dataset. Clearly, the RMSEP decreases with the increase of population number before 30. Above 30, the RMSEP value has an increasing trend. The lowest RMSEP is located at 30. Thus, the optimal population number is 30 for wheat dataset. Because FA is swarm intelligence, it is easy to understand that too few and too many fireflies are not good for population performance. Similar analysis can be performed for blood and diesel fuel datasets and the optimal population number is set as 30 and 40, respectively.

Environmental absorbance γ is determined subsequently. To determine the optimal environmental absorbance, interval number and population number are used the optimal values determined above, the constant is used the default value 0.5. Environmental absorbance is investigated in the range of 0.1–1.2 with interval of 0.1. Figure 4(b) depicts the variation of RMSEP with environmental absorbance for wheat dataset. It can be seen that the RMSEP is comparatively large at the beginning. With the increasing of environmental absorbance, although it has some fluctuation, the overall trend of RMSEP is decreasing. When the environmental absorbance is 0.5, RMSEP reaches the lowest value. Accordingly, 0.5 is used as the optimal environmental absorbance for wheat dataset. For blood and diesel fuel datasets, the optimal environmental absorbance is 0.8 and 0.9, respectively.

The optimal constant α is the last parameter need to be optimized. The interval number, population number, environmental absorbance are set to 10, 30 and 0.5 determined above for wheat dataset, respectively. The constant is investigated in the range of 0.1–1 with an interval 0.1. From Fig. 4(c), it is obvious that the RMSEP tends to decrease with the increase of constant before 0.3. When it is above 0.3, the overall trend in RMSEP is increasing. Thus, the optimal constant is set as 0.3 for wheat dataset. Similarly, 0.1 and 0.5 are the optimal α for blood and diesel fuel dataset, respectively.

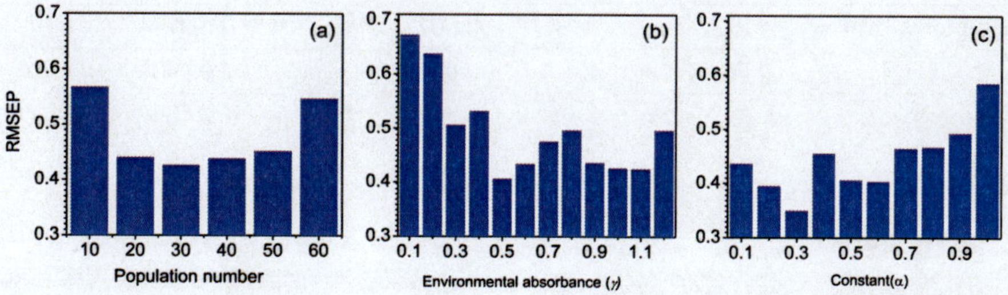

Fig. 4. Variation of RMSEPs with population number (a), environmental absorbance γ (b) and constant α (c) for wheat dataset.

4.3 Prediction Results

The optimal interval number, population number n, environmental absorbance γ and constant α for the three datasets are summarized in Table 1. With the optimal parameters, FA is used to select the spectral intervals of the samples in training set and then built PLS model. To validate the efficiency of FA method, the same spectral intervals of the samples in prediction set are selected and input into the model. The predicted values are obtained and used to calculate RMSEP and R for the three datasets, which are shown in Table 2. For comparison, the full-spectrum PLS results for the three datasets are also listed in Table 2. Obviously, a better model should have a lower RMSEP and a larger R. As shown in Table 2, with FA, the RMSEP of PLS decrease from 0.7763, 0.4310, 0.0023 to 0.3498, 0.3308, 0.0014 for wheat, blood and diesel fuel datasets, respectively. The R values of the three datasets all increase after FA interval selection. The results show that FA spectral interval selection can improve the prediction accuary of PLS obviously. To sum up, FA-PLS is an efficient method for NIR spectral interval selection and NIR spectral quantitative analysis.

Table 1. The parameter optimization results for different datasets

Datasets	Interval number	n	γ	α
Wheat	10	30	0.5	0.3
Blood	20	30	0.8	0.1
Diesel oil	20	40	0.9	0.5

Table 2. Prediction results of PLS before and after interval selection by FA

Datasets	Methods	RMSEP	R
Wheat	PLS	0.7763	0.9365
	FA-PLS	0.3498	0.9762
Blood	PLS	0.4310	0.9613
	FA-PLS	0.3308	0.9777
Diesel oil	PLS	0.0023	0.9744
	FA-PLS	0.0014	0.9901

5 Conclusion

FA interval selection coupled with PLS is used for NIR spectroscopic quantification of complex samples. In this approach, the parameters are firstly optimized and then wavelength intervals are selected by FA. With the optimal parameters, FA-PLS model is established and applied to predict unknown samples. In order to verify the validity of the method, the contents of protein in wheat, hemoglobin in blood and cetane number in diesel fuel samples are predicted, respectively. The RMSEP and R of FA-PLS are compared with those of PLS on the three datasets. Result shows that FA can effectively improve the prediction performance of PLS.

Acknowledgments. This study is supported by Key Lab of Process Analysis and Control of Sichuan Universities (No. 2020001) and Opening Foundation of State Key Laboratory of Plateau Ecology and Agriculture (No. 2021-KF-07).

References

1. Wei, J.L., Huang, D.Y., Chen, Y.C.: Using gadolinium ions as affinity probes to selectively enrich and magnetically isolate bacteria from complex samples. Anal. Chim. Acta **1113**, 18–25 (2020)
2. Offermans, T., et al.: ENDBOSS: industrial endpoint detection using batch-specific control spaces of spectroscopic data. Chemometr. Intell. Lab. Syst. **209**, 104229 (2021)
3. Bian, X.H., et al.: Rapid identification of milk samples by high and low frequency unfolded partial least squares discriminant analysis combined with near infrared spectroscopy. Chemometr. Intell. Lab. Syst. **170**, 96–101 (2017)
4. Su, T., Sun, Y., Han, L., Cai, W.S., Shao, X.G.: Revealing the interactions of water with cryoprotectant and protein by near-infrared spectroscopy. Spectrochim. Acta A **266**, 120417 (2022)
5. Li, J.Y., Chu, X.L., Tian, S.B., Lu, W.Z.: The identification of highly similar crude oils by infrared spectroscopy combined with pattern recognition method. Spectrochim. Acta A **112**, 457–462 (2013)
6. Appell, M., Compton, D.L., Bosma, W.B.: Raman spectral analysis for rapid determination of zearalenone and alpha-zearalanol. Spectrochim. Acta A **270**, 120842 (2022)
7. Lin, L., et al.: A rapid analysis method of safflower (Carthamus tinctorius L.) using combination of computer vision and near infrared. Spectrochim. Acta A **236**, 118360 (2020)
8. Bian, X.H., Lu, Z.K., van Kollenburg, G.H.: Ultraviolet-visible diffuse reflectance spectroscopy combined with chemometrics for rapid discrimination of Angelicae Sinensis Radix from its four similar herbs. Anal. Methods **12**, 3499–3507 (2020)
9. Urickova, V., Sadecka, J.: Determination of geographical origin of alcoholic beverages using ultraviolet, visible and infrared spectroscopy: a review. Spectrochim. Acta A **148**, 131–137 (2015)
10. Ma, Y.J., et al.: Rapid determination of four tobacco specific nitrosamines in burley tobacco by near-infrared spectroscopy. Anal. Methods **4**, 1371–1376 (2012)
11. Lemos, T., Kalivas, J.H.: Leveraging multiple linear regression for wavelength selection. Chemometr. Intell. Lab. Syst. **168**, 121–127 (2017)
12. Gemperline, P.J., Salt, A.: Principal components regression for routine multicomponent UV determinations: a validation protocol. J. Chemom. **3**, 343–357 (1989)
13. Wold, S., Sjostrom, M., Eriksson, L.: PLS-regression: a basic tool of chemometrics. Chemometr. Intell. Lab. Syst. **58**, 109–130 (2001)
14. Zhang, H., Hu, X.Y., Liu, L.M., Wei, J.F., Bian, X.H.: Near infrared spectroscopy combined with chemometrics for quantitative analysis of corn oil in edible blend oil. Spectrochim. Acta A **270**, 120841 (2022)
15. Fei, Q., Li, M., Wang, B., Huan, Y.F., Feng, G.D., Ren, Y.L.: Analysis of cefalexin with NIR spectrometry coupled to artificial neural networks with modified genetic algorithm for wavelength selection. Chemometr. Intell. Lab. Syst. **97**, 127–131 (2009)

16. Bian, X.H., Li, S.J., Fan, M.R., Guo, Y.G., Chang, N., Wang, J.J.: Spectral quantitative analysis of complex samples based on extreme learning machine. Anal. Methods **8**, 4674–4679 (2016)
17. Wu, W., Walczak, B., Massart, D.L., Prebble, K.A., Last, I.R.: Spectral transformation and wavelength selection in near-infrared spectra classification. Anal. Chim. Acta **315**, 243–255 (1995)
18. Xu, Z.C., Liu, W.S., Cai, X.G.: Shao, A wavelength selection method based on randomization test for near-infrared spectral analysis. Chemometr. Intell. Lab. Syst. **97**, 189–193 (2009)
19. Han, Q.J., Wu, H.L., Cai, C.B., Xu, L., Yu, R.Q.: An ensemble of monte carlo uninformative variable elimination for wavelength selection. Anal. Chim. Acta **612**, 121–125 (2008)
20. Shamsipur, M., Zare-Shahabadi, V., Hemmateenejad, B., Akhond, M.: Ant colony optimisation: a powerful tool for wavelength selection. J. Chemom. **20**, 146–157 (2006)
21. Wu, X.Y., Bian, X.H., Yang, S., Xu, P., Wang, H.T.: A variable selection method for near infrared spectroscopy based on gray wolf optimizer algorithm. J. Instrum. Anal. **39**, 1288–1292 (2020)
22. Mehmood, T., Liland, K.H., Snipen, L., Saebo, S.: A review of variable selection methods in partial least squares regression. Chemometr. Intell. Lab. Syst. **118**, 62–69 (2012)
23. Mohammadi, M., et al.: Genetic algorithm based support vector machine regression for prediction of SARA analysis in crude oil samples using ATR-FTIR spectroscopy. Spectrochim. Acta A **245**, 118945 (2021)
24. Yang, X.S.: Firefly algorithms for multimodal optimization. Lect. Notes Comput. Sci. **5792**, 169–178 (2009)
25. Goodarzi, M., Coelho, L.D.: Firefly as a novel swarm intelligence variable selection method in spectroscopy. Anal. Chim. Acta **852**, 20–27 (2014)
26. Attia, K.A.M., Nassar, M.W.I., El-Zeiny, M.B., Serag, A.: Firefly algorithm versus genetic algorithm as powerful variable selection tools and their effect on different multivariate calibration models in spectroscopy: a comparative study. Spectrochim. Acta A **170**, 117–123 (2017)
27. Bian, X.H., et al.: Robust boosting neural networks with random weights for multivariate calibration of complex samples. Anal. Chim. Acta **1009**, 20–26 (2018)
28. Kuenstner, J.T., Norris, K.H., McCarthy, W.F.: Measurement of hemoglobin in unlysed blood by near-infrared spectroscopy. Appl. Spectrosc. **48**, 484–488 (1994)
29. Soyemi, O.O., Busch, M.A., Busch, K.W.: Multivariate analysis of near-infrared spectra using the G-programming language. J. Chem. Inf. Comput. Sci. **40**, 1093–1100 (2000)

Application of Convolutional Neural Network Model Based on Combined NIR-Raman Spectra in Feed Composition Analysis

Wenjie Zhang[1], Yihao Liang[1], Gongyi Cheng[1], Chao Dong[3], Bin Wang[1,3], Jing Xu[2], and Xiaoxuan Xu[1](✉)

[1] The Key Laboratory of Weak-Light Nonlinear Photonics, Ministry of Education, School of Physics, Nankai University, Tianjin 300071, China
xuxx@nankai.edu.cn
[2] College of Artificial Intelligence, Nankai University, Tianjin 300350, China
[3] Changzhou Hengxian Sensing Technology Co., Ltd., 213172 Changzhou, China

Abstract. The maturity of machine learning algorithms such as convolutional neural networks have made it possible to use them in feed spectroscopy. This paper compares convolutional neural network (CNN) and multivariate scattering processing support vector machine (MSC-SVM) modeling, including NIR spectroscopy, Raman spectroscopy modeling, and NIR-Raman spectroscopy modeling, to predict the protein content in feed. The experiments were based on measured NIR (wave number 4000–12000 wavenumbers) and Raman spectral (500–3000 wavenumbers) data due to the complementary roles of NIR and Raman spectroscopy techniques. The organic combination of the two spectral data adds useful information in model building. The CNN and MSC-SVM models based on NIR-Raman spectra, the predictions are better than single spectra.

Keywords: Feed · Convolutional neural network · Support vector machine · Raman spectroscopy

1 Introduction

Feed is the basis for the development of China's animal husbandry industry. With the booming development of feed industry and animal husbandry, it becomes especially important to identify the quality of feed quickly. Because of the imperfection of the feed testing system and the high price of high-quality feed in the market, resulting in a part of the unscrupulous elements in the feed adulterated with some other substances. Protein feed ingredients are again the cornerstone of the feed industry, and protein, as one of the most important material bases for life activities, plays a huge role in the farming industry and is one of the important qualities in analyzing feed quality. Common feed protein testing methods include Kjeldahl nitrogen determination, Dumas combustion, near infrared spectrometry, nano colorimetric method, double condensation urea method, atomic fluorescence spectrophotometer method, etc. [1].

X. Chu et al. (Eds.): ICNIR 2021, *Sense the Real Change: Proceedings of the 20th International Conference on Near Infrared Spectroscopy*, pp. 283–290, 2022.
https://doi.org/10.1007/978-981-19-4884-8_31

Near-infrared spectroscopy is widely used because of its fast and non-destructive characteristics. NIR spectra mainly consist of the ensemble and octave frequencies of hydrogen-containing group vibrations, which contain rich information such as intensity and position differences of absorption peaks [24]. Since most feed can provide characteristic information in the NIR spectral region, NIR spectroscopy techniques combined with traditional machine learning algorithms such as KNN, PCA, CART, etc. They are suitable for application in feed protein detection.

Support vector machines (SVM) are supervised learning models for analyzing data in classification and regression analysis with associated learning algorithms. Given a set of training instances, each labeled as belonging to one or the other of two classes, the SVM training algorithm creates a model that assigns new instances to one of the two classes, making it a non-probabilistic binary linear classifier. The SVM model represents instances as points in space, so that the mapping allows instances of separate classes to be separated by as wide an apparent interval as possible. The new instances are then mapped into the same space and the categories to which they belong are predicted based on which side of the interval they fall.

Convolutional Neural Networks (CNN), one of the hot topics in deep learning for several years, has excellent performance in image classification and other areas [3]. Fully connected artificial neural network methods require too many parameters to train and have a large time cost. In contrast, the neurons of CNN perceive only local information, and the number of parameters to be trained is greatly reduced by sharing the parameters of the convolution kernel used in the same layer. NIR spectral data are spatially correlated to form features, and the sampling principle is consistent on each space, which is suitable for learning with CNN.

2 Materials and Methods

2.1 Sample and Spectral Acquisition

We have 25 groups of samples and each group of feed samples has protein content arrange of 14% to 20%. The spectrometer used is a Fourier NIR spectrometer with a scanning band of 4000 wave numbers to 12000 wave numbers, with 4001 spectral points. The spectrum consists of a laser at 532 nm as the excitation light and a scanning band of 100–3200 nm with 2295 data points.

The NIR and Raman spectra measured of samples are shown in Fig. 1. 20% of the samples and 80% of the samples are classified as the training and validation set.

2.2 Data Pre-processing

The measurement process of sample spectra may be affected by noise, sample background and measurement instruments. The first derivative method (1st Der), the second derivative method (2nd Der), centralization, standardization, Savitzky-Golay smoothing method (SG) [11], multiplicative scatter correction, MSC) [12], and standard normal variate (SNV) [13] method to preprocess the spectral data. Based on the results under subsequent modeling, multiplicative scatter correction was selected as the data preprocessing method.

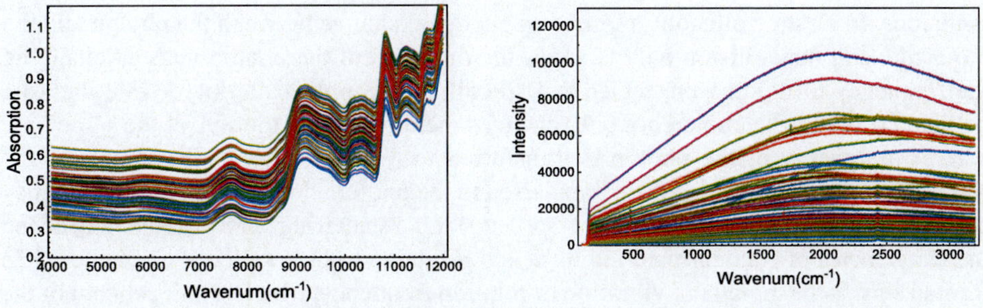

Fig. 1. Raw NIR and Raman spectra

2.3 One-Dimensional Convolutional Neural Network

Convolutional neural network is a nonlinear model. The convolutional neural network model contains an input layer, a convolutional layer, a pooling layer, a fully connected layer and an output layer, and all layers can be set up in multiple layers except for the input and output layers. After inputting data and labels, the convolutional layer uses multiple one-dimensional convolutional kernels with set size and step size to obtain feature maps after convolutional operations. The number of convolutional layers can be one, two or more, but too many layers and kernels may lead to overfitting. The pooling layer is usually used after the convolution layer to extract the local features of the data. Pooling methods include maximum pooling and average pooling, with the former focusing on extracting feature information. After one or more fully connected layers, the features are mapped to the sample space for classification. In order to build a nonlinear model, an activation function needs to be added, and the gradient disappearance problem can be avoided by using the ReLU function.

The model is trained by first initializing the weights, inputting the feed training set NIR spectral data and category labels, and passing through each layer of the neural network to obtain the final output. The model loss function value is calculated, and the loss function value is passed from the final layer to each layer of the network by back propagation, and the weights are updated in the direction of minimizing the loss function value to continue the training.

In order to avoid overfitting phenomenon as much as possible, regular terms and random deactivation (Dropout) are added to the neural network. And to reduce the model complexity, the convolutional layer uses 8 convolutional kernels with a size of 1×3. The pooling layer uses the maximum pooling method with kernel size of 1×2 and step size of 2. The final fully-connected layer of the model uses Softmax activation function, AdamOptimizer optimizer, learning rate of 0.001, number of convolutional kernels of 3, and iteration count of 10000.

2.4 Combined NIR-Raman Spectroscopy

The near-infrared spectra of the samples measured in this study were coupled with Raman spectra, which are produced by inelastic scattering of light irradiated onto matter. Incident photons of monochromatic light can interact with molecules in elastic and inelastic collisions. In elastic collisions, there is no energy exchange between the photon and the molecule, and the collision only changes the direction of the photon without changing the frequency of the photon, which is also called Rayleigh scattering. While inelastic collision process, the photon not only changes the direction of motion, at the same time part of the energy of the photon is transferred to the molecule, or the vibration and rotation energy of the molecule is transferred to the photon, thus changing the frequency of the photon, this scattering process is called Raman scattering. The difference between the frequencies of Raman scattered light and Rayleigh light is called Raman shift. The Raman shift is the molecular vibration or rotation frequency, which is independent of the incident frequency and is related to the molecular structure. Raman spectrum is similar to infrared spectrum, which is one of the scattering spectra, and its signal comes from the vibration and rotation of molecules. However, IR spectra are related to the change of dipole moment when the molecule vibrates, while Raman scattering is the result of the change of polarization rate of the molecule. Raman spectroscopy and IR spectroscopy are complementary to each other in molecular structure analysis.

3 Results and Discussion

3.1 Pre-treatment Method

For the different pre-processing methods, the RMSEs of the processed spectra after 10 regressions using the support vector machine model according to the divided training and prediction sets are shown in Table 1.

Table 1. Comparison of RMSE based on different preprocessing methods

Pretreatment methods	Validation RMSE
Raw NIR spectra	1.32
MSC	0.84
Centralization	1.63
Standardization	1.65
1 st Der	1.72
2 nd Der	1.85
SG	2.52

From Table 1, it can be seen that after comparing the spectra without any treatment with the other treatments, the spectra treated with MSC have the smallest prediction error.

3.2 NIR Spectra Combined with Raman Spectra for Stitching

The results of the sample NIR and Raman spectra after MSC processing are shown in Fig. 2 and Fig. 3.

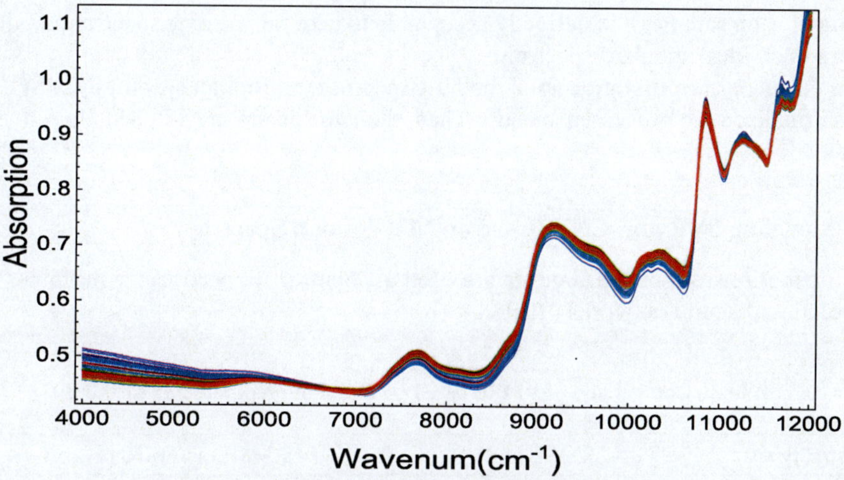

Fig. 2. NIR spectra after MSC treatment

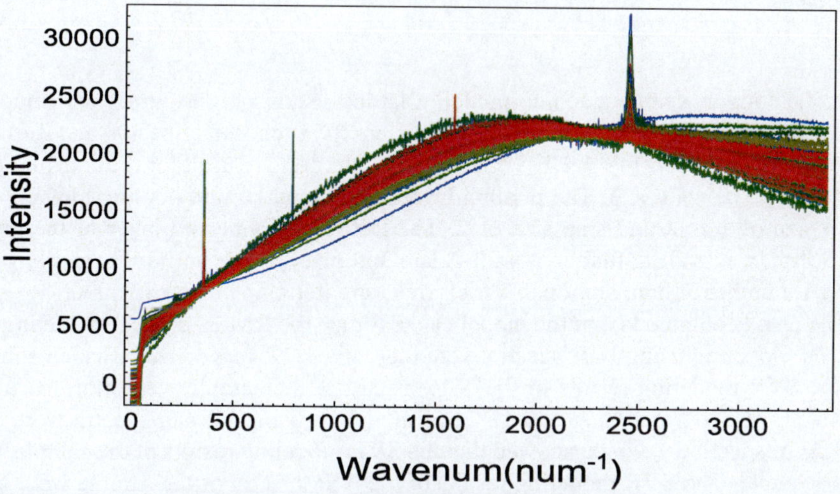

Fig. 3. NIR and Raman spectra after MSC treatment

The multivariate scattering correction method is a common data processing method for multi-wavelength calibration modeling. The scattering correction can effectively eliminate the scattering effect and enhance the absorption information related to the content of the components. The use of this method first requires the establishment of an "ideal spectrum" of the sample to be measured. There is a direct linear relationship between the change in the spectrum and the content of the components in the sample. This spectrum was used to correct the NIR spectra of all other samples, including baseline shifts and offset corrections. In practice, the "ideal spectrum" is difficult to obtain. since the method is only used to correct the relative baseline shifts and offsets in the NIR spectra of each sample, it is perfectly acceptable to take the average spectrum of all the spectra as an ideal standard spectrum.

In this paper, we first processed the NIR spectra with Raman spectra by MSC, and then normalized the processed spectra. Then, the two spectra are stitched close to each other.

3.3 Modeling SVR and CNN Based on NIR-Raman Spectra

In our experiments, spectral images are used as input to support vector machines and convolutional neural networks (Table 2).

Table 2. Comparison of RMSE based on different preprocessing methods

Spectral type	RMSE of SVR	RMSE of 1DCNN
NIR	0.84	1.22
Raman	0.92	1.53
NIR-Raman	0.12	0.35

In the support vector machine model, a lattice-seeking algorithm is performed to find the best parameters, where Penalty coefficient (C) is chosen to be 400 and Kernel is chosen to be Poly, Gamma = 2.1. 8 convolutional kernels are used for the convolutional layer, with a size of 1×3. The pooling layer uses the maximum pooling method with kernel size of 1×2 and step size of 2. The last fully connected layer of the model uses Softmax activation function, with AdamOptimizer as the optimizer, learning rate of 0.001, number of convolutional kernels of 3, and iteration count of 10000.

The results obtained from the model calculations: the RMSE of SVR modeling and 1DCNN modeling using NIR spectra were 0.84 and 1.22, respectively, which showed that the SVR modeling results of the NIR spectra of the sample were more accurate; the RMSE of SVR modeling and 1DCNN modeling using Raman spectra were 0.92 and 1.53, respectively, which showed that the SVR modeling results of the sample were more accurate. The SVR modeling results of the NIR spectra of the sample were more accurate; the RMSE of the results of SVR modeling and 1DCNN modeling using NIR-Raman spectroscopy were 0.12 and 0.35, respectively, and the results surface that the modeling accuracy of the sample using NIR-Raman spectroscopy was greatly improved. For such small sample data, the SVR results are better than 1DCNN.

4 Conclusion

To address the difficulty of detecting nutrient content in farm feeds nowadays, this study measured NIR and Raman spectra of feed samples with different protein contents, and selected the MSC method with the highest accuracy by comparing various pre-processing methods. In this study, samples were modeled by NIR spectroscopy alone, Raman spectroscopy alone and NIR-Raman spectroscopy. The support vector machine regression and 1DCNN modeling were performed for each of the three types of data to compare the different results. It was found that both modeling methods had higher accuracy of NIR spectra results than Raman spectra; support vector machine regression models had higher accuracy of results than 1DCNN model. For both models, the NIR-Raman spectral accuracy has a significant improvement.

Acknowledgements. This work is supported by Jiangsu Province: Program for High-Level Entrepreneurial and Innovation Talents Introduction.

References

1. Li, B., Yang, W.: Research progress on detection methods of protein in feed. Anim. Husb. Feed Sci. **38**, 38–40 (2017)
2. Liu, Y., Xie, Y., Li, Z.: Research progress of deep learning target detection algorithm in security field. Com. Tech. **54**, 2063–2073 (2021)
3. He, R., Wu, X., Sun, Z.: Wasserstein CNN: learning invariant features for nir-vis face recognition. IEEE Trans. Pattern Anal. Mach. Intell. **41**, 1761–1773 (2018)
4. Lee, Y.W., Kim, K.W., Hoang, T.M.: Deep residual CNN-based ocular recognition based on rough pupil detection in the images by NIR camera sensor. Sensors. **19**, 842 (2019)
5. Yu, G., Ma, B., Chen, J.: Nondestructive identification of pesticide residues on the Hami melon surface using deep feature fusion by Vis/NIR spectroscopy and 1D-CNN. J. Food Process Eng. **44**, e13602 (2021)
6. Tazim, R.J., Miah, M.M.M., Surma, S.S.: Biometric authentication using CNN features of dorsal vein pattern extracted from NIR image. In: TENCON 2018–2018 IEEE Region 10 Conference, pp. 1923–1927. IEEE (2018)
7. Cui, C., Fearn, T.: Modern practical convolutional neural networks for multivariate regression: applications to NIR calibration. Chemom. Intell. Lab. Syst. **182**, 9–20 (2018)
8. Chakravartula, S.S.N., Moscetti, R., Bedini, G.: Use of convolutional neural network (CNN) combined with FT NIR spectroscopy to predict food adulteration: a case study on coffee. Food Control **135**, 108816 (2022)
9. Murray, I.: Value of traditional analytical methods and near-infrared (NIR) spectroscopy to the feed industry, Ve of traditional analytical methods and near-infrared (NIR) spectroscopy to the feed industry, pp. 87–108 (1996)
10. Wu, Y.F., Peng, S.L., Xie, Q.: An improved weighted multiplicative scatter correction algorithm with the use of variable selection: application to near-infrared spectra. Chemometr. Intel. Lab. Sys. **185**, 114–121 (2019)
11. Bi, Y.M., Yuan, K.L., Xiao, W.Q., Wu, J.Z.: A local pre-processing method for near-infrared spectra, combined with spectral segmentation and standard normal variate transformation. Anal. Chim. Acta. **909**, 30–40 (2016)
12. Zhang, H., Yang, S., Guo, L.: Comparisons of isomiR patterns and classification performance using the rank-based MANOVA and 10-fold cross-validation. Gene **569**, 21–26 (2015)
13. Undersander, D.: Uses and abuses of NIR for feed analysis. In: Florida Ruminant Nutrition Symposium, Gainseville (2006). http://dairy.ifas.ufl.edu/rns/2006/Undersander.pdf
14. Fernandez-Ahumada, E., Garrido-Varo, A., Guerrero, J.E.: Taking NIR calibrations of feed compounds from the laboratory to the process: calibration transfer between predispersive and postdispersive instruments. J. Agr. Food Chem. **56**, 10135–10141 (2008)
15. Peeters, M., Peeters, E., Van Hauwermeiren, D.: Determination and understanding of lead-lag between in-line NIR tablet press feed frame and off-line NIR tablet measurements. Int. J. Pharm. **611**, 121328 (2021)

16. Yi, C.: Discrimination of fresh tobacco leaves with different maturity levels by near-infrared (NIR) spectroscopy and deep learning. J. Anal. Meth. Chem. **2021** (2021)
17. Balabin, R.M., Lomakina, E.I.: Support vector machine regression (SVR/LS-SVM)-an alternative to neural networks (ANN) for analytical chemistry? Comparison of nonlinear methods on near infrared (NIR) spectroscopy data. Analyst **136**, 1703–1712 (2011)
18. Liu, X., Chen, H.C., Liu, T.A.: Application of PCA-SVR to NIR prediction model for tobacco chemical composition. Spectrosc. Spect. Anal. **27**, 2460–2463 (2007)
19. Xu, S., Lu, B., Baldea, M.: An improved variable selection method for support vector regression in NIR spectral modeling. J. Process Control **67**, 83–93 (2018)
20. Liu, J., Pu-mei, G., Hou, Z.: CWT-SVR model and its application in NIR analysis of corn. J. Southwest Univ. National. (Nat. Sci. Edn.) (2008)
21. Argyri, A.A., Jarvis, R.M., Wedge, D.: A comparison of Raman and FT-IR spectroscopy for the prediction of meat spoilage. Food Control **29**, 461–470 (2013)
22. Zhang, Y., Zhang, F., Fu, X.: Detection of fatty acid content in mixed oil by Raman spectroscopy based on ABC-SVR algorithm. Spectros. Spectr. Anal. **39**, 2147 (2019)
23. Fan, Y., Lai, K., Rasco, B.A.: Determination of carbaryl pesticide in Fuji apples using surface-enhanced Raman spectroscopy coupled with multivariate analysis. LWT-Food Sci. Tech. **60**, 352–357 (2015)
24. Han, X., Yang, W.: Characteristic spectrum response of folic acid solution on two-dimensional visible-near infrared transmitted spectroscopy. Agri. Eng. **9**, 74–78 (2019)
25. Zhang, H.: Comparisons of isomiR patterns and classification performance using the rank-based MANOVA and 10-fold cross-validation. Gene **569**, 21–26 (2015)

LASSO Based Extreme Learning Machine for Spectral Multivariate Calibration of Complex Samples

Zizhen Zhao[1], Kaiyi Wang[1], Shuyu Wang[1], Yang Xiang[2], and Xihui Bian[1,2,3](✉)

[1] State Key Laboratory of Separation Membranes and Membrane Processes, School of Chemical Engineering and Technology, Tiangong University, Tianjin 300387, China
bianxihui@163.com

[2] State Key Laboratory of Plateau Ecology and Agriculture, Qinghai University, Xining 810016, China

[3] Key Lab of Process Analysis and Control of Sichuan Universities, Yibin University, Yibin 644000, Sichuan, China

Abstract. Extreme learning machine (ELM) has received increasing attention in multivariate calibration of complex samples due to its advantages of fast learning speed and good generalization ability. However, irrelevant variables in spectral matrix to target can interfere the quality of ELM modeling. Therefore, variable selection is required before multivariate calibration. In this study, least absolute shrinkage and selection operator (LASSO) combined with ELM (LASSO-ELM) is used for spectral quantitative analysis of complex samples. In the method, LASSO is firstly used to selected variables by shrinking regression coefficients of unselected variables to zero. The optimal model position *s* of LASSO is determined by S*p* criterion. Then ELM model is built between the selected variables and analyzed target with the optimal activation function and hidden node number determined by the ratio of mean to standard deviation of correlation coefficients (MSR). Near infrared (NIR) spectra of tobacco lamina and ultraviolet (UV) spectra of fuel oil samples are used to evaluate the prediction performance of LASSO-ELM. Results show that only with tens of variables, LASSO-ELM achieves the lowest root mean square error of prediction (RMSEP) and highest correlation coefficient (R) compared with full-spectrum partial least squares (PLS) and ELM. Thus, LASSO-ELM is an effective variable selection and multivariate calibration method for quanatitive analysis of complex samples.

Keywords: Least absolute shrinkage and selection operator · Extreme learning machine · Variable selection · Spectral analysis · Quantification

1 Introduction

Quantitative analysis of complex samples is a challenging task in the field of analytical chemistry. Chromatographic analysis offers high sensitivity and selectivity in analyzing complex samples but the whole process is time-consuming and laborious. As fast, green

X. Chu et al. (Eds.): ICNIR 2021, *Sense the Real Change: Proceedings of the 20th International Conference on Near Infrared Spectroscopy*, pp. 291–300, 2022.
https://doi.org/10.1007/978-981-19-4884-8_32

and nondestructive technology, spectral analysis [1, 2] has been increasingly used to analyze complex samples. However, hundreds of components in complex samples make it difficult to quantitative analysis directly by univariate analysis. Therefore, multivariate calibration methods, including linear and nonlinear methods, are widely used in the quantitative analysis of complex samples [3–5]. The linear methods, such as principal component regression (PCR) [6, 7], multiple linear regression (MLR) [8] and partial least square (PLS) [9–11] have simple structure, good stability and few parameters involved. However, when nonlinear behavior appears, the prediction performances of linear methods may decrease. Nolinear methods, such as artificial neural network (ANN) [12], support vector regression (SVR) [13, 14], can solve the nonlinear problems. However, they usually need more parameters and easily fall into local minimum.

Extreme learning machine (ELM) firstly proposed by Huang [15] and has been applied in spectral quantitative analysis of complex samples in recent years [16–18]. It can model nonlinear data well with two parameters [19, 20]. However, when the number of variable in spectra is much larger than that of samples, it is likely to cause dimensional disaster. Furthermore, only some variables related to the analyzed target, irrelevant variables involved in the model could reduce the prediction ability of ELM model. Therefore, it is necessary to select variables [21, 22] before ELM modeling.

The least absolute shrinkage and selection operator (LASSO) is a shrinkage estimation method proposed by Tibshirani [23], which has been used in medicine [24, 25], food [26, 27] and wastewater treatment [28], *etc.* LASSO can shrink regression coefficients to zero by adding a penalty term L1 to the objective function. Thus, it not only can be used to solve dimensional disaster problem but also can be used for variable selection.

Inspired by the attractive advantage of LASSO and ELM, LASSO-ELM is developed for spectral quantitative analysis. In this approach, the model position of LASSO, the activation function and hidden node number of ELM are investigated first. With the optimal parameters, the related variables are selected by LASSO and then ELM is built between the selected variables and target in training set. To validate the prediction performance, the method is used to predict reducing sugar contents of tobacco lamina samples and monoaromatics contents of fuel oil samples in prediction set, respectively. Compared with full-spectrum ELM and PLS, the result shows that LASSO-ELM is better than PLS and ELM in terms of root mean square error of prediction (RMSEP) and correlation coefficient (R).

2 Theory and Algorithm

2.1 Extreme Learning Machine (ELM)

The ELM algorithm is a novel single-hidden layer feed forward neural networks (SLFNs), which includes input layer, hidden layer and output layer. The input weights and hidden layer biases of ELM are randomly assigned. Moreover, the output weights are realized by Moore-Penrose generalized inverse method. Supposing that $\mathbf{X} = [\mathbf{x}_1, \ldots, \mathbf{x}_m]^{\mathrm{T}} \in R^{m\times n}$ and $\mathbf{y} = [\mathrm{y}_1, \ldots, \mathrm{y}_m]^{\mathrm{T}} \in R^{m\times 1}$ are the spectra and targets, respectively. m is the number of samples, n denotes the variables in training set. For the jth training sample $(\mathrm{x}_j, \mathrm{y}_j)$, the SLFNs model can be represented as follows.

$$\sum\nolimits_{i=1}^{L} \beta_i g(w_i \cdot x_j + b_i), j = 1, 2, \cdots m \quad (1)$$

where g is an activation function, L is the number of hidden layer nodes, $i = 1, 2, \cdots, L$. β_i is the output weights. w and b represent the input weight vector and the hidden layer biases, respectively. The above formula can also be written as

$$\mathbf{H\beta} = \mathbf{y} \tag{2}$$

where $\boldsymbol{\beta}$ represents the output weight vector, which is a $L \times 1$ matrix. H is a $m \times L$ hidden layer output matrix. The optimal solution of β can be obtained by

$$\boldsymbol{\beta} = \mathbf{H}^{+}\mathbf{y} \tag{3}$$

where $\mathbf{H}^{+}$ is the Moore-Penrose generalized inverse of matrix $\mathbf{H}$.

Before building the ELM model, two parameters including activation functions and the number of hidden nodes should be optimized. The optimal parameters of ELM are determined by the highest MSR [20], which is the ratio of mean value to standard deviation of Rs.

2.2 Least Absolute Shrinkage and Selection Operator (LASSO)

LASSO is an efficient technique to eliminate redundant variables. The principle of LASSO is to minimize the residual sum of squares when the sum of the absolute values of regression coefficients is less than a constant. The LASSO can be defined as

$$\hat{\beta}_{LASSO} = \arg\min_{\beta} \sum\nolimits_{i=1}^{m} \left(y_i - \sum\nolimits_{b=1}^{n} x_{ib}\beta_b \right)^2, \ s.t. \sum\nolimits_{b=1}^{\mathrm{n}} |\beta_b| \le t \tag{4}$$

where t is tuning parameter, x_i and y_i is the spectrum and target value of the ith sample in training set. Equation (4) can also be written as

$$\hat{\beta}_{LASSO} = \arg\min_{\beta} \left\{ \sum\nolimits_{i=1}^{m} \left(y_i - \sum\nolimits_{b=1}^{n} x_{ib} \right)^2 + \lambda \sum\nolimits_{b=1}^{n} |\beta_b| \right\} \tag{5}$$

Assuming that $\mathbf{X}^T\mathbf{X} + \lambda\Omega^-$ is nonsingular. The solution of the equation is as follows.

$$\hat{\beta}_{LASSO} = \left(\mathbf{X}^{\mathrm{T}}\mathbf{X} + \lambda\Omega^- \right)^{-1} \mathbf{X}^{\mathrm{T}}\mathbf{y} \tag{6}$$

The parameter λ is predetermined, which determines the number of zeros in the regression coefficients. In this paper, the forward stagewise method is used to find the optimal variable number of LASSO. Before iteration, the regression coefficients are all zero. In the iteration process, some new variables are increasing or decreasing accordingly. When the number of iterations reaches the maximum, it means that the number of variables is the largest. However, a large number of variables do not represent that an optimal model can be formed. Therefore, in the whole iterative process, it is necessary to find an optimal position s. In this study, the optimal model position s of LASSO is obtained by residual sum of squares (RSS) of 10-fold cross-validation combined with Sp criterion.

3 Experimental

Two spectral datasets were used to evaluate the prediction performance of LASSO-ELM. The tobacco lamina dataset, provided by tobacco corporation, consists of NIR spectra and reducing sugar of 271 tobacco samples. The NIR spectra and reducing sugar contents were used in this study. The spectra were measured with MPA FT-NIR spectrometer (Bruker, Germany). Each spectrum is composed of 1296 variables recorded in the wavelength range of 9000–4000 cm^{-1} (1111–2500 nm) with the digitization interval ca. 4 cm^{-1}. The reducing sugar contents were measured on an AA III auto analyzer (Bran Luebbe, Germany). Figure 1(a) displays the NIR spectra of 271 samples.

Fuel oil dataset was provided by Wentzell *et al.* [29] It includes ultraviolet (UV) spectra and saturates, monoaromatics, diaromatics and polyaromatics percentage contents of 115 fuel oil samples. The UV spectra and monoaromatics percentage contents were investigated for this study. The spectra were measured on a Cary 3 UV-visible spectrophotometer (Varian Instruments, San Fernando, Calif.). Each spectrum is composed of 572 variables recorded in the wavelength range of 200–400 nm with an interval of 0.35 nm. Being an outlier, the sample No. 115 was not used. Figure 1(b) shows the measured UV spectra of the 114 samples.

Before calculation, the two datasets were divided into the training and prediction set for modeling and performance validation, respectively. For tobacco lamina dataset, 181 samples were taken as the training set, and then 90 samples were selected as the prediction set by using the Kennard-Stone (KS) algorithm. For fuel oil dataset, 70 and 44 samples were taken as training and prediction set, respectively.

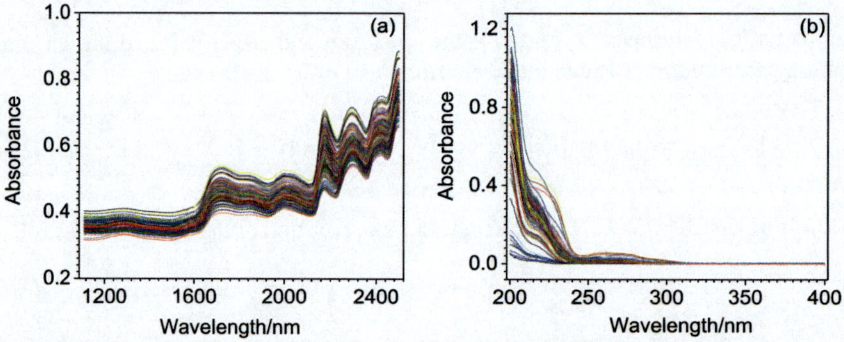

Fig. 1. NIR spectra for tobacco lamina (a) and UV spectra for fuel oil (b) datasets.

4 Results and Discussion

4.1 Determination of the Optimal Model Position for LASSO

In LASSO process, the optimal model position *s* should be determined. The value range of *s* is in the range of 0–1. This range is divided into 1000 segments with an interval of 0.001. Each s can get 10 RSS by 10-fold cross-validation. The average and standard deviation (SD) of 10 RSS can be calculated accordingly. Figure 2(a) and (b) show the variations of mean RSS with 1000 s for tobacco lamina and fuel oil dataset, respectively. Since 1000 SD values could be overlapped, only the SD of RSS at every 50 s values are displayed as vertical line in the sub-figures. Clearly, in the beginning, the mean and SD of RSS is relative large. Then with the increasing of s values, the mean and SD of RSS begins to decrease and finally tends to be unchanged. As the red dash dot represented in the sub-figures, the optimal model position s is obtained according to the mean and SD of RSS by the Sp criterion, which is 0.103 and 0.308 for tobacco lamina and fuel oil datasets, respectively.

Figure 2(c) and (d) represent coefficient changes of all variables with the increase of s for tobacco lamina and fuel oil datasets, respectively. The coefficient variation of a variable during the whole iterative process is presented as a line. When s is zero, which represents the variable is not selected. On the first iteration, it can be clearly seen that some coefficients of variables start to change. With the increase of s, more and more variables are added. When s is 1, the number of variables reaches the largest, and all iterations are completed. The red dash dot represents the optimal model position. The variables correspond to non-zero coefficients are selected at the red dash dot position.

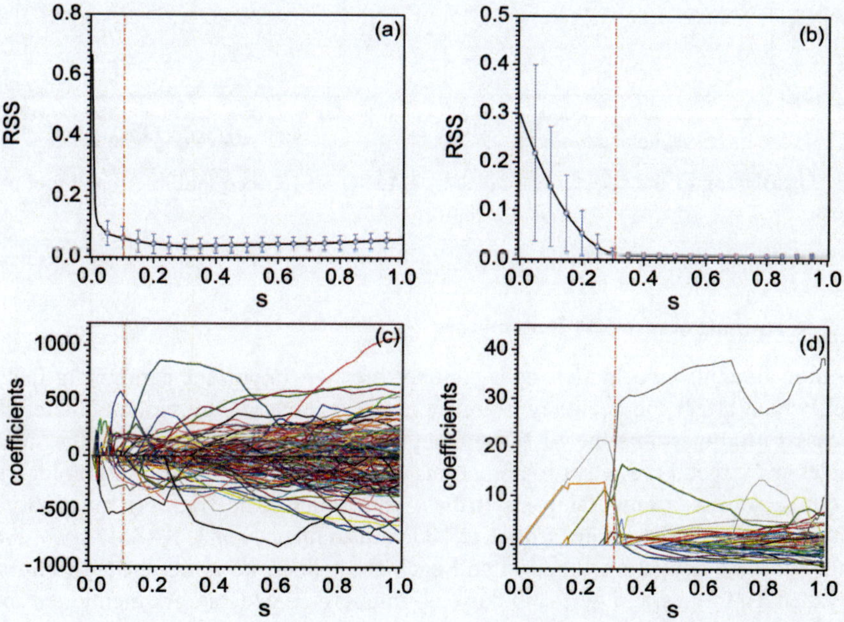

Fig. 2. Variation of RSS for tobacco lamina (a) and fuel oil (b) and coefficients of each variable for tobacco lamina (c) and fuel oil (d) datasets with the increasing of s. In plot (a) and (b), the black solid line represents the average value of 10 RSS obtained by 10-fold cross-validation for each s value and the circle and vertical line indicate the average and standard deviation of RSS at every 50 s values.

4.2 Distribution of the Selected Variables

To clearly demonstrate the selected variables, the distribution of the selected variables is shown in Fig. 3(a) and (b) for the tobacco lamina and fuel oil dataset, respectively. The longitudinal coordinate is the β coefficients at the optimal positions in Fig. 2(c) and (d) and horizontal ordinate is the wavelength. There are a total of 1296 variables in Fig. 3(a), of which the non-zero coefficients represent 27 variables are selected by LASSO. The number of 27 variables is much smaller than the original number of variables, which reduces the input dimension. A total of 7 variables are selected for fuel oil dataset as shown in Fig. 3(b), which are mainly distributed in the 235, 288 and 280–400 nm. Therefore, LASSO solves the high dimensional problem and eliminates redundant variables. Furthermore, the number of reserved variable for both the datasets indicates that LASSO has good sparsity.

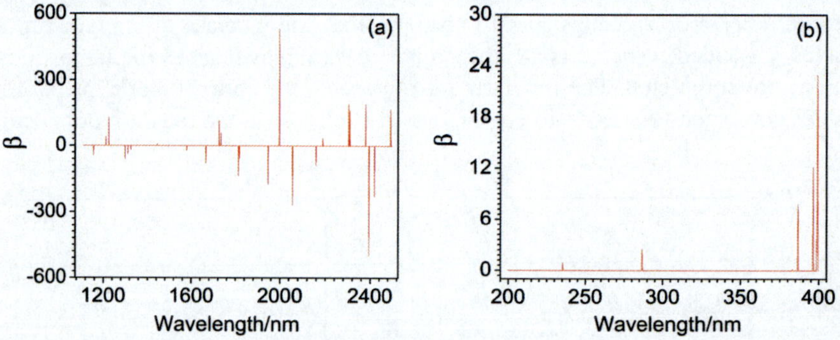

Fig. 3. Distribution of the selected variables by LASSO for tobacco lamina (a) and fuel oil (b) datasets.

4.3 Determination of ELM Parameters

Activation function and hidden node number are two important parameters in ELM model, which affect the accuracy, stability and efficiency of the model. Therefore, it is necessary to determine the ELM parameters before modeling between the selected variables and target. Five activation functions are investigated, including sigmoidal (sig), sine (sin), hardlim, triangular basis (tribas) and radial basis (radbas) functions. The number of hidden nodes ranges from 1 to 100 with an interval of 1. LASSO-ELM model is built for each activation function and each the hidden node number. This process is repeated 100 times and gets 100 correlation coefficients (Rs). As mentioned in our previous research [20], the ratio of mean value to standard deviation of Rs, termed as MSR, used to evaluate the LASSO-ELM prediction performance, which considers the accucary and stability of the model simultaneously. Figure 4(a) and (b) show the variation of MSR values with the activation function and the number of hidden nodes for tobacco lamina and fuel oil datasets, respectively.

From Fig. 4(a), MSR values of sig and sin functions are significantly larger than those of hardlim, tribas and radbas functions. By further comparison, the prediction

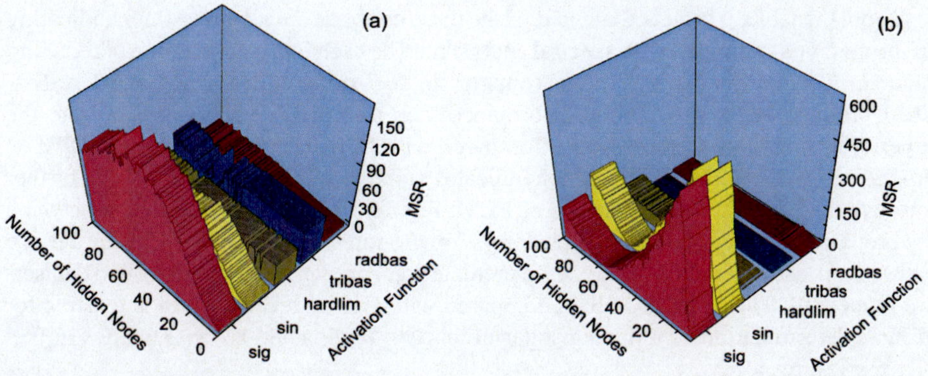

Fig. 4. Variation of MSR with activation function and hidden node number by LASSO-ELM for tobacco lamina (a) and fuel oil (b) datasets.

performance of the sig function is better than that of sin function for the same hidden node. MSR values of the sig functions increase as the number of hidden nodes increases at first, then reach the highest points at 61. After that, the MSR begins to decrease. Therefore, the optimal activation function and hidden node number are sig and 61 for the tobacco dataset, respectively. From Fig. 4(b), the MSR values of hardlim, tribas and radbas functions hardly change with the increase of hidden node number. Although sig and sin functions have similar performance, sig function has the best performance for further comparison. Thus, the optimal activation function and hidden node number are sig and 10 for fuel oil dataset, respectively.

4.4 Comparison of the Prediction Results

With the optimal model position, activation function and hidden node number determined above, LASSO-ELM can be developed and used to predict the contents of unknown samples. The prediction performance of LASSO-ELM is evaluated by RMSEP and R for the samples in prediction set. A good model should have a low RMSEP and a high R. Table 1 shows RMSEP and R for LASSO-ELM for the two datasets. As comparison, the prediction results of full-spectrum ELM and PLS are also summarized in Table 1.

Table 1. Prediction results of different methods for the two datasets

	Tobacco lamina		Fuel oil	
	RMSEP	R	RMSEP	R
PLS	1.5875	0.9648	0.9059	0.9907
ELM	1.3224	0.9763	0.7465	0.9935
LASSO-ELM	1.1965	0.9808	0.7072	0.9943

From the table, it is clear that the R of the three methods are all above 0.98, indicating all the methods combined with spectral analysis can be used for quantification of reducing sugar in tobacco lamina and monoaromatics in fuel oil samples. ELM and LASSO-ELM produce better prediction performance than that of PLS. This demonstrate the superiority of ELM based methods. Compared with full-spectrum ELM, LASSO-ELM produces better RMSEP and R, which indicates variable selected by LASSO can further improve the prediction performance of ELM. Figure 5 shows the relationship between the prepared and the predicted values for the prediction set of LASSO-ELM for the two datasets. As can be seen from Fig. 5, the predicted and prepared values of the two datasets are fitted well. Thus, LASSO-ELM compared with NIR and UV spectra is feasible for accuate determination of reducing sugar in tobacco lamina and monoaromatics in fuel oil samples, respectively.

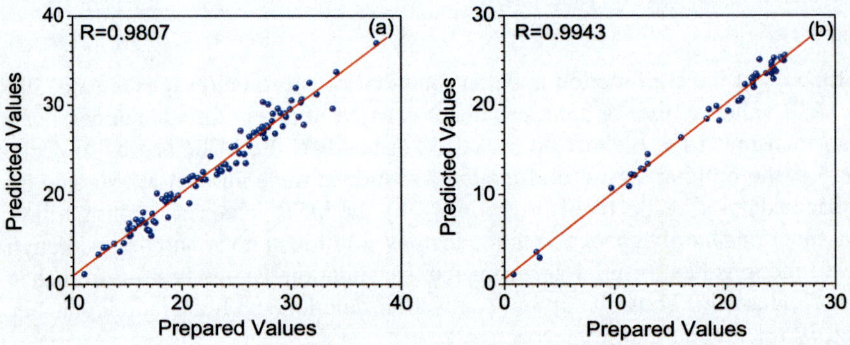

Fig. 5. The relationship between the prepared and the predicted values for the prediction set of LASSO-ELM for tobacco lamina (a) and fuel oil (b) datasets.

5 Conclusion

The feasibility of LASSO-ELM for variable selection and quantification of complex samples are investigated. The model position, the activation function and the number of hidden nodes are optimized before modeling and then built LASSO-ELM model to predict the samples in prediction set. In order to verify the validity of the method, the contents of reducing sugar in tobacco lamina samples and monoaromatics in fuel oil samples are predicted. The RMSEP and R values of LASSO-ELM are compared with those of ELM and PLS on the two datasets. The result shows that LASSO-ELM has the best predictive performance with only tens of variables. Therefore, LASSO-ELM is a feasible method for quantitative analysis of complex samples.

Acknowledgments. This study is supported by Opening Foundation of State Key Laboratory of Plateau Ecology and Agriculture (No. 2021-KF-07) and Key Lab of Process Analysis and Control of Sichuan Universities (No. 2020001).

References

1. Beganovic, A., Bec, K.B., Grabska, J., Stanzl, M.T., Brunner, M.E., Huck, C.W.: Vibrational coupling to hydration shell-mechanism to performance enhancement of qualitative analysis in NIR spectroscopy of carbohydrates in aqueous environment. Spectrochim. Acta A **237**, 118359 (2020)
2. Ma, B., Wang, L., Han, L., Cai, W.S., Shao, X.G.: Understanding the effect of urea on the phase transition of poly (N-isopropylacrylamide) in aqueous solution by temperature-dependent near-infrared spectroscopy. Spectrochim. Acta A **253**, 119573 (2021)
3. Li, X., et al.: Review of NIR spectroscopy methods for nondestructive quality analysis of oilseeds and edible oils. Trends Food Sci. Technol. **101**, 172–181 (2020)
4. Bian, X.H., Lu, Z.K., van Kollenburg, G.H.: Ultraviolet-visible diffuse reflectance spectroscopy combined with chemometrics for rapid discrimination of Angelicae Sinensis Radix from its four similar herbs. Anal. Methods **12**, 3499–3507 (2020)
5. Zhang, H., Hu, X.Y., Liu, L.M., Wei, J.F., Bian, X.H.: Near infrared spectroscopy combined with chemometrics for quantitative analysis of corn oil in edible blend oil. Spectrochim. Acta A **270**, 120841 (2022)
6. Gemperline, P.J., Salt, A.: Principal components regression for routine multicomponent UV determinations: a validation protocol. J. Chemom. **3**, 343–357 (1989)
7. Lin, Y.W., Deng, B.C., Xu, Q.S., Yun, Y.H., Liang, Y.Z.: The equivalence of partial least squares and principal component regression in the sufficient dimension reduction framework. Chemometr. Intell. Lab. Syst. **150**, 58–64 (2016)
8. Lemos, T., Kalivas, J.H.: Leveraging multiple linear regression for wavelength selection. Chemometr. Intell. Lab. Syst. **168**, 121–127 (2017)
9. Li, J.Y., Chu, X.L.: Rapid determination of physical and chemical parameters of reformed gasoline by near-infrared (NIR) spectroscopy combined with the Monte Carlo virtual spectrum identification method. Energy Fuel **32**, 12013–12020 (2018)
10. Li, Q., Huang, Y., Tian, K., Min, S., Hao, C.: Rapid quantification of analog complex using partial least squares regression on mass spectrum. Chem. Pap. **73**, 1003–1012 (2019)
11. Song, X.Z., Du, G.R., Li, Q.Q., Tang, G., Huang, Y.: Rapid spectral analysis of agro-products using an optimal strategy: dynamic backward interval PLS-competitive adaptive reweighted sampling. Anal. Bioanal. Chem. **412**, 2795–2804 (2020)
12. Marini, F., Bucci, R., Magri, A.L., Magri, A.D.: Artificial neural networks in chemometrics: history, examples and perspectives. Microchem. J. **88**, 178–185 (2008)
13. Thissen, U., Ustun, B., Melssen, W.J., Buydens, L.M.C.: Multivariate calibration with least-squares support vector machines. Anal. Chem. **76**, 3099–3105 (2004)
14. Algamal, Z.Y., Qasim, M.K., Lee, M.H., Ali, H.T.M.: Improving grasshopper optimization algorithm for hyperparameters estimation and feature selection in support vector regression. Chemometr. Intell. Lab. Syst. **208**, 104196 (2021)
15. Huang, G.B., Zhu, Q.Y., Siew, C.K.: Extreme learning machine: theory and applications. Neurocomputing **70**, 489–501 (2006)
16. Jiang, H., Liu, G.H., Mei, C.L., Chen, Q.S.: Qualitative and quantitative analysis in solid-state fermentation of protein feed by FT-NIR spectroscopy integrated with multivariate data analysis. Anal. Methods **5**, 1872–1880 (2013)
17. Zhang, C.X., et al.: Subagging for the improvement of predictive stability of extreme learning machine for spectral quantitative analysis of complex samples. Chemometr. Intell. Lab. Syst. **161**, 43–48 (2017)
18. Bian, X.H., et al.: Robust boosting neural networks with random weights for multivariate calibration of complex samples. Anal. Chim. Acta **1009**, 20–26 (2018)

19. Chen, H., Tan, C., Lin, Z.: Ensemble of extreme learning machines for multivariate calibration of near-infrared spectroscopy. Spectrochim. Acta A **229**, 117982 (2020)
20. Bian, X.H., Li, S.J., Fan, M.G., Guo, Y.G., Chang, N., Wang, J.J.: Spectral quantitative analysis of complex samples based on extreme learning machine. Anal. Methods **8**, 4674–4679 (2016)
21. Yun, Y.H., Li, H.D., Deng, B.C., Cao, D.S.: An overview of variable selection methods in multivariate analysis of near-infrared spectra. ATrends Analyt. Chem. **113**, 102–115 (2019)
22. Zhang, J., Cui, X., Cai, W., Shao, X.: A variable importance criterion for variable selection in near-infrared spectral analysis. Sci. China Chem. **62**, 271–279 (2019)
23. Robert, T.: Regression shrinkage and selection via the LASSO. J. R. Statist. Soc. B **58**, 267–288 (1996)
24. Mozafari, Z., Chamjangali, M.A., Arashi, M.: Combination of least absolute shrinkage and selection operator with Bayesian regularization artificial neural network (LASSO-BR-ANN) for QSAR studies using functional group and molecular docking mixed descriptors. Chemometr. Intell. Lab. Syst. **200**, 103998 (2020)
25. Cui, X.C., et al.: Adaptive LASSO logistic regression based on particle swarm optimization for Alzheimer's disease early diagnosis. Chemometr. Intell. Lab. Syst. **215**, 104316 (2021)
26. Higashi, H., ElMasry, G.M., Nakauchi, S.: Sparse regression for selecting fluorescence wavelengths for accurate prediction of food properties. Chemometr. Intell. Lab. Syst. **154**, 29–37 (2016)
27. Wang, Y., Bian, X.H., Tan, X.Y., Wang, H.T., Li, Y.K.: A new ensemble modeling method for multivariate calibration of near infrared spectroscopy. Anal. Methods **13**, 1374–1380 (2021)
28. Xiao, H.J., Ba, B.X., Li, X.X., Liu, J., Liu, Y.Q., Huang, D.P.: Interval multiple-output soft sensors development with capacity control for wastewater treatment applications: a comparative study. Chemometr. Intell. Lab. Syst. **184**, 82–93 (2019)
29. Wentzell, P.D., Andrews, D.T., Walsh, J.M., Cooley, J.M., Spencer, P.: Estimation of hydrocarbon types in light gas oils and diesel fuels by ultraviolet absorption spectroscopy and multivariate calibration. Can. J. Chem. **77**, 391–400 (1999)

Others

Prediction of Rubber Leaf Nitrogen Content Based on Fractional-Order GWO-SVR

Rongnian Tang, Xiaowei Li, Chuang Li, Kaixuan Jiang, and Jingjin Wu(✉)

College of Mechanical and Electrical Engineering, Hainan University, Haikou 570228, China
jingjin.wu@hainanu.edu.cn

Abstract. Grey Wolf Optimizer (GWO) algorithm, based on swarm intelligence, is easy to implement due to its few parameters and simple structure characteristics. However, few studies employ GWO for the spectral analysis to our knowledge. In this study, the GWO algorithm is introduced into the detection of nitrogen content to provide a reference for detecting nitrogen content in rubber leaves, which further facilitated the rubber yield of the rubber tree. GWO is used to optimize the SVR model of support vector regression, and 11 GWO-SVR models are established by taking spectral data of different fractional-order as input consecutively. The results show that the GWO-SVR model is superior to the SVR model as GWO-SVR optimizes SVR parameters by GWO algorithm, penalty factor c, and kernel function parameter g. The prediction correlation coefficient (Rp) improves by 10.88%, and the root mean square error (RMSEP) reduces by 9.15%. The GWO-SVR model base on fractional-order is optimized at 0.6 order, and the correlative Rp and RMSEP are 0.907 and 0.213, respectively. Compared with the GWO-SVR model based on the original spectrum, the Rp increases by 2.96%, and the RMSEP is reduced by 3.15%. Hence, it is feasible to predict rubber tree nitrogen content based on fractional GWO-SVR, which provides technical support for variable rate fertilization in precision rubber tree agriculture.

Keywords: Rubber tree leaves · Nitrogen · Near infrared spectral · Fractional-order · SVR · GWO

1 Introduction

Natural rubber is an important industrial raw material and strategic material, and plays an essential role in developing the national economy and society [1, 2]. Literature shows that rubber tree leaves are sensitive to nutrient abundance and deficiency. Their nutrient content strongly correlates with latex yield, which can more accurately reflect the effect of fertilization and is suitable as a sampling site for nutritional diagnosis [3, 4]. Among the many nutrients, nitrogen plays a vital role in the growth, rubber production, and latex stability of rubber trees [5]. In order to make rubber trees thrive and improve the yield and quality of rubber, adequate fertilization is essential [6]. However, traditional nitrogen content diagnostic methods are time-consuming, labor-intensive, expensive, and cannot

X. Chu et al. (Eds.): ICNIR 2021, *Sense the Real Change: Proceedings of the 20th International Conference on Near Infrared Spectroscopy*, pp. 303–315, 2022.
https://doi.org/10.1007/978-981-19-4884-8_33

obtain data on a large scale. Therefore, it is imperative to find an accurate and efficient method for measuring leaf nitrogen content.

As a non-destructive testing method, near infrared reflectance (NIR) spectroscopy has been widely used to detect nitrogen content in crops or agricultural products [7–9]. Studies have shown that choosing an appropriate pre-processing method can effectively improve the accuracy of the model. Ana Belén et al. tested several standard data pre-processing methods to analyze the impact on PLSR prediction leaf water potential precision, finding that derivative pre-processing increased the determination coefficients of our models and reduced root mean squared error (RMSE) [10]; based on baseline correction; Xu et al. have developed a novel baseline correction procedure, through extensive numerical experiments on a wide variety of spectra including simulated spectra, mineral spectra, and dialysate spectra, that show that is simple, fast, and can yield consistent and accurate baselines that preserve all the meaningful Raman peaks [11]; Xu et al. analysis of tumor functional characteristics using Savitzky-Golay filter (S-G filter) based CEUS quantification (SGCQ) software, that study proves SGCQ software to be a valid, sensitive and repeatable method for therapeutic evaluation [12]; Li et al. used a series of pre-processing methods, such as first-order derivative and second-order derivative, to model and predict the potassium content of rubber tree leaves, which improved the prediction ability of the model [13].

The swarm intelligence optimization algorithm has had excellent potential in parameter selection and variable screening due to its powerful global search ability. Studies have shown that the gray wolf algorithm (GWO) quantitative analysis has fewer input parameters, faster convergence speed, and easier to find the optimal global solution, and its optimization effect is much more significant than that of intelligent optimization algorithms such as particle swarm optimization algorithm and evolution strategy [14]. Zhang et al. established a soil liquefaction prediction model based on standard penetration test data and the grey wolf optimizer algorithm. The results indicated that the GWO algorithm could not only improve the accuracy of SVM fitting and optimize the performance of the prediction but also can fasten the operation [15]. Basak Hritam et al. used the gray wolf optimization algorithm to optimize a fully automated cytology image classification framework based on deep learning and feature selection. They selected a non-redundant and optimal feature subset from this feature space to improve the classification performance [16]. Sharma Sharad et al. improves the routing process using the algorithm grey wolf optimizer (GWO), which represents a semantic form of optimization that typically reduces the drop, time delay, and energy [17]. In recent years, spectral technology has been widely used in plant nutrient content detection. However, swarm intelligence optimization algorithms in the nutritional diagnosis model of rubber tree leaves are still rare. The research to explore the applicability of the GWO algorithm in the nitrogen diagnosis model of rubber tree leaves is still blank.

In this study, the SVR algorithm was used as the optimization object to explore the applicability of the gray wolf algorithm in the diagnosis model of nitrogen nutrients in rubber tree leaves. At the same time, we introduce fractional derivatives and compare the fractional-order with the traditional first-order and second-order derivatives to obtain more profound information on the leaf spectrum. Therefore, this study will use the fractional derivative combined with the GWO-SVR model to predict the nitrogen content of rubber tree leaves. The optimal parameters in the SVR are found through gray wolf optimization, and on this basis, the spectral data is processed by fractional-order to model again to find the optimal order. This study aims to provide a new method for the nutritional detection of other agricultural products and provide technical guidance for the adequate fertilization of rubber trees.

2 Materials and Methods

2.1 Sample Acquisition

Danzhou city, is the region with the most extensive rubber tree planting area in Hainan, and its rubber output ranks first in the province. In order to make the study more representative, we chose the experimental field in Danzhou city, Hainan province, southern China, as the study area. Figure 1 shows the sample collection area. In this study, samples were collected according to the following principles: 1. Mature, healthy, and intact leaves were randomly selected, 2. The picked leaves belonged to the rubber tree canopy. In July 2019, a total of 147 samples were collected. In order to reduce the experimental error, all collected samples were sealed and stored at low temperatures for subsequent physical and chemical analysis.

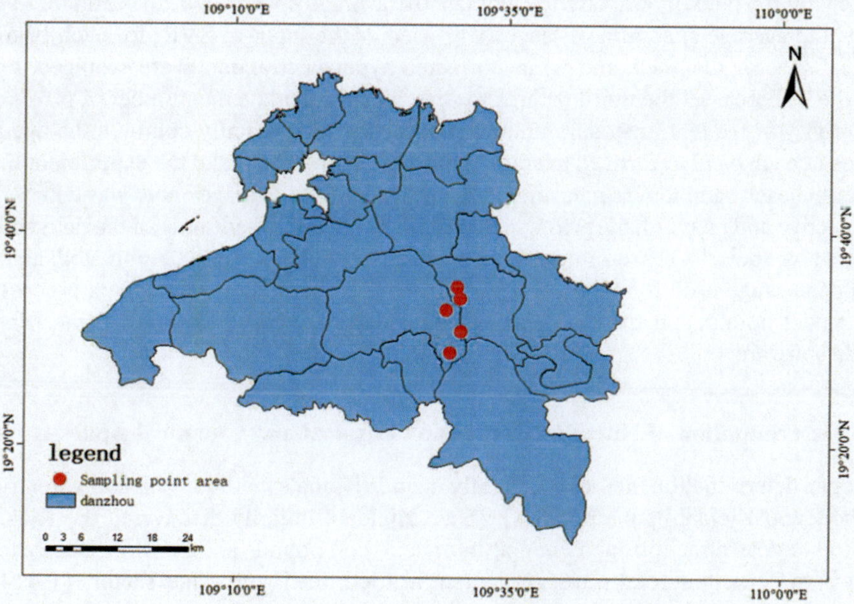

Fig. 1. Location of the study area in Hainan province

2.2 Spectral Data Acquisition

Spectral data were acquired by a hyperspectral imaging system (GaiaField-F-N17E) with a spectral resolution of 4 nm, uniformly collected at an acquisition interval of 3.28 nm in the wavelength range 841.44–1678.32 nm. Its working principle is to illuminate the sample to be tested placed on the electronically controlled moving platform through the light source. The 200 W halogen lamp provides the light source in the sample area with an incident angle of 45° about 0.8 m away from the soil sample. The latent optical camera captures the emitted light of the sample through the lens, and one-dimensional image and spectral information are obtained. With the electronically controlled moving platform driving the sample to run continuously, continuous one-dimensional images and real-time spectral information can be obtained. A three-dimensional data cube containing image information and spectral information is finally obtained recorded by computer software.

Before acquiring hyperspectral data, we need to perform initial calibration of the instrument, adjust the height of the lens, and correct the position of the sample, so that the whole leaf sample can be taken and the spectral image taken is clear. The collected leaf samples were brought back to the laboratory on the same day for a simple surface cleaning treatment. Put the rubber tree leaves on the designated area of the pad, which is placed on the electronically controlled mobile platform. Finally, start the device and obtain leaf spectral data. The acquired sample data needs to be corrected for black and white with the help of software to calculate the average spectrum of the sample. In this study, the average spectrum of leaves was used as the input to SVR. for each band of the leaf spectra, the black-and-white corrected hyperspectral data were averaged. First, find the leaf area, set the non-leaf area to zero. Then, calculate the number of pixels and reflectivity in the leaf area, sum up and take the average. Finally combine the average values of each band to form an average spectrum. In order to make the experiment more scientific, each sample was measured three times, and the average value was taken. Due to objective and unavoidable errors in the measurement, the beginning of the data will be affected by "noise". Therefore, the data of the first 26 bands will be eliminated, and the final band range used for analysis is 926.06–1678.32 nm. The spectral data processing and model building in the entire experimental process was performed using Matlab 2018a software.

2.3 Determination of Nitrogen Content by Physical and Chemical Analysis

Nitrogen determination methods generally include Duma's method, Will and Varentrap's method, and Kjeldahl's method [18]. Johan Kjeldahl initially discovered the Kjeldahl method to determine nitrogen content in organic compounds and has now been widely used in to determine total nitrogen content in food, feed, and other sectors [19]. This method is divided into three types: constant, semi-micro, and micro. Among them, the semi-micro method is one-time digestion and multiple distillations, which is conducive to parallel determination and facilitates the comparison of the accuracy of the determination results. Therefore, this study adopts the semi-micro Kjeldahl method for nitrogen determination The nitrogen content in rubber tree leaves was detected [20, 21]. After the spectral data of the leaves are collected, a quantitative amount of leaves is randomly collected. Then the leaf samples are subjected to greening, drying, and grinding treatments and the treated samples are weighed and placed in a Kjeldahl flask, followed by a series of digestion, distillation, Absorption, titration process, and finally measure the nitrogen content of the sample. Each leaf sample was analyzed twice, and the average value was taken as the leaf nitrogen content.

2.4 Support Vector Machines

Support vector machine (SVM) was first proposed by Vapnik and Chervonenkis in 1963. The core idea of SVM is to find the optimal separation hyperplane with the most considerable geometric interval in the feature space [22, 23]. Regardless of the specific relationship between the independent and dependent variables, the SVM is used to convert nonlinear and low-dimensional problems into linear and high-dimensional problems, and solve the optimal regression hyperplane [24, 25]. As a commonly used classification and recognition method, SVM is more suitable for occasions with few samples and high generalization ability than algorithms such as neural network and fuzzy recognition [26]. In the environment of small samples, SVM can realize spectral recognition and classification more quickly and efficiently than the enormous computational load of a neural network [27]. Support vector regression (SVR) is a generalized model of support vector machine (SVM), which inherits the advantages of support vector machine, such as structural risk minimization strong data generalization ability, and it is convenient to use the Kernel method to deal with nonlinear problems [28]. The choice of SVR in this study is due to its better predictive ability for small sample spectral data sets, and the more commonly used RBF is selected as the kernel function.

2.5 Modeling Based on GWO-SVR

The gray wolf optimization algorithm is a swarm intelligence evolutionary algorithm proposed by Seyedali Mirjalili et al. in 2014 to simulate the social class and predation behavior of gray wolves. It has been used to solve many practical optimization problems [29, 30]. Gray wolves have a pyramid-like social hierarchy in the hunt. The head wolf is α, and the remaining sub-level gray wolf individuals are labeled as β, δ, and ω in turn, and they cooperate to prey. The algorithm is an optimization search method developed according to the predation activities of gray wolves. It has the characteristics of solid convergence performance, few parameters, and easy implementation. In this study, the GWO algorithm is used to search for the optimal parameters c and g in SVR. The specific process is as follows:

First, we use the concentration gradient method to divide the spectral data of rubber tree leaves into a calibration set and a prediction set. The division principle is 3:2, and the calibration set is used as a training sample. Then, the initial value of GWO is set, the number of wolves H, the maximum number of iterations N, c and g are: 40, 60, 0.01–100000, respectively, and the positions of α, β and δ wolves are initialized. Next, use c and g to randomly generate the initial position of the gray wolf, update the position of each individual in the wolf pack according to the fitness value, and save the position of the best fitness value obtained. The optimized objective function value of this study is the mean squared error (MSE). Finally, when the number of iterations N exceeds 40, the training is terminated, the optimal global position is output, that is, the optimal values of *c* and *g*, and the optimal SVR model is established by using the optimal *c* and *g*, and the prediction set samples are predicted and analyzed.

2.6 Fractional-Order

The Grünwald-Letnikov (G-L) definition derives from the classical fractional differential definition. Extending the differential order from integers to fractions makes it easier to obtain numerical solutions, and it is widely used in the numerical analysis [31]. Therefore, this study is based on the definition of GL. By considering the independent variable x as the wavelength, the fractional step size is 1, and in the range from 0 to 2, with an interval of 0.2, a total of 11 spectral data are obtained. In this study, the 11 orders are used as the input of the SVR-GWO modeling method, and the optimal *c* and *g* are obtained in turn, and the optimal order SVR-GWO model is established.

2.7 Model Evaluation Indicators

In this study, the evaluation indicators of the rubber tree leaf nitrogen content model are mainly determined by the coefficient of determination R^2, root mean square error (RMSE), correlation coefficient of calibration (Rc), correlation coefficient of prediction (Rp), root mean square error of calibration (RMSEC), root mean square error of prediction (RMSEP) and other parameters. The higher the values of the first two parameters, the lower the value of the last two parameters, the better the accuracy and stability of the established model are.

The closer the R^2 score is to 1, the better the fit of the model, and the value range of R^2 is [0,1]. The smaller the value of RMSEC and RMSEP, the more accurate the model, the closer the values are, the more stable the model, and the stronger the prediction ability [32].

3 Results and Discussion

3.1 Nitrogen Content of Rubber Tree Leaves

In order to establish a more stable and explanatory model, we use the concentration gradient method to divide the sample set. The 147 data were divided into calibration set and prediction set according to the ratio of 3:2 by concentration gradient method. Table 1 shows the statistical analysis of nitrogen content in rubber tree leaves. It can be seen from Table 1 that the maximum value of all the data is 4.709, the minimum value is 1.951, the mean value is 3.577, the coefficient of variation is 19.67%, the calibration set is 19.78%, and the prediction set is 19.56%. The coefficients of variation of the calibration and prediction sets were generally similar, which indicates that the data are feasible.

Table 1. Descriptive statistics of nitrogen content (NT, %).

Sample set	n	Minimum	Mean	Median	Maximum	SD	CV
Entire	147	1.951	3.577	3.808	4.709	0.703	19.67%
Calibration	98	1.951	3.568	3.803	4.442	0.705	19.78%
Prediction	49	2.046	3.595	3.824	4.709	0.706	19.65%

Notes: *n*, sample number; SD, standard deviation; CV, coefficient of variation

3.2 Spectral Data Preprocessing

Figure 2 shows the original spectral reflectance of rubber tree leaves. It can be seen from Fig. 2 that the general trend of spectral reflectance is first to decrease and then increase, and there is a strong absorption valley at 1440 nm due to the existence of water molecules, which is in line with the spectral characteristics of plants. The leaf reflectance has two small absorption valleys around 970 and 1200 nm, and the reflectivity is higher at 900–1300 nm. The reason is that the internal cell structure of plant leaves is loose, the internal cavity of the sponge tissue increases the reflection area, the light is scattered in the leaves multiple times, and the cellulose and pigment have no absorption effect on this band, the resulting spectral reflectance in this range is relatively high. The reflectivity of leaves at 1440 nm–1678.32 nm showed a monotonically increasing trend.

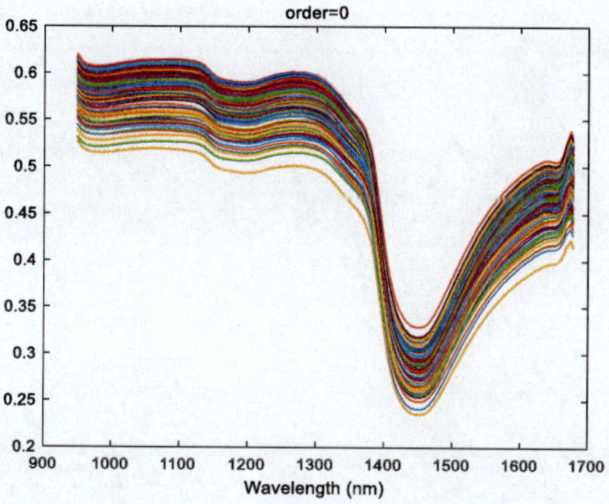

Fig. 2. Original spectral reflectance of rubber tree leaves

3.3 Establishment and Analysis of SVR and GWO-SVR Models

In order to explore the effect of the GWO algorithm on SVR, we used the full-wavelength spectral data as the input variables of the SVR and GWO-SVR regression models and the nitrogen content as the output variable. Select the optimal parameters *c* and *g* as the input parameters of the SVR prediction model to predict the prediction set. The model evaluation index judges the robustness and stability of the established model. Figure 3 is a comparison chart of the prediction results of the SVR and GWO-SVR models. It can be seen from Fig. 3 that the GWO-SVR model can fit the actual value-line better than the SVR model, and the GWO-SVR can better predict the low and high nitrogen content of leaves and also accurately predict the rubber tree globally. Nitrogen content of leaves. Table 2 shows the SVR and GWO-SVR model results. It can be seen from Table 2 that the prediction correlation coefficient Rp of the GWO-SVR model is increased by 10.88% compared with the SVR model, and the prediction root mean square error RMSEP is reduced by 9.15%, and the accuracy of the model has been dramatically improved. The results show that the gray wolf optimization algorithm has indeed optimized the performance of the model and can find the optimal *c* and *g*. This result is consistent with Gholamhossein Sodeifian and Nedasadat Saadati Ardestani [33], which also confirms the use of GWO optimization algorithm for the rubber tree leaves SVR model plays the role of optimization.

Table 2. SVR and GWO-SVR model results.

Model	Rc	RMSEC	Rp	RMSEP
SVR	0.814	0.303	0.768	0.336
GWO-SVR	0.910	0.211	0.877	0.245

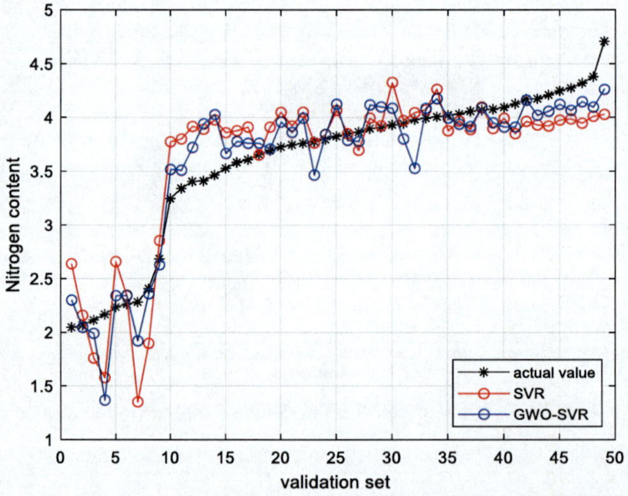

Fig. 3. Comparison of prediction results of SVR and GWO-SVR models

3.4 Establishment and Analysis of Fractional-Order GWO-SVR Model

In this paper, after the spectral data is subjected to corresponding mathematical transformation, fractional differential calculation with 0.2 intervals is performed, and a total

Table 3. Results of GWO-SVR model with different orders.

Order	Rc	RMSEC	Rp	RMSEP
0.0	0.910	0.211	0.877	0.245
0.2	0.953	0.153	0.892	0.230
0.4	0.917	0.203	0.898	0.224
0.6	0.916	0.203	0.907	0.213
0.8	0.920	0.198	0.907	0.213
1.0	0.938	0.175	0.902	0.219
1.2	0.930	0.186	0.887	0.235
1.4	0.969	0.124	0.872	0.250
1.6	0.978	0.104	0.834	0.285
1.8	0.895	0.227	0.829	0.289
2.0	0.892	0.231	0.815	0.301

of 11 spectral data are obtained in the range of 0 to 2 orders. Figure 4 shows the spectral reflectance of rubber tree leaves under fractional-order = 0.0, 0.6, 0.8, and 1.6 treatments. It can be seen from Fig. 4 that after fractional processing, more information in the original spectral data is mined. As shown in the box marked in Fig. 4(b), compared to the original spectral data in Fig. 4(a), it can be seen that some detailed features are more pronounced. However, as shown in Figs. 4(c) and (d), with the increase of the fractional-order, part of the noise information is amplified, and the noise is more obvious. After the spectrum undergoes fractional-order 0–2 transformation, the tiny details of some wavelength points will be highlighted, and it will also increase the noise, which will make the fractional-order spectrum look rough. Fractional-order aims to find a balanced order between display details and noise, and then build a more accurate model.

Table 3 shows the GWO-SVR model results under different orders. As shown in Table 3, as the fractional-order increases, the value of Rp gradually increases, reaching a maximum at the 0.8 order, and then the value of Rp gradually decreases. Derivatives, the second-order derivatives have better effects, and the Rp of the first-order and second-order derivatives are more prominent in the 0.6-order and 0.8-order, indicating that the fractional-order can be used to mine some more detailed features and improve the accuracy of the model. Figure 5 compares the prediction results of the GWO-SVR model under the fractional-order of order = 0.0 and 0.6. From Fig. 5, the fitting ability of the 0.6-order GWO-SVR model is better than the original spectral data. Combined with the spectral modeling effects of different orders in Table 3, it can be seen that fractional calculus improves the effect of the GWO-SVR model. It can be seen from a comprehensive comparison of various orders that when the Rp value is not much different, in the 0.6 order, at the same time, the difference between RMSEP and RMSEC is more negligible, and it can also accurately predict the nitrogen content of rubber tree leaves.

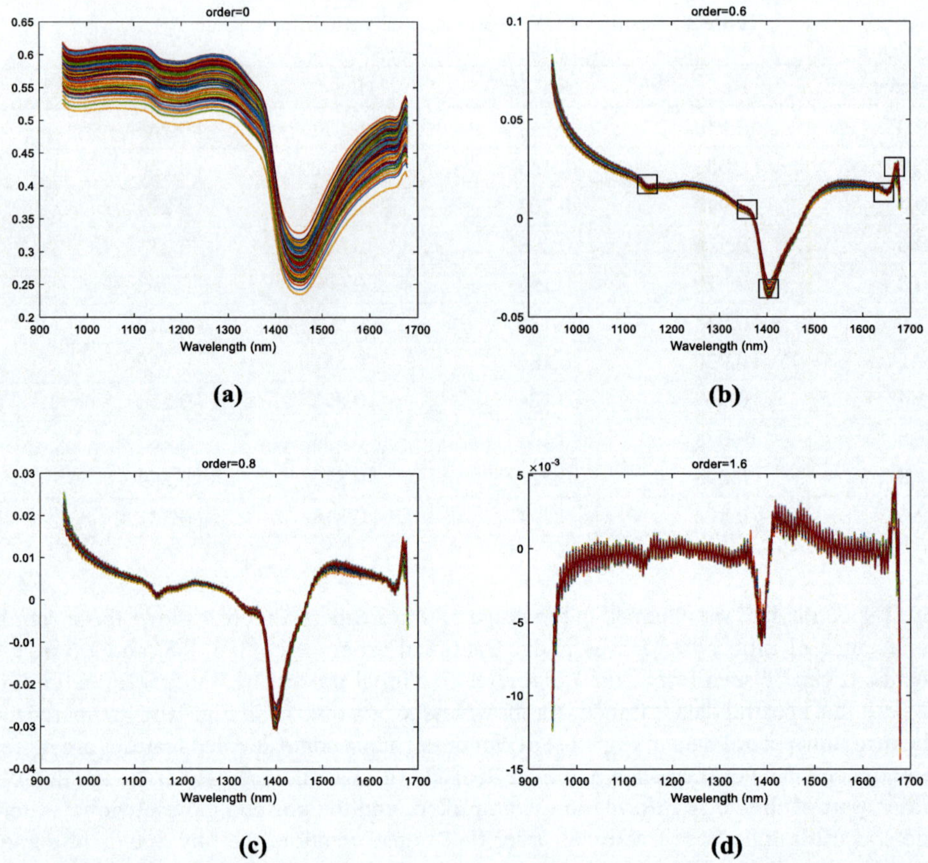

Fig. 4. Spectral reflectance of rubber leaves under fractional-order (a) order = 0.0, (b) order = 0.6, (c) order = 0.8 and (d) order = 1.6

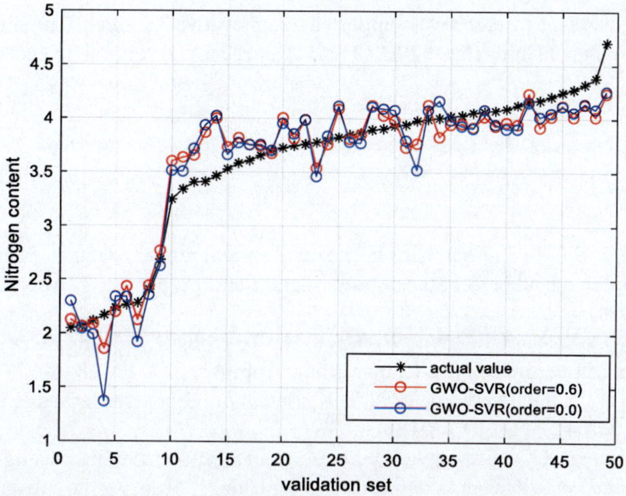

Fig. 5. Comparison of GWO-SVR model prediction results under fractional-order = 0.0 and 0.6

4 Conclusion

In this study, rubber tree leaves were taken as the experimental object, the near-infrared spectral data of the samples were collected. The SVR and GWO-SVR prediction models were established with the original spectra as input. The 11 spectral data as input and 11 GWO-SVR prediction models are established respectively. The results show that the performance of the GWO-SVR model is better than that of the SVR model. The introduction of the GWO algorithm can find the optimal *c* and *g* in the SVR model. The prediction correlation coefficient Rp is 0.877, and the prediction root mean square error RMSEP is 0.245. Rp increased by 10.88% and RMSEP decreased by 9.15% compared with the SVR model.

The GWO-SVR modeling effect of spectral data processed under different fractional-orders is different. Compared with the GWO-SVR modeling of the original spectrum, the model under fractional-order processing has a better effect, and the accuracy is greatly improved, in which the 0.6-order spectral modeling is the optimal model.

These results show that it is feasible to use the swarm intelligence optimization GWO algorithm to optimize *c* and *g* in the SVR model, and at the same time, the fractional-order-based GWO-SVR modeling method is better than the original spectrum-based GWO-SVR modeling method. The robustness and accuracy of the model are significantly improved, which provides a feasible idea for further research on the spectral diagnostic model of nitrogen content in rubber trees. However, since the spectrum characteristics have not been discussed in this study, in future work, it is possible to use the characteristic band as the input and combine the fractional-order to construct the GWO-SVR model to achieve a better prediction effect and improve work efficiency. Satisfactory fertilization provides theoretical support.

Acknowledgments. This research was supported by Innovative Research Team Project of Hainan Natural Science Found of China (No. 320CXTD431) and National Natural Science Found of China (No. 32060413).

References

1. Salomez, M., et al.: Micro-organisms in latex and natural rubber coagula of Hevea brasiliensis and their impact on rubber composition, structure and properties. J. Appl. Microbiol. **117**, 921–929 (2014)
2. Liu, R.J., Mo, Y.Y., Yang, L., Wu, W., He, C.H.: Re-recognition and advice on the strategic role of natural rubber industry in China. China Trop. Agric. **1**, 13–18 (2022)
3. Lin, Q.H., et al.: Annual variation of N, P, K content of rubber tree leaves in Hainan. Chin. J. Trop. Crops **33**(04), 595–601 (2012)
4. Wang, D.P., Wang, X.Q., Cheng, J., Tan, H.D., He, P.: Problems in increasing yield of natural rubber in Hainan province and countermeasures. Chin. J. Trop. Agric. **33**, 66–70 (2013)
5. Chen, Y.B., Zhang, Y.F., Wang, W.B., Xue, X.X., Luo, X.H.: Nitrogen nutrition characteristics of rubber tree leaves and its response to nitrogen application rate. Chin. J. Trop. Crops **40**, 831–838 (2019)
6. Zhang, X.C., Xie, G.H.: Production status of high-yield rubber plantations in rubber planting areas in my country and cultivation measures. China Trop. Agric. **6**, 6–9 (2018)
7. Fernandez, C.I., Leblon, B., Haddadi, A., Wang, K.R., Wang, J.F.: Potato late blight detection at the leaf and canopy levels based in the red and red-edge spectral regions. Remote Sens. **12**, 1292 (2020)
8. Sterling, A., Melgarejo, L.M.: Leaf spectral reflectance of Hevea brasiliensis in response to Pseudocercospora ulei. Eur. J. Plant Pathol. **156**, 1063–1076 (2020)
9. Hu, W.F., Tang, R.N., Li, C., Zhou, T., Chen, J., Chen, K.: Fractional order modeling and recognition of nitrogen content level of rubber tree foliage. J. Near Infrared Spectrosc. **29**, 42–52 (2021)
10. González-Fernández, A.B., Sanz-Ablanedo, E., Gabella, V.M., Garcia-Fernandez, M., Rodriguez-Perez, J.R.: Field spectroscopy: a non-destructive technique for estimating water status in vineyards. Agronomy **9**, 427 (2019)
11. Xu, Y., Du, P., Senger, R., Robertson, J., Pirkle, J.L.: ISREA: an efficient peak-preserving baseline correction algorithm for Raman spectra. Appl. Spectrosc. **75**, 34–45 (2021)
12. Xu, Z.T., et al.: Savitzky-Golay filter based quantitative dynamic contrast-enhanced ultrasound on assessing therapeutic response in mice with hepatocellular carcinoma. J. Signal Process Syst. **92**, 315–323 (2020)
13. Li, X.Q., Chen, G.L., Xu, M.G., Ding, H.P., Liu, Z.M.: Study on hyperspectral estimation models for potassium content of rubber tree leaves, Southwest China. J. Agric. Sci. **33**, 769–774 (2020)
14. Chen, W., Li, C., Tang, R.N.: Application of interval random frog combined with successive projections algorithm to detecting nitrogen content in rubber tree leaves. J. Henan Univ. Sci. Technol. (Nat. Sci. Ed.) **40**, 51–56 (2019)
15. Zhang, Y., Qiu, J., Zhang, Y., Xie, Y.L.: The adoption of a support vector machine optimized by GWO to the prediction of soil liquefaction. Environ. Earth Sci. **80**, 1–9 (2021)
16. Basak, H., Kundu, R., Chakraborty, S., Das, N.: Cervical cytology classification using PCA and GWO enhanced deep features selection. SN Comput. Sci. **2**, 1–17 (2021)
17. Sharma, S., Kapoor, A.: An efficient routing algorithm for IoT using GWO approach. Int. J. Appl. Metaheur. **12**, 67–84 (2021)

18. Li, W.T., Zhou, Y., Hu, J., Wang, L.: Determination of the total nitrogen content in some unknown compound by the Kjeldahl method. China Health Stand Manage **7**, 110–112 (2016)
19. Zeng, Q.T.: The influence factors and elimination methods of the determination of nitrogen in geochemical samples by SEI-micro Kjeldahl method. Guangdong Chem. Ind. **45**, 225–226 (2018)
20. Wang, X.: Application of dumas combustion and Kjeldahl methods in the determination of nitrogen concentration in wheat. Agric. Tech. Equip. **11**, 7–8 (2020)
21. Jia, X.Y.: A comparative study of Kjeldahl nitrogen determination and near-infrared spectroscopy in the detection of protein raw materials. Hunan Agric. Univ. (2011)
22. Liu, Q.Y.: Research on rapid detection of flammable liquid based on Raman spectroscopy technology. Yantai University (2021)
23. Zhu, W.B., Wen, Y., Ma, L., Chu, W.T., Li, C.N., Sheng, Q.P.: Predicting strength of poured cement mortar in semi-flexible pavement based on grey relational-support vector machine. Concrete **11**, 126–129 (2021)
24. Jiang, C.C., Tang, R.N.: Detection of nitrogen in mature rubber tree leaves using near infrared spectroscopy. J. Anhui Agric. Univ. **44**, 429–433 (2017)
25. Li, X.Q., Chen, G.L., Xu, M.G., Liu, Z.M., Deng, Y.Y.: Effects of variety and tapping age on estimation of rubber leaf nitrogen content based on hyper-spectral. Southwest China J. Agric. Sci. **30**, 2497–2505 (2017)
26. Du, J., Gu, J.W., Qiu, S.K.: Prediction of total nitrogen by using hyperspectral data based on support vector regression. Henan Sci. **38**, 1585–1590 (2020)
27. Li, H.B., Li, H., Lou, X.P., Meng, F.Y., Zhu, L.Q.: FBG reflelection spectrum type recognition based on support vector machine. Comput. Appl. Software **38**, 159–163 (2021)
28. Du, P.J.: Terahertz spectroscopic analysis of single - component and multi - component substances based on support vector regression. Shenzhen University (2018)
29. Mirjalili, S.: How effective is the grey wolf optimizer in training multi-layer perceptrons. Appl. Intell **43**, 150–161 (2015)
30. Li, Z.W.: Research on the grey wolf optimizer and its application. Hebei GEO University (2019)
31. Garg, V., Singh, K.: An improved Grunwald-Letnikov fractional differential mask for image texture enhancement. Int. J. Adv. Comput. Sci. Appl. **3**(3), 130–135 (2012)
32. Guo, P.T., Shi, Z., Li, M.F., Luo, W., Cha, Z.Z.: A robust method to estimate foliar phosphorus of rubber trees with hyperspectral reflectance. Ind. Crops Prod. **126**, 1–12 (2018)
33. Sodeifian, G., Ardestani, N.S., Sajadian, S.A., Ghorbandoost, S.: Application of supercritical carbon dioxide to extract essential oil from Cleome coluteoides Boiss: experimental, response surface and grey wolf optimization methodology. J. Supercrit. Fluids **114**, 55–63 (2016)

Feature Recognition of Tobacco by Independent Component Analysis - Back Propagation Neural Network

Jia Duan[1], Yue Huang[1(✉)], Yizhi Shi[1], Rui Chen[1], Guorong Du[2], Yitong Dong[1], and Shungeng Min[1]

[1] China Agricultural University, Beijing 100193, People's Republic of China
huangyue@cau.edu.cn

[2] Beijing Third Class Tobacco Supervision Station, Beijing 101121, People's Republic of China

Abstract. As the most important base for studying the quality stability of tobacco products, the characteristics of flue-cured tobacco are of great significance for both cigarette enterprises and producing areas. In this study, the characteristics of tobacco were qualitatively recognized based on the directly acquired gas chromatography-mass spectrometry (GC-MS) accumulative data via the pattern recognition, in an effort to rapidly and conveniently identify the grades information of tobacco. Specifically, an independent component analysis -back propagation neural network (ICA-BPNN) method was proposed to process the GC-MS ion accumulation. First, independent components were extracted after cumulating all the spectrum peaks of the acquired mass data. Next, a BPNN recognition model was established for the obtained independent components and then used to qualitatively discriminate four tobacco grades from Yunnan Province in China. Finally, a comparison was made between ICA-BPNN and principal component analysis (PCA)-BPNN models in the qualitative effect. Given nodes of 40, the ICA-BPNN model achieved the best accuracy, with the accuracy of calibration and prediction set of 72.86% and 80.82%, respectively. Results revealed that the proposed pattern recognition method, where mass data of tobacco were directly accumulated as overall information, is of certain potentials in the fast discrimination of agricultural product quality.

Keywords: Flue-cured tobacco · Gas chromatography-mass spectrometry · Pattern recognition · Independent component analysis · Back propagation neural network

1 Introduction

As an important analytical means of complex samples, the GC-MS technology has been extensively applied to the qualitative and quantitative analysis of sample components. In the current GC-MS studies, particular emphasis has been laid on the separation of sample components, which facilitates qualitative and quantitative analysis. For instance, all kinds of tedious separation and concentration operations are adopted in sample analysis

X. Chu et al. (Eds.): ICNIR 2021, *Sense the Real Change: Proceedings of the 20th International Conference on Near Infrared Spectroscopy*, pp. 316–324, 2022.
https://doi.org/10.1007/978-981-19-4884-8_34

[1–4]. The spectrum peaks of overlapping signals are separated using chemometrics methods [5, 6]. However, the components with approximate properties are usually prone to overlapping spectrum peaks due to the characteristics of chromatographic separation. In addition, the chromatographic peaks of low-content components are very approximate to baseline, making the qualitative and quantitative component analysis very difficult.

In the current pattern recognition of samples via GC-MS, sample components are generally qualitatively and quantitatively analyzed first, followed by the recognition and classification of the qualitative and quantitative analysis results [7]. However, for some complex systems with unavoidable overlapping peaks owing to the complicated chemical components [8, 9], the pattern recognition will be time-consuming and laborious if the traditional method is used. In fact, not the specific quantitative results of all components are required for the recognition and classification of samples, but instead, it is only necessary to extract complex data directly using chemometrics method and capture some connection between the application requirement goal and the obtained data. Or, mass information can be processed under the influence of many factors to find one or several chemical components related to the pattern recognition goal [10, 11].

Tobacco, a Solanaceae plant and important economic crop, contains very complex chemical components, which constitute the material basis for the quality of tobacco. Moreover, the chemical components, which are important factors determining the quality of tobacco, are closely related to the appearance quality and sensory quality. So far, there are about over 4,000 known chemical components in tobacco, along with many substances with approximate properties [12]. When the chemical components of tobacco are analytically determined via GC-MS, a complex system featured with serious overlapping chromatographic peaks is obtained [13, 14], which brings about great difficulties to the analytical comparison of different sample sources. Hence, how to extract single-component information from mixed signals is a hot chemometrics issue in complex sample analysis, especially for "black" systems [15–17].

ICA, a statistical signal processing method, has unique advantages in processing complex mixed systems. With the potential of processing blind source signals, ICA has been mainly applied to the separation and feature extraction of blind source signals [18, 19]. Besides, it is capable of separating independent components from measured overlapping mixed signals to obtain the latent variables with chemical significance. Moreover, it has already been applied to the chemical field, e.g., interpreting the pure-component spectrum in spectral measurement data in the event of the lack of prior information [20] and separating the peaks of overlapping mass spectrum information [21].

Back propagation neural network (BPNN) has received much attention since it has many advantages. It is increasingly widely used to solve a variety of analysis issues [22]. Compared with the previous neural networks, it has better classification ability. In the optimization of the new-generation neural network, it also has the ability of multidimensional function mapping. Compared with the simple perceptron, it expands the scope of solving problems, which cannot be solved in many previous studies and have been broken through the limitation of the algorithm [23, 24].

In this study, the pattern recognition of tobacco samples was conducted directly using the data of GC-MS. All the chromatographic peaks were directly overlapped into one peak along with all the mass data cumulated, followed by the feature extraction of independent components. Subsequently, the tobacco grade was rapidly recognized by combining BPNN classification model based on independent components. It would provide an approach for mining and analyzing chromatographic-mass data.

2 Material and Methods

2.1 Sample Preparation

All samples were oven dried under normal pressure at 60 °C for 24 h and then ground to certain granularity through the whirlwind grinding instrument. A total of 20 g of tobacco powders were weighed and placed in a 1000 mL round-bottom flask, and then nitrobenzene (internal standard) and 350 mL of NaCl solution were added. After blending, the mixture was heated by an electric jacket at one end of simultaneous distillation and extraction (SDE) device. The other end of this device was connected to a round-bottom flask containing 40 mL of dichloromethane for heating in water bath at 60 °C. After SDE for 2 h, the extract liquor was collected, dried with anhydrous sodium sulfate and then concentrated to 1 mL using a rotary evaporator for the subsequent sample injection. Next, the collected GC-MS data were superposed as the sample data and processed via MATLAB (Version 2016a, Mathworks, USA) software.

2.2 GC-MS Analysis

GC-MS was used to obtain chromatograms of the extract, which consisted of a Clarus 600 gas chromatograph equipped with a TurboMass 5.0 mass spectrometer (Perkin Elmer Corporation of America). Chromatographic separations were optimized and performed on a DB-5MS column (30 m × 0.25 mm, film thickness 0.25 μm). Ultrahigh purity helium (99.999%) was used as the carrier gas and the flow rate was 1.0 mL/min. Injector temperature was set at 280 °C. Heating program was as follows: Column was held at 60 °C for 2 min; subsequently, the temperature was raised to 200 °C at 5 °C/min and held for 5 min, finally increasing to 250 °C at 5 °C/min and held for 5 min, totally took 50 min for one process. Accurate 1 μL sample was injected using split mode with a split ratio of 10:1.

An Ion Trap Mass Spectrometer was then operated. Ionization voltage was set to 70 eV, the ion trap temperature to 150 °C, and mass transmission line temperature to 250 °C. Mass ranged from 50.0–350.0 m/z, with scanning time of 0.15 s and interval of 0.05 s. Electron multiplier voltage was set using its in-built autotune facility. Solvent delay was about 2 min.

3 Results and Discussion

3.1 Data Preprocessing

The total ion current diagram of Hongda flue-cured tobacco samples in Dali is displayed in Fig. 1(A). It could be seen that a lot of chemical components were detected, which differed greatly in the response value, with the difference value reaching 3 orders of magnitude. Compared with the maximum peak, many peaks were processed as noises. The data were processed using 10-point averaging method and logarithmic transformation-detrending algorithm to equally treat each peak value and eliminate the peak difference in the retention time. The peak difference was smaller after processing, and the total ion current of preprocessed samples is shown in Fig. 1(B). Following the preprocessing of the original total ion current diagram, the number of data points was changed into 715. A total of 135 information variables were selected through uninformative variable elimination (UVE), accounting for 15.8% of total information variables. In Fig. 1(C), the red dots represent the information variables selected through UVE of the preprocessed total ion current data. It could be seen that the target information obtained by selecting information variables after preprocessing was relatively scattered at each peak position. While the data were further reduced, the chemical information of the whole concentrated sample could also be covered. The signal-to-noise ratio turned poor after 40 min, so the variables selected also decreased.

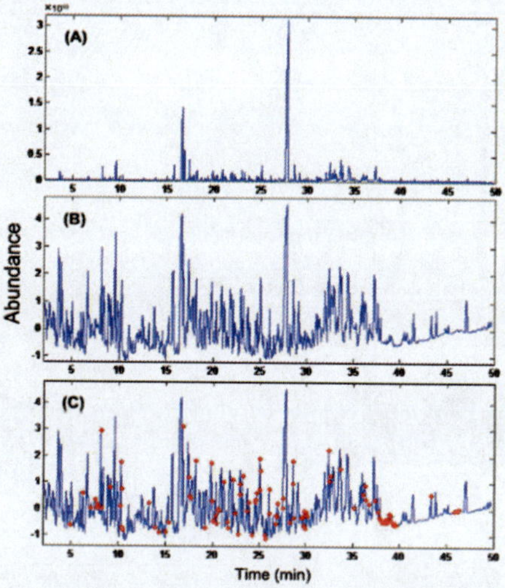

Fig. 1. Total ionic current (TIC) chromatogram of the extracted components in Hongda flue-cured tobacco (A) raw TIC, (B) preprocessed TIC, (C) informative variables after spectral preprocess and UVE selection

3.2 ICA on Mass Data

The accumulation of all mass ions of one tobacco sample is presented in Fig. 2(A), where x-coordinate denotes the mass-to-charge ratio (range: 51–350) and y-coordinate represents the ion abundance. First, the principal components were extracted from the accumulated mass data. The loadings of the first two principal components, PC1 and PC2, are shown in Fig. 2(B). It could be observed that a large number of signals far exceeding the accumulated mass spectrum appeared in the obtained loadings, and many mass-to-charge ratio signals are negative in PC2, so the principal component loadings could not be directly used to analyze the chemical components.

Subsequently, independent components were extracted by Fast ICA. Given that all the mass-to-charge ratios in the mass spectrum were non-negative, absolute values were taken from the extracted independent component diagram, where the first six independent components are shown in Fig. 3. It could be seen that the characteristic ions 57, 71, 85 and 99 of IC1 were similar to those of alkane, and the characteristic ions 207 and 281 of IC3 were similar to the ion information generated by column bleeding, indicating, to some extent, that the extracted independent components presented corresponding relations with chemical substances. Considering the complex chemical components of tobacco, however, the independent components extracted from the full-spectrum accumulated mass data were the most statistically independent components, which just reflected one kind of substances or its rough information to a certain degree, being not completely identical with actual chemical components, so they could not be accurate to some specific chemical components.

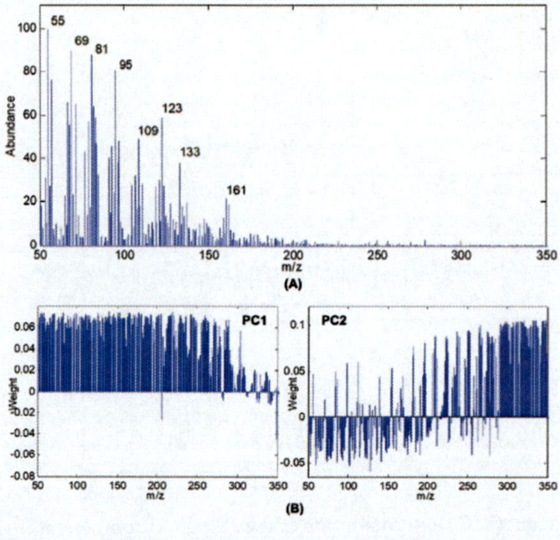

Fig. 2. Spectrum of ion accumulation of Hongda flue-cured tobacco (A), and loading diagram of the first two extracted principal components (B)

3.3 ICA-BPNN Modeling

In this study, a three-layer error BPNN model was used, with the first 12 independent components taken as the input nodes. The established model was utilized to identify the four grades of tobacco samples, where 0 0 0 1, 0 0 1 0, 0 1 0 0 and 1 0 0 0 represent grades B2F, C2L, C3F and X2L, respectively, so the output-layer node count of this network was set to 4. Actually, the neural network model was also influenced by the HLC. If the node count was too small, the BPNN failed to establish complex mapping relations,

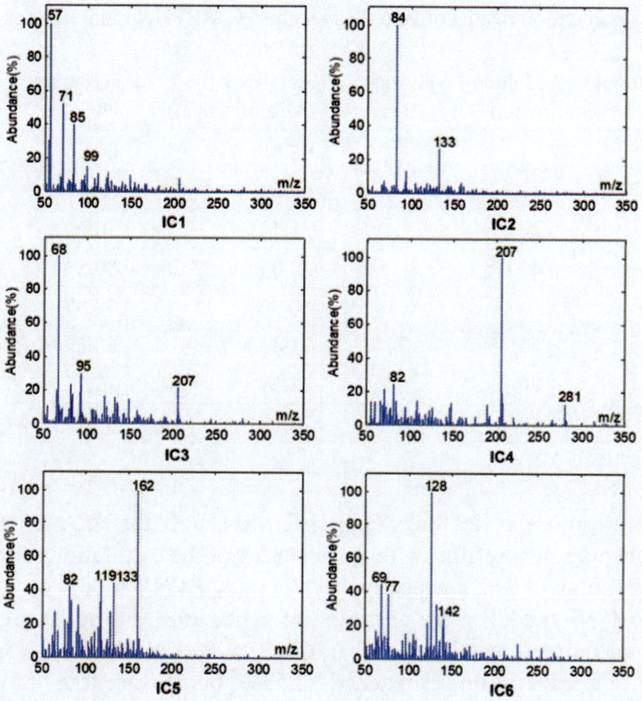

Fig. 3. Loading images of the first six independent components of mass accumulation

accompanied by the network prediction accuracy which was not high. Nevertheless, if the node count was too large, the network learning time was lengthened, and "overfitting" might occur. In consideration of the influence of HLC on the neural network model, the discrimination results of neural networks with different HLCs were experimentally investigated, as seen in Table 1. When HLC equaled 40, the ICA-BPNN model achieved the highest accuracy, where the accuracy of the calibration set and prediction set was 72.86% and 80.82%, respectively.

3.4 PCA-BPNN Model

PCA has been commonly used to process complex data systems. For comparison, the PCA was used in this study to process the accumulated mass data. Thus, a classification model was also established for the grading of Hongda tobacco by combining PCA and neural network method and then compared with the method proposed in this study.

To parallel the comparison, the first 12 principal components were also taken as the inputs of neural network model, which accounted for 99.99% of the gross information content. The PCA-BPNN model shared the same neural network structure with the ICA-BPNN model. Similarly, the influence of HLC on the discrimination result of neural network model was compared. The main parameters and discrimination results are listed in Table 1. It could be seen that with the increase in HLC, the discrimination accuracy of the calibration set was gradually elevated, so was the model calculated quantity and

Table 1. Classification results of ICA-BPNN and PCA-BPNN modeling for mass data

Pattern recognitions	Hidden-layer node count	Accuracy of calibration (%)	Accuracy of prediction (%)
ICA-BPNN	4	62.86	60.00
	10	64.29	65.71
	20	71.43	70.00
	40	**72.86**	**80.82**
PCA-BPNN	4	60.00	58.57
	10	71.43	68.57
	20	**74.29**	**75.71**
	40	78.57	68.57

operating time. Results showed that when HLC was set 20, the PCA-BPNN model harvested favorable judgment results, with the accuracy of the calibration set and prediction set being 74.29% and 75.71%, respectively. The PCA-BPNN model was slightly better than the ICA-BPNN model in the accuracy of calibration set, but the accuracy of its prediction set was much lower than that in the ICA-BPNN model. Given HLC of 40, the accuracy of the calibration set reached 78.57%, but the accuracy of the prediction set was far lower than that in the ICA-PBNN model. To balance the calibration set and prediction set and avoid the overfitting risk, the reasonable HLC was still taken as 20 in the case of PCA-BPNN model. The comparison between the two methods revealed that the ICA-BPNN method showed better effect on processing the accumulated mass data of tobacco. In other words, the ICA-BPNN model could be applied to the grading of raw tobacco materials very well by virtue of its convenient, fast and effective qualitative discrimination ability for the accumulated mass data of tobacco.

4 Conclusion

In this study, an ICA-BPNN pattern recognition applied to GC-MS accumulation data was proposed. After accumulating all the acquired spectral peaks, the extracted independent components were of certain chemical significance. Subsequently, a BPNN model was established based on the obtained independent components, followed by the qualitative discrimination of tobacco grade. Moreover, the ICA-BPNN model was compared with the PCA-BPNN model, and the results indicated that it is feasible to accumulate the mass data as the sample information for the pattern recognition.

Furthermore, it could be found from the results that the prediction accuracy of the PCA-BPNN model can be more optimized, manifesting that it is far from enough to process the mass data just through the direct accumulation. For instance, the data can be further refined before accumulation, as specifically, ICA or other pattern recognition are implemented after preprocessing the mass spectrum (e.g., baseline elimination) in the future study. As well, different from PCA, the information and sequence of characteristic components extracted by ICA are not fixed, so a further study can be done with regard to the selection of ICs for analysis.

Acknowledgements. This work was supported by the Beijing Natural Science Foundation (No. 8222070); Health Food Industry Research Institute (Xinghua), China Agricultural University (No. 201905). The authors would like to express their gratitude to Dali branch of China Tobacco, Inc. for the financial and the technical support.

References

1. Cabanes, A., Valdés, F.J., Fullana, A.: A review on VOCs from recycled plastics. Sustain. Mater. Tech. **25**, e00179 (2020)
2. Zou, W., Gao, B., Ok, Y.S., Dong, L.: Integrated adsorption and photocatalytic degradation of volatile organic compounds (VOCs) using carbon-based nanocomposites: a critical review. Chemosphere **218**, 845–859 (2019)
3. Ning, M., Jun, Y., Meseret, A., et al.: Accelerated solvent extraction combined with GC–MS: a convenient technique for the determination and compound-specific stable isotope analysis of phthalates in mine tailings. Microchem. J. **153**, 104366 (2020)
4. Liew, C.S., Li, X., Zhang, H., Lee, H.K.: A fully automated analytical platform integrating water sampling-miniscale-liquid-liquid extraction-full evaporation dynamic headspace concentration-gas chromatography-mass spectrometry for the analysis of ultraviolet filters. Anal. Chim. Acta **1006**, 33–41 (2018)
5. Lebanov, L., Ghiasvand, A., Paul, B.: Data handling and data analysis in metabolomic studies of essential oils using GC-MS. J. Chromatogr. A **1640**, 461896 (2021)
6. Duan, L., Ma, A., Meng, X., Shen, G., Qi, X.: QPMASS: a parallel peak alignment and quantification software for the analysis of large-scale gas chromatography-mass spectrometry (GC-MS)-based metabolomics datasets. J. Chromatogr. A **1620**, 460999 (2020)
7. Ebrahimabadi, E.H., Ghoreishi, S.M., Masoum, S., Ebrahimabadi, A.H.: Combination of GC/FID/Mass spectrometry fingerprints and multivariate calibration techniques for recognition of antimicrobial constituents of Myrtus communis L. essential oil. J. Chromatography B **1008**, 50–57 (2016)
8. Shao, X., Liu, Z., Cai, W.: Resolving multi-component overlapping GC-MS signals by immune algorithms, TrAC-Trend. Anal. Chem. **28**, 1312–1321 (2009)
9. Zeng, Z.D., Hugel, H.M., Marriott, P.J.: Simultaneous estimation of retention times of overlapping primary peaks in comprehensive two-dimensional GC. J. Sep. Sci. **36**, 2728–2737 (2013)
10. Hoggard, J.C., Siegler, W.C., Synovec, R.E.: Toward automated peak resolution in complete GC × GC-TOFMS chromatograms by PARAFAC. J. Chemometr. **23**, 421–431 (2009)
11. Duarte, L.M., Amorim, T.L., Grazul, R.M., Oliveira, M.A.: Differentiation of aromatic, bittering and dual-purpose commercial hops from their terpenic profiles: an approach involving batch extraction, GC–MS and multivariate analysis. Food Res. Int. **138**, 109768 (2020)
12. Savareear, B., Escobar-Arnanz, J., Brokl, M., et al.: Comprehensive comparative compositional study of the vapour phase of cigarette mainstream tobacco smoke and tobacco heating product aerosol. J. Chromatogr. A **1581**, 105–115 (2018)
13. Lim, H.H., Choi, K.Y., Shin, H.S.: Qualitative and quantitative comparison of flavor chemicals in tobacco heating products, traditional tobacco products and flavoring capsules. J. Pharm. Biomed. Anal. **207**, 114397 (2022)

14. Huang, L.F., Zhong, K.J., Shun, X.J.: Comparative analysis of the volatile components in cut tobacco from different locations with gas chromatography–mass spectrometry (GC-MS) and combined chemometric methods. Anal. Chim. Acta **575**, 236–245 (2006)
15. Medina, S., Perestrelo, R., Silva, P., Pereira, J.A., Carama, J.S.: Current trends and recent advances on food authenticity technologies and chemometric approaches. Trend Food Sci. Tech. **85**, 163–176 (2019)
16. Paul, A., Harrington, P.B.: Chemometric applications in metabolomic studies using chromatography-mass spectrometry. TrAC-Trend Anal. Chem. **135**, 116165 (2021)
17. Bovens, M., Ahrens, B., Alberink, I., Nordgaard, A., Salonen, T., Huhtala, S.: Chemometrics in forensic chemistry-Part I: implications to the forensic workflow. Forensic Sci. Int. **301**, 82–90 (2019)
18. Babu, P.R., Narasimhan, S.: Multivariate techniques for preprocessing noisy data for source separation using ICA. Int. J. Adv. Eng. Sci. App. Math. **4**, 32–40 (2012)
19. Monakhova, Y.B., Rutledge, D.N.: Independent components analysis (ICA) at the "cocktail-party" in analytical chemistry. Talanta **208**, 120451 (2020)
20. Monakhova, Y.B., Tsikin, A.M., Kuballa, T., et al.: Independent component analysis (ICA) algorithms for improved spectral deconvolution of overlapped signals in 1H NMR analysis: application to foods and related products. Magn. Reson. Chem. **52**, 231–240 (2014)
21. Habchi, B., Kassouf, A., Padellec, Y., et al.: An untargeted evaluation of food contact materials by flow injection analysis-mass spectrometry (FIA-MS) combined with independent components analysis (ICA). Anal. Chim. Acta **1022**, 81–88 (2018)
22. Wu, D., Shen, Y.: English feature recognition based on GA-BP Neural Network algorithm and data mining. Comput. Intel. Neurosci. 1890120 (2021)
23. Huang, X., Jin, H., Zhang, Y.: Risk assessment of earthquake network public opinion based on global search BP neural network. PLoS ONE **14**, e0212839 (2019)
24. Low, C.Y., Park, J., Teoh, A.B.J.: Stacking-based deep neural network: deep analytic network for pattern classification. IEEE Trans. Cybern. **50**, 5021–5034 (2020)

Insight into Hydration Behavior of Poly(Hydroxypropyl Acrylate) Block Copolymer by Temperature-Dependent Infrared Spectroscopy

C. Xiong[1], S. Han[1], Y. Guo[1], and L. Guo[1,2](✉)

[1] State Key Laboratory of Organic-Inorganic Composites, Beijing University of Chemical Technology, Beijing 100029, People's Republic of China
guolh@mail.buct.edu.cn

[2] Beijing Engineering Research Center of Synthesis and Application of Waterborne Polymer, Beijing University of Chemical Technology, Beijing 100029, People's Republic of China

Abstract. Dynamic light scattering (DLS), microscope, temperature-dependent infrared (IR) spectroscopy with perturbation correlation moving window (PCMW) were used to investigate the phase transition of a thermo-sensitive copolymer poly(*N,N*-dimethylacrylamide)-*b*-poly(hydroxypropyl acrylate)-*b*-poly(*N,N*-dimethylacrylamide) (PDMAA-*b*-PHPA-*b*-PDMAA). The copolymer was synthesized via reversible addition-fragmentation chain transfer (RAFT) polymerization. In the DLS analysis, the sudden increase in particle size at 49 °C and the polymer droplets appearing at 49 °C in the microscope photos indicate that the lower critical solution temperature (LCST) of the copolymer is around 49 °C. IR spectra show that the OH groups of PHPA change from hydration states to free ones. The redshift of the spectra in C-H bands indicates the dehydration of CH groups. The hydration C = O bands transform into free states or the other part, which owns stronger intermolecular interactions. Moreover, the O-H bands show a more responsive to temperature during the phase transition by PCMW analysis. The reduced hydrophilicity of O-H bands is not enough to stabilize the polymer in water, which leads to a phase transition.

Keywords: Thermo-sensitive polymer · Temperature-dependent IR spectra · Phase transition · Hydroxypropyl acrylate

1 Introduction

Thermo-sensitive polymers, that change their physical properties in response to temperature, have wide studied in the last few decades. They are applied to fields such as biosensors [1], material for tissue engineering [3, 4], and drug delivery carriers [2] due to their potential biomedical applications and easy to control stimulus. When the temperature is over the lower critical solution temperature (LCST), phase separation will

X. Chu et al. (Eds.): ICNIR 2021, *Sense the Real Change: Proceedings of the 20th International Conference on Near Infrared Spectroscopy*, pp. 325–335, 2022.
https://doi.org/10.1007/978-981-19-4884-8_35

occur. High concentration polymer droplets or polymer precipitation will lead to clouding of the solution [5]. At the LCST, the enthalpy gain produced by the interaction of the hydrophilic part of the polymer with water does not compensate for the entropy loss produced by the hydrophobic part. Thus, the balance of hydrophilicity and hydrophobicity within the polymer usually determines the LCST. Poly(*N*-isopropylacrylamide) (PNIPAM), which shows a coil-to-globule phase transition during heating and cooling, is the most studied temperature-sensitive polymer for its LCST (~32 °C) close to physiological temperature and quick response to temperature [6]. However, potential alternatives for PNIPAM are being sought due to some disadvantages such as difficulty to adjust LCST, producing toxic low-molecular weight amines during hydrolysis, etc. [7].

An emerging class of thermo-sensitive polymers are a family of acrylic monomers, which own the better biocompatibility and adjustable LCST that may compete with or even surpass PNIPAM [8, 38]. Hydroxypropyl acrylate (HPA) is one of the family and was first reported to be thermo-sensitive by Taylor and Cerankowski in 1975, which is still the lowest LCST (~16 °C, 10 wt%) found so far [9]. Subsequently, the temperature-sensitive nature of HPA has been reported [10, 11]. Schubert and co-workers investigate the effect of HPA polymer solution concentration and the ratio of hydrophilic block and HPA to LCST. HPA is polymerized with *N*-acryloylmorpholine (Amor) or *N,N*-dimethylacrylamide (DMAA) in different proportions, and a wide range of LCST (21.4 to 88 °C) is obtained [12]. Meanwhile, the PHPA homopolymer concentration was adjusted from 1.5 to 0.25 wt%, and the LCST increased from 18.3 to 33.3 °C. The large hysteresis and width of the transition at low concentration were also observed. High polymer concentration facilitates hydrophobic interaction between polymers to a low LCST and the transition with a narrow temperature range. Homopolymer of PHPA was synthesized via reversible addition-fragmentation chain transfer (RAFT) polymerization, and its LCST was significantly increased due to the hydrophilic COOH end-group [13]. PHPA with COOH end-groups on both ends also shows a phase transition at LCST. However, they will lose their temperature sensitivity at neutral pH, which may cause by the ionization of COOH end-groups. Although the temperature-sensitive properties of HPA have been generally studied, the phase transition process of HPA at the molecular level is still limited understood.

Recently, infrared (IR) spectroscopy has been a powerful tool to investigate the transition process of thermo-sensitive polymers [14–17]. When the temperature is below the LCST, the ordered water structure will appear around C-H bands of the polymer, which will increase the frequency of C-H bands in the spectrum [18]. The interaction with the hydrogen bonds reduces the frequency of the group vibration, which appears as a peak shift to a low wavenumber on the spectrum [19]. Thus, by observing the change of wavenumber shift of different polar groups such as C-H and C = O, the change of hydration and intramolecular hydrogen bonds can be completely extracted. For instance, the phase transition process of PNIPAM was obtained by temperature-dependent IR spectra. There is a redshift of spectra in C-H bands, which indicates the dehydration of C-H groups.

In this work, a triblock thermo-sensitive copolymer PDMAA-*b*-PHPA-*b*-PDMAA was synthesized by RAFT polymerization. Dynamic light scattering (DLS) and microscope analysis were used to observe the particle size and macro changes. The phase change process is manifested as liquid-liquid separation in the solution, and the polymer forms small droplets. Moreover, temperature-dependent IR spectra with PCMW2D analysis were performed to obtain the phase transition point and phase transition interval. The results show that the phase transition process of thermo-sensitive polymers can be studied by temperature-dependent infrared.

2 Materials and Methods

2.1 Materials

RAFT agent *S,S′*-bis (*R,R′*-dimethyl-*R″*-acetic acid)-trithiocarbonate (BDAAT) was self-synthesized. Hydroxypropyl acrylate (DAAM, Shanghai Macklin Biochemical Co., Ltd), *N, N*-Dimethylacrylamide (DMAA, Shanghai Macklin Biochemical Co., Ltd). 4,4′-azobis (4-cyanovaleric acid) (ACVA, Shanghai Macklin Biochemical Co., Ltd), 1,4-dioxane (>99%, Beijing Chemical Works), n-hexane (>99%, Beijing Chemical Works), Deuterium Oxide (>99.9%, Energy Chemical). Deionized (DI) water was self-prepared in our laboratory.

2.2 Synthesis of PDMAA-*B*-pHPA-*B*-pDMAA Triblock Copolymers

The ABA-type triblock copolymer of PDMAA-*b*-PHPA-*b*-PDMAA with a designed chain structure was synthesized via RAFT polymerization, where the PDMAA block and the PHPA block have a degree of polymerization (DP) at 40 and 100, respectively. The hydrophilic chain transfer agent of PDMAA-CTA was firstly synthesized by using ACVA and BDAAT as initiator and RAFT agent, respectively. The solution of ACVA, BDAAT, DMAA, and 1, 3, 5-trioxane (as an internal standard) in 1,4-dioxane (20 wt%) with a molar ratio at 0.2:1:40:1 was added into a glass tube with a magnetic stir bar. RAFT polymerization was carried out at 70 °C for 3 h after refrigerating and filling with nitrogen for three cycles. Then, the solution was precipitated using *n*-hexane and further dried in a vacuum oven at 45 °C for overnight to obtain yellow PDMAA-CTA powder. Furthermore, the solution of ACVA, PDMAA-CTA, and HPA in 1,4-dioxane (20 wt%) with a molar ratio at 0.2:1:100 was added into a glass tube. And the reaction was carried out for 4 h at 70 °C after deoxygenation. Finally, the yellow powder of PDMAA-*b*-PHPA-*b*-PDMAA was obtained after precipitation purification. The aqueous solution containing 10 wt% of the synthesized product was prepared using redistilled water and D_2O as the solvent. The ^{1}H-NMR of PDMAA-CTA and PDMAA-*b*-PHPA-*b*-PDMAA is shown in Fig. 1.

2.3 Instruments and Measurements

Temperature-dependent DLS of copolymer particles (10 wt% aqueous dispersion) was analyzed from 35 to 65 °C using a Malvern Zetasizer Nano-ZS. The sample temperatures were stepwise raised with an increment of 1 °C at a rate of 1 °C/min and kept at each temperature for 2 min before repeated measurement for 3 times. The results were determined by the average value.

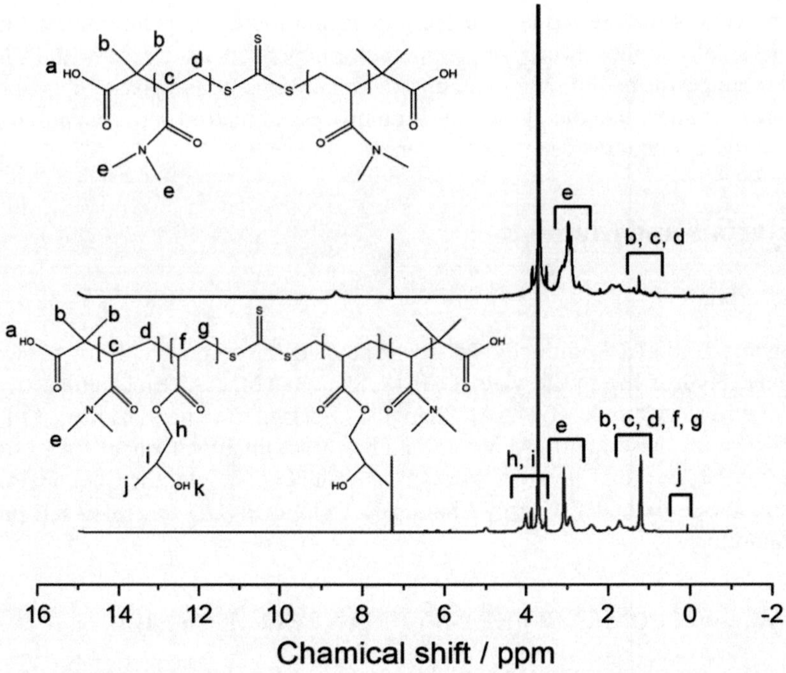

Fig. 1. The ^{1}H-NMR curve of PDMAA-CTA and PDMAA-*b*-PHPA-*b*-PDMAA

Microscope with temperature control accessories (Instec, HCS621GXY) was used to observe the phase transition of the copolymer. Observing the change of the aqueous solution (10 wt%) at a heating rate of 1°C/min from 40 to 55 °C.

For FTIR measurements, PDMAA-*b*-PHPA-*b*-PDMAA D_2O solution with a concentration of 10 wt% was sealed between two CaF_2 tablets for 15 μm. All the temperature resolved FTIR spectra were recorded utilizing Spectrum-100 (PerkinElmer, Germany) FTIR spectrometer equipped with a DTGS detector. In order to obtain an acceptable signal-to-noise ratio, 20 scans at a resolution of 4 cm^{-1} were accumulated. Linkam THMS600 was used to control the temperature from 35 to 60 °C (The interval was 1 °C). The heating rate was 1 °C/min. Stay 2 min before the test to stabilize the temperature.

Perturbation correlation moving window (PCMW) was processed, calculated, and plotted using two-dimensional correlation spectroscopy software [20, 21].

3 Results and Discussion

3.1 DLS and Microscope Analysis

Firstly, DLS measurement is used to analyze the phase transition behavior of the copolymer in water via the change of particle sizes. Below LCST, the small particle sizes (around 30 nm) indicate that the block copolymer is fully extended in water, which makes the solution appear transparent. The curve in Fig. 2a rises slightly at 45 °C and undergoes a dramatic increase at 49 °C. Then, the changing trend of the particle size becomes gentle. The change of particle sizes indicate that the copolymer solution undergoes a sharp phase change process. More details can be found in microscope photos in Fig. 2b. The LCST observed from the microscope is 49 °C. However, it is still transparent at 48 °C and a sudden liquid-liquid separation occurred in the process of rising to 49 °C, which indicates the copolymers aggregate due to hydrophobic interaction. Strong intermolecular hydrogen bonds will lead to coil conformation, but weak hydrogen bonds between the polymer and water allow a small amount of water to enter between the polymers to form a polymer-rich phase [22]. The polymer-rich domains show that the copolymers own weak interaction with water after phase separation. The small droplets that appear within a short temperature interval indicate that the block copolymer PDMAA-*b*-PHPA-*b*-PDMAA undergoes sharp dehydration during heating, which seems to be similar to the phase transition process of PNIPAM [23] instead of the gradual dehydration of PVCL in a wide temperature range [24]. The sharp dehydration may be also related to high concentration, which increases the chance of polymer contact with each other [25].

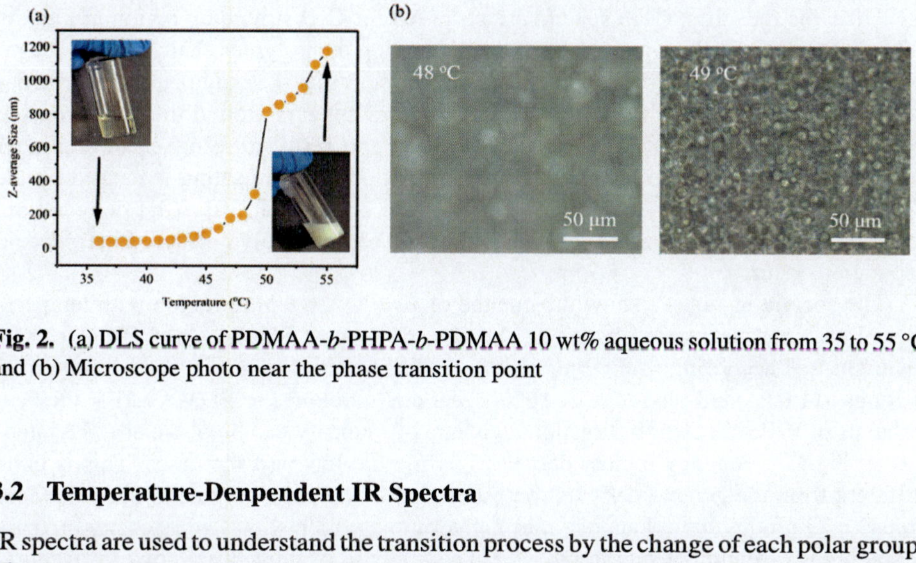

Fig. 2. (a) DLS curve of PDMAA-*b*-PHPA-*b*-PDMAA 10 wt% aqueous solution from 35 to 55 °C and (b) Microscope photo near the phase transition point

3.2 Temperature-Denpendent IR Spectra

IR spectra are used to understand the transition process by the change of each polar group. The temperature range of IR spectra measurement is from 35 to 60 °C for covering the LCST of the copolymer solution. D_2O is a solvent instead of H_2O to dissolve copolymers for eliminating the band of O-H stretching vibration around 3300 cm^{-1} and O-H bending vibration around 1640 cm^{-1}. Three main regions are focused here at 3700–3100 cm^{-1} for OH groups, 3060–2840 cm^{-1} for C-H groups, and 1860–1540 cm^{-1} for C = O groups. Then, we can monitor the changes of almost all groups of PDMAA-*b*-PHPA-*b*-PDMAA copolymer in the water.

Firstly, blue shifts can be observed clearly in O-H stretching region in Fig. 3a. O-H bands can be roughly divided into two parts, the hydration O-H bands and the free O-H bands. Intensity increases at high wavenumber and that decreases at low wavenumber, which indicates the hydration O-H states transform into free O-H states [26]. HPA is perhaps a kind of unique temperature-sensitive monomer since it results in rather

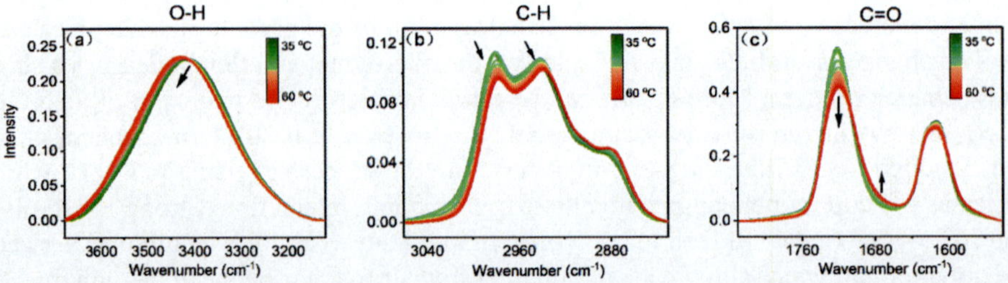

Fig. 3. Temperature-dependent IR spectra of PDMAA-*b*-PHPA-*b*-PDMAA in 10 wt% D_2O at (a) 3700–3100 cm-1 (b) 3060–2840 cm^{-1} (c) 1860–1540 cm^{-1}.

hydrophobic polymers while containing a hydroxyl group. Hydrophilic OH groups may play an important role to make the polymer chain extend in the water. Heating may weaken the hydrogen bonds between the O-H bands and water. The decrease in the hydrophilicity of the O-H bands may lead to a decrease in the overall hydrophilicity of the polymer.

Then, the redshifts of the spectra are observed for C-H stretching region in Fig. 3b. When below LCST, the water molecules surrounding the polymer chain form the cage-like structure with hydrogen bonds, which makes sure that the polymer chain stretches in the water [27]. Once the ordered water structure appears around the C-H group, it will increase the vibration frequency of the C-H and result in a higher wavenumber [28]. Therefore, the redshift in C-H stretching region during heating is caused by the destruction of the order water structure around CH groups [29, 30]. After dehydration, the hydrophobic interaction causes the aggregation of the copolymers and Liquid-liquid separation occurs.

The spectra in Fig. 3c show the change of C = O stretching region with temperature. Two broad peaks are observed. The C = O band of PDMAA and PHPA can be distinguished according to the literature [31]. The spectral range of 1760–1680 cm^{-1} belongs to PHPA and the range of 1650–1580 cm^{-1} belongs to PDMAA. For the C = O bands of PHPA, a clear bidirectional change of intensity can be obtained. The intensity of higher frequency moiety decreases, while the low part increases. This is quite different from the general dehydration process of LCST-type polymers [32, 33]. More free C = O groups should appear after dehydration, which shows an intensity increase in the high wavenumber region [34, 35]. The increase of intensity in lower wavenumber implies stronger interaction within the groups. Thus, there is a stronger interaction between copolymers after the phase transition.

More critical information about changes between copolymers and water molecules can be found in the second derivative curves in Fig. 4, which shows the difference before and after the phase transition. In Fig. 4a, the redshift of the spectra in C-H band is more obvious. The ordered water structure is destroyed and copolymers undergo a dehydration process. It is noticed that a broadening of C = O bands occurs after heating above the LCST in Fig. 3c, which can be further elucidated by the related second derivative curves in C = O bands of PHPA and two transition trends can be found in Fig. 4b. The high wavenumber frequency part shifts to a higher frequency, indicating that the hydrated C = O bands of PHPA become freer. Meanwhile, the hydrated C = O band changes into low frequency part, suggesting that the hydrated C = O forms a stronger intramolecular interaction. The C = O bands of PDMAA show a clear change on second derivative curve in Fig. 4b. The peak at low wavenumber shows a decrease, which considered the hydrated C = O of PDMAA transform into free ones.

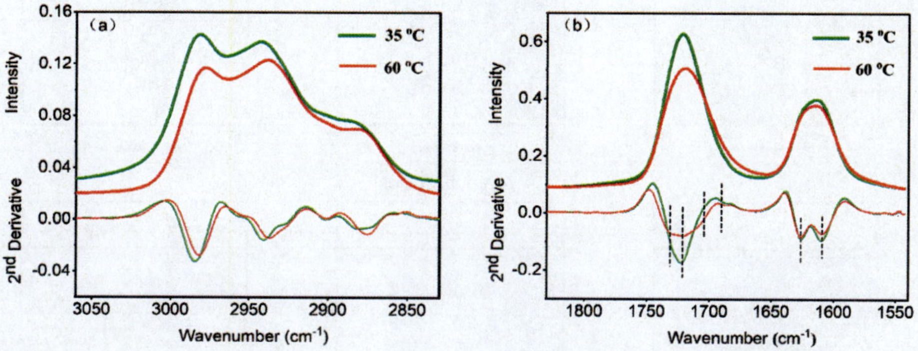

Fig. 4. IR and second derivative curve spectra at 35 and 60 °C in (a) 3060–2840 cm^{-1} (b) 1860–1540 cm^{-1}.

3.3 Perturbation Correlation Moving Window (PCMW) Analysis

In order to determine the LCST of PDMAA-*b*-PHPA-*b*-PDMAA solution more accurately, PCMW was employed. PCMW is a technique with basic principles dating back to the conventional moving window provided by Thomas et al. [36] Then, Morita improved it in 2006 [37]. PCMW contains the synchronous and asynchronous correlation spectra. The transition point and the transition interval can be presented intuitively via PCMW. The PCMW synchronous and asynchronous spectra from 35 to 60 °C of O-H, C-H, and C = O regions are shown in Fig. 5.

Via the PCMW synchronous spectra, the transition points of the O-H, C-H, and C = O bands are roughly 45, 46, and 47 °C, respectively. The O-H bands have an earlier response during the entire transition process. This also proves that the terminal OH groups of PHPA provide the necessary hydrophilic effect for the copolymer. When the temperature is higher than the LCST of the copolymer, the interaction between the hydroxyl group and water is weakened, which results in an aggregation of the copolymer. It is worth noting that the transition point from the PCMW analysis is advanced than that from the DLS analysis. This may be due to D_2O as a solvent to slightly weaken the hydrogen bond. Moreover, the transition interval can be obtained by the PCMW asynchronous spectra. All groups contain the same transition interval from 43 to 49 °C.

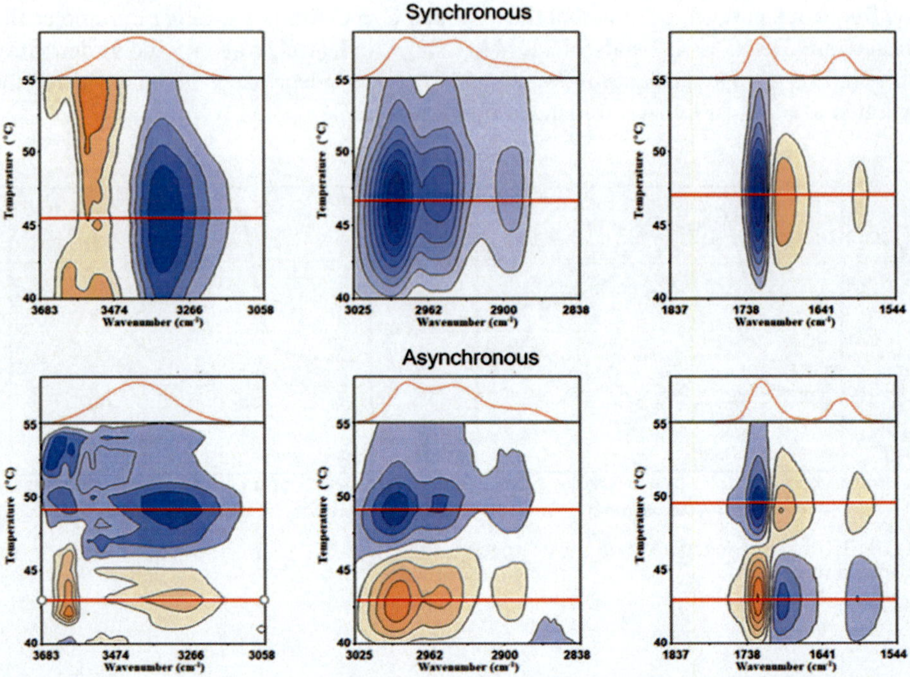

Fig. 5. PCMW synchronous and asynchronous spectra of O-H, C-H, and C = O regions. The orange represents the positive intensity and the blue represents the negative intensity.

4 Conclusion

In this work, a temperature-sensitive triblock copolymer PDMAA-*b*-PHPA-*b*-PDMAA was synthesized by RAFT polymerization. A sharp transition point at 49 °C was obtained by DLS analysis. The microscope photo shows that the solution changes from transparent to opaque is due to the formation of small copolymer droplets during the heating, which may cause by that the hydrophobic interaction between coplomer. The IR spectra with proves it. The hydrated O-H bands transform into free states during the transition process. The redshift of IR spectra in C-H bands indicates that the destruction of the order water structure around the methyl, which leads to a hydrophobic interaction between the copolymers. The C = O bands of PHPA shows a bidirectional trend of change to more free states and more hydrated states. Determination of phase transition point and phase transition interval of thermally responsive polymers by PCMW2D and the results correspond to the DLS analysis. Temperature-dependent infrared is a powerful tool for analyzing thermo-responsive polymers.

Acknowledgments. This study is financially supported by the National Key Research and Development Program of China (Grant No. 2020YFE0100300).

References

1. Hao, F., Wang, L., Chen, B., Qiu, L., Nie, J., Ma, G.: Interfaces, bifunctional smart hydrogel dressing with strain sensitivity and NIR-responsive performance. ACS Appl. Mater. Interfaces **13**, 46938–46950 (2021)
2. Cui, S., et al.: Interfaces, injectable thermogel generated by the "Block Blend" strategy as a biomaterial for endoscopic submucosal dissection. ACS Appl. Mater. Interfaces **13**, 19778–19792 (2021)
3. Parchehbaf-Kashani, M., et al.: Heart repair induced by cardiac progenitor cell delivery within polypyrrole-loaded cardiogel post-ischemia. ACS Appl. Bio Mater. **4**, 4849–4861 (2021)
4. Liu, F., Urban, M.: Recent advances and challenges in designing stimuli-responsive polymers. Prog. Polym. Sci. **35**, 3–23 (2010)
5. Xia, M., Cheng, Y., Theato, P.M.: Physics, thermo-induced double phase transition behavior of physically cross-linked hydrogels based on oligo (ethylene glycol) methacrylates. Macromol. Rapid Commun. **216**, 2230–2240 (2015)
6. Okada, Y., Tanaka, F.: Cooperative hydration, chain collapse, and flat LCST behavior in aqueous poly (N-isopropylacrylamide) solutions. Macromolecules **38**, 4465–4471 (2005)
7. Akdemir, Ö., et al.: Washington, Controlled/Living Radical Polymerization: Progress in ATRP. American Chemical Society, Washington (2009)
8. Vancoillie, G., Frank, D., Hoogenboom, R.: Thermoresponsive poly(oligo ethylene glycol acrylates). Prog. Polym. Sci. **39**, 1074–1095 (2014)
9. Taylor, L.D., Cerankowski, L.: Preparation of films exhibiting a balanced temperature dependence to permeation by aqueous solutions—a study of lower consolute behaviour. Science **13**, 2551–2570 (1975)
10. Christova, D., Velichkova, R., Loos, W., Goethals, E.J., Du Prez, F.: New thermo-responsive polymer materials based on poly (2-ethyl-2-oxazoline) segments. Polymer **44**, 2255–2261 (2003)
11. Sugihara, S., Yoshida, A., Fujita, S., Maeda, Y.: Design of hydroxy-functionalized thermoresponsive copolymers: improved direct radical polymerization of hydroxy-functional vinyl ethers. Macromolecules **50**, 8346–8356 (2017)
12. Eggenhuisen, T.M., Becer, C.R., Fijten, M.W., Eckardt, R., Hoogenboom, R., Schubert, U.: Libraries of statistical hydroxypropyl acrylate containing copolymers with LCST properties prepared by NMP. Macromolecules **41**, 5132–5140 (2008)
13. Vo, C.-D., Rosselgong, J., Armes, S.: RAFT synthesis of branched acrylic copolymers. Macromolecules **40**, 7119–7125 (2007)
14. Park, Y., Jin, S., Noda, I., Jung, Y.: Recent progresses in two-dimensional correlation spectroscopy (2D-COS). J. Mol. Struct. **1168**, 1–21 (2018)
15. Wang, G., Wu, P.: Toward the dynamic phase transition mechanism of a thermoresponsive ionic liquid in the presence of different thermoresponsive polymers. Soft Matter **12**, 925–933 (2016)
16. Wen, L., Zhang, J., Zhou, T., Zhang, A.: Hydrogen bonding in micro-phase separation of poly (polyamide 12-block-polytetrahydrofuran) alternating block copolymer: enthalpies and molecular movements. Vib. Spectrosc. **86**, 160–172 (2016)
17. Dai, Y., Wu, P.: Exploring the influence of the poly (4-vinyl pyridine) segment on the solution properties and thermal phase behaviours of oligo (ethylene glycol) methacrylate-based block copolymers: the different aggregation processes with various morphologies. Phys. Chem. Chem. Phys. **18**, 21360–21370 (2016)
18. Schmidt, P., Dybal, J., Trchová, M.: Investigations of the hydrophobic and hydrophilic interactions in polymer–water systems by ATR FTIR and Raman spectroscopy. Vib. Spectrosc. **42**, 278–283 (2006)

19. Maeda, Y., Yamauchi, H., Kubota, T.: Confocal micro-raman and infrared spectroscopic study on the phase separation of aqueous poly (2-(2-methoxyethoxy) ethyl (meth) acrylate) solutions. Langmuir **25**, 479–482 (2009)
20. Zhou, T., et al.: Identification of weak transitions using moving-window two-dimensional correlation analysis: treatment with scaling techniques. Anal. Bioanal. Chem. **406**(17), 4157–4172 (2014). https://doi.org/10.1007/s00216-014-7788-6
21. Su, G., Zhou, T., Zhang, Y., Liu, X., Zhang, A.: Microdynamics mechanism of D2O absorption of the poly (2-hydroxyethyl methacrylate)-based contact lens hydrogel studied by two-dimensional correlation ATR-FTIR spectroscopy. Soft Matter **12**, 1145–1157 (2016)
22. Su, G., Zhou, T., Liu, X., Zhang, J., Bao, J., Zhang, A.: Two-dimensional correlation infrared spectroscopy reveals the detailed molecular movements during the crystallization of poly (ethylene-co-vinyl alcohol). RSC Adv. **5**, 84729–84745 (2015)
23. Li, T., Tang, H., Wu, P.: Molecular evolution of poly(2-isopropyl-2-oxazoline) aqueous solution during the liquid–liquid phase separation and phase transition process. Langmuir **31**, 6870–6878 (2015)
24. Wang, X., Qiu, X., Wu, C.: Comparison of the coil-to-globule and the globule-to-coil transitions of a single poly (N-isopropylacrylamide) homopolymer chain in water. Macromolecules **31**, 2972–2976 (1998)
25. Sun, S., Wu, P.: Infrared spectroscopic insight into hydration behavior of poly(N-vinylcaprolactam) in water. J. Phys. Chem. B **115**(40), 11609–11618 (2011). https://doi.org/10.1021/jp2071056
26. Hou, L., Wu, P.: On the abnormal "forced hydration" behavior of P (MEA-co-OEGA) aqueous solutions during phase transition from infrared spectroscopic insights. Phys. Chem. Chem. Phys. **18**, 15593–15601 (2016)
27. Deshmukh, S.A., Sankaranarayanan, S.K., Suthar, K., Mancini, D.C.: Role of solvation dynamics and local ordering of water in inducing conformational transitions in poly(N-isopropylacrylamide) oligomers through the LCST. J. Phys. Chem. B. **116**, 2651–2663 (2012)
28. Maeda, Y., Takaku, S.: Lower critical solution temperature behavior of poly (N-tetrahydrofurfuryl (meth) acrylamide) in water and alcohol– water mixtures. J. Phys. Chem. B **114**, 13110–13115 (2010)
29. Su, G., Zhou, T., Liu, X., Ma, Y.: Micro-dynamics mechanism of the phase transition behavior of poly(N-isopropylacrylamide-co-2-hydroxyethyl methacrylate) hydrogels revealed by two-dimensional correlation spectroscopy. Polym. Chem. **8**, 865–878 (2017)
30. Sun, S., Wu, P.: On the thermally reversible dynamic hydration behavior of oligo(ethylene glycol) methacrylate-based polymers in water. Macromolecules **46**, 236–246 (2012)
31. Byard, S.J., Williams, M., McKenzie, B.E., Blanazs, A., Armes, S.P.: Preparation and cross-linking of all-acrylamide diblock copolymer nano-objects via polymerization-induced self-assembly in aqueous solution. Macromolecules **50**, 1482–1493 (2017)
32. Ye, Z., Li, Y., An, Z., Wu, P.: Exploration of doubly thermal phase transition process of PDEGA-b-PDMA-b-PVCL in water. Langmuir **32**, 6691–6700 (2016)
33. Park, Y., Hashimoto, C., Ozaki, Y., Jung, Y.: Understanding the phase transition of linear poly (N-isopropylacrylamide) gel under the heating and cooling processes. J. Mol. Struct. **1124**, 144–150 (2016)
34. Wang, Q., Tang, H., Wu, P.: Dynamic phase transition behavior and unusual hydration process in poly(ethylene oxide)-b-poly(N-vinylcaprolactam) aqueous solution. J. Polym. Sci. Part B: Polym. Phys. **54**, 385–396 (2016)
35. Wang, G., Wu, P.: Unusual phase transition behavior of poly(N-isopropylacrylamide)-co-Poly(tetrabutylphosphonium styrenesulfonate) in water: mild and linear changes in the poly(N-isopropylacrylamide) part. Langmuir **32**, 3728–3736 (2016)

36. Thomas, A.C., Richardson, H.: 2D-IR correlation analysis of thin film water adsorbed on α-Al2O3 (0001). J. Mol. Struct. **799**, 158–162 (2006)
37. Morita, S., Shinzawa, H., Noda, I., Ozaki, Y.: Perturbation-correlation moving-window two-dimensional correlation spectroscopy. Appl. Spectrosc. **60**, 398–406 (2006)
38. Hu, Z., Cai, T., Chi, C.: Thermoresponsive oligo (ethylene glycol)-methacrylate-based polymers and microgels. Soft Matter **6**, 2115–2123 (2010)

Author Index

X. Chu et al. (Eds.): ICNIR 2021, *Sense the Real Change: Proceedings of the 20th International Conference on Near Infrared Spectroscopy*, pp. 337–339, 2022.
https://doi.org/10.1007/978-981-19-4884-8